Ecotones Between Forest and Grassland

Randall W. Myster
Editor

Ecotones Between Forest and Grassland

Springer

Editor
Randall W. Myster
Department of Biology
Oklahoma State University
Oklahoma City, OK, USA

ISBN 978-1-4614-3796-3 ISBN 978-1-4614-3797-0 (eBook)
DOI 10.1007/978-1-4614-3797-0
Springer New York Heidelberg Dordrecht London

Library of Congress Control Number: 2012941221

Printed on acid-free paper

Springer is part of Springer Science+Business Media (www.springer.com)

For my mother, Dorothy, on her 90th birthday.
Many more and much love, always.

Preface

I have lived within the Cross Timbers for 10 years, observing its many changes. It did not take me long to realize that many of the processes maintaining this ecotone—between eastern deciduous forest and tallgrass prairie—may also control how trees invaded the ecotone defined after abandonment from agriculture in New Jersey, USA which was the subject of my Ph.D dissertation and included in my first book. Not only were many of the dominant species and genera the same (e.g., *Quercus* spp., *Ulmus* spp., *Rubus* spp., *Juniperius virginiana*) but the ecotonal dynamics itself could be looked at as an interplay between the process of tree invasion on the forest side of the ecotone and those processes that present barriers to that invasion on the prairie side. Consequently I could combine my previous knowledge of the biology and ecology of many of the trees, shrubs, forbs, and grasses with refined field experimentation, to investigate this Cross Timbers ecotone and set the stage for this new book. I have been fortunate to work with these generous colleagues at Oklahoma State University (OSU): Drs. John Weir, Stephen Hallgren, Edwardo Lorenzi, and Chris Stansberry at OSU—Stillwater, and Drs. Edward Vezey and Armando Cruz-Rodz at OSU—Oklahoma City.

Oklahoma City, OK, USA Randall W. Myster

Contents

Contributors

William B. Batista IFEVA–CONICET and Departamento de Métodos Cuantitativos, Universidad de Buenos Aires, 1417 Buenos Aires, Argentina

Enric Batllori Departament d'Ecologia, Universitat de Barcelona, 08028 Barcelona, Spain

Department of Environmental Science, Policy and Management, University of California, Berkeley, USA

Juan Carlos Berrio Department of Geography, University of Leicester, Leicester, LE1 7RH, UK

Carolina C. Blanco Departamento de Ecologia, Universidade Federal do Rio Grande do Sul, 91501-970 Porto Alegre, RS, Brazil

Raul G. Bogotá-A Department Paleoecology and Landscape Ecology, University of Amsterdam, 1098 XH, Amsterdam, The Netherlands

Universidad Distrital Francisco José de Caldas, Bogotá, Colombia

Arnoud Boom Department of Geography, University of Leicester, Leicester, LE1 7RH, UK

J. Julio Camarero ARAID, Instituto Pirenaico de Ecología, CSIC, 50080 Zaragoza, Spain

Andrew S. Carr Department of Geography, University of Leicester, Leicester, LE1 7RH, UK

Enrique J. Chaneton IFEVA–CONICET, Universidad de Buenos Aires, 1417 Buenos Aires, Argentina

Claudio M. Ghersa IFEVA–CONICET, Universidad de Buenos Aires, 1417 Buenos Aires, Argentina

Zaire González-Carranza Department Paleoecology and Landscape Ecology, University of Amsterdam, 1098 XH Amsterdam, The Netherlands

EcoPetrol, Bucaramanga, Colombia

Mirella H. M. Groot Department Paleoecology and Landscape Ecology, University of Amsterdam, 1098 XH, Amsterdam The Netherlands

Sándor Gulyás Department of Geology and Landscape, University of Szeged, 6722 Szeged, Egyetem u.2-6, Hungary

Emilia Gutiérrez Departament d'Ecologia, Universitat de Barcelona, 08028 Barcelona, Spain

Henry Hooghiemstra Department of Paleoecology and Landscape Ecology, University of Amsterdam, Amsterdam, The Netherlands

Thomas Kitzberger Laboratorio Ecotono, Universidad Nacional del Comahue, INIBIOMA-CONICET, 8400 Bariloche, Argentina

Leif Kullman Department of Ecology and Environmental Science, Umeå University, 901 87 Umeå, Sweden

Guillaume de Lafontaine Département de biologie, Conservateur/Curator Herbier Louis-Marie, Québec, Canada

Francis E. Mayle School of GeoSciences, University of Edinburgh, Edinburgh, EH89XP, UK

Noemí Mazía Cátedra de Dasonomía, Universidad de Buenos Aires, 1417 Buenos Aires, Argentina

Juliano M. de Oliveira Programa de Pós-Graduação em Biologia, Universidade do Vale do Rio dos Sinos—UNISINOS, 93022-000 São Leopoldo, RS, Brazil

Marcela Moscol Olivera Department Paleoecology and Landscape Ecology, University of Amsterdam, 1098 XH Amsterdam, The Netherlands

Sandra C. Müller Departamento de Ecologia, Universidade Federal do Rio Grande do Sul, 91501-970 Porto Alegre, RS, Brazil

Randall W. Myster Department of Biology, Oklahoma State University, Oklahoma City, OK, USA

Gerhard E. Overbeck Departamento de Botânica, Universidade Federal do Rio Grande do Sul, 91501-970 Porto Alegre, RS, Brazil

Serge Payette Département de biologie, Université Laval, Québec, Canada

Gergő Persaits Department of Geology and Paleontology, University of Szeged, 6722 Szeged, Egyetem u.2-6, Hungary

Valério D. Pillar Departamento de Ecologia, Universidade Federal do Rio Grande do Sul, 91501-970 Porto Alegre, RS, Brazil

Andrés G. Rolhauser Departamento de Biología, Facultad de Ciencias Exactas, Físicas y Naturales, Universidad Nacional de San Juan, Av. Ignacio de la Roza San Juan, Av., Ignacio de la Roza (Oeste) 599, Rivadavia, San Juan, Argentina

Temuulen Tsagaan Sankey Idaho State University, Boise, ID, USA

Pál Sümegi Department of Geology and Paleontology, University of Szeged, 6722 Szeged, Egyetem u.2-6, Hungary

Bronwen S. Whitney School of GeoSciences, University of Edinburgh, Edinburgh EH89XP, UK

Hanne Wouters Department of Geography, University of Leicester, Leicester, LE1 7RH, UK

Chapter 1
Introduction

Randall W. Myster

1.1 Rationale

The Oxford English Dictionary (OED: www.oed.com) defines an ecotone as "A transitional area between two or more distinct ecological communities" and gives the etymology of both "eco" and "tone" as deriving from ancient Greek. The OED goes on to give this first usage by Clements (1904):

> Zonation in a habitat... The line that connects the points of accumulated or abrupt change in the symmetry is a stress line or *ecotone*... Ecotones are well marked between formations, particularly where the medium changes: they are less distinct within formations. It is obvious that an ecotone separates two different series of zones in the one case, and merely two distinct zones in the other.

Whereas terrestrial vegetation can be organized into large-scale biomes composed of species having common adaptations to conditions within each biome (Walter 1979), ecotones are still today thought of as dynamic border regions of abrupt transition between communities, ecosystems or biomes (Holland et al. 1991; Gosz 1993). Ecotones can have a major influence on those bordering ecosystems (e.g., material flow: Cadenasso et al. 2003) and indicate both local and global changes (Hufkens et al. 2009). As major components of landscapes, ecotones can (1) move unidirectionally through time although the rate may be variable (directional ecotones), (2) have no net change over relatively long periods but with movement of patches across the ecotone for a period of time (shifting ecotones), or (3) be relatively stable over scales of decades with little movement from one biome into another (stationary ecotones: Peters et al. 2006).

The dynamics of all three of these kinds of ecotones are controlled by the relative importance of abiotic drivers (e.g., climate), positive (e.g., facilitation) and negative

R.W. Myster (✉)
Department of Biology, Oklahoma State University, Oklahoma State University, Oklahoma City, OK 73107, USA
e-mail: rwmyster@gmail.com

R.W. Myster (ed.), *Ecotones Between Forest and Grassland*,
DOI 10.1007/978-1-4614-3797-0_1, © Springer Science+Business Media New York 2012

(e.g., competition) biotic feedback mechanisms, inherent abiotic constraints (e.g., parent material), and abiotic feedback mechanisms (e.g., organic matter accumulation: Peters et al. 2006). Ecotones can arise naturally, such as at a lakeshore, or they can be the product of human activity, such as clearing a forest to create an agricultural field (Myster 2007a). While ecotones can retain some characteristics of the ecosystems on either side, they often have their own unique properties such as physiognomic change, the existence of a vegetational mosaic, ecotonal species not found in the adjacent communities, and a spatial mass effect (Walker et al. 2003). Among these traits, the pronounced, sharp spatial change in vegetation compared to the bordering plant communities is often of greatest interest to ecologists.

Just as there are many kinds of ecosystems and communities, there are many kinds of ecotones such as forest/grassland, grassland/desert, tundra/forest, forest/marsh, land/fresh water, land/salt water, ground water/surface water, and forest riparian zones. As humans increasingly disturb and fragment the landscape in the future, ecotones will become even more common and important to the dynamics of the ecosystems on either side, redefining their boundaries and influencing their structure and function.

As past studies have shown, and as the chapters in this book will illustrate, an ecotone's structure, size, and scope can change considerable over the millennia, expanding and shrinking as other conditions also change. For example, today's ecotones are changing at a rate not seen for many years with woody plants increasingly invading grasslands worldwide (Scholes and Archer 1997). Indeed many believe that current climate change and temperature increases are major "drivers" determining those changes because ecotones are more sensitive to climate change than the biomes on either side, perhaps due to them being located at sharp climatic and/or species range boundaries. If so then ecotones can serve as early indicators of pending climate change before it is detected elsewhere (Kupfer and Cairns 1996; Allen and Breshears 1998). In this regard, the middle Holocene (approximately 8,000–4,000 year BP) may be a useful analog for current and future global warming. In addition to these direct effects of warming, climate change may also affect ecotones by altering the frequency of disturbances, e.g., fire and the strong wind events of Hurricanes and Tornados (Myster and Malahy 2010). Finally ecotones are important because they direct evolution and speciation (Simonetta 1990), helping to generate biodiversity worldwide (Smith et al. 1997).

Various techniques have been developed for the analysis of spatial and temporal patterns of ecotones (Hufkens et al. 2009) which will also be explored in this book. In addition to the importance of the patterns themselves, patterns are of interest because they may suggest the underlying mechanisms, tolerances, and processes that cause them, for example, fire, precipitation, grazing, solar radiation, temperature, wind disturbance, soil resources, canopy cover, litter, pathogens, crop signatures (Myster 1993, 2007a), livestock effects, seed rain, and species invasion which produce both the ecotone itself and those alterations in species distributions and other structures that define the ecosystems on either side. Furthermore, any of these ecotonal mechanisms, tolerances, or processes can involve negative feedbacks within the ecotone and positive feedbacks in the ecosystems on both sides (Stevens and Fox 1991). For example, plant-soil feedbacks can be either positive or negative, mediated by nematods and fungal pathogens (Peltzer 2001). Such information can have great utility for mangers who wish to restore and/or maintain biomes and their ecotones.

Many of the most common and most studied ecotones involve transitions between woody and nonwoody vegetation, often between forests and various kinds of grasslands (Butler et al. 2009; Holtmeier 2009) where tree–grass interactions are key to their dynamics (Scholes and Archer 1997). Whether the forests are tropical, temperate, or boreal, grasslands are often found naturally at their borders due to, for example, precipitation levels where trees can no longer maintain a positive carbon balance (Cairns and Malanson 1998). Tropical forests are often bordered by savannas (Hoffmann et al. 2004) and llanos (Marchant et al. 2006), temperate forests by prairies (Nelson and Hu 2008), steppes (Dulamsuren et al. 2008) and pampas (Mazia et al. 2001), and boreal forests by mires/bogs (Anschlag et al. 2008), tundra (Jia et al. 2006), and both alpine and temperate grasslands (Bergengren et al. 2001).

To fully understand forest-grassland ecotone dynamics, many approaches are needed. As the chapters in the book will illustrate, these include (1) forest and grassland sampling of both the current vegetation, soils and other ecosystem parameters but also those of the past, (2) field experimentation, and (3) modeling. In all these approaches, I suggest that the focus be on the process of tree invasion (Myster 1993, 2010) rather than the process of grass invasion because investigations need to be individualistic and it is difficult to identify individual grass plants in the field. This basic structural issue places strong limits on what can be easily known in other grassland studies, often forcing them to focus on the grassland community and ecosystem rather than on individual grass plants and grass populations (see chapters in Knapp et al. 1998).

Over the past few decades, a number of conceptual models have been proposed for plant community dynamics. These have been reviewed (e.g., Czaran and Bartha 1992; McCook 1994; Chave et al. 2002; Lortie et al. 2004; Myster 2012a) and various shortcomings have been discussed. I suggest these additional issues need to be incorporated into future plant community models: while (1) some models have included a spatial context, the real spaces that plants occupy and influence—and/or are influenced by—need to be added, (2) some models have included one-to-one plant replacements, the actual replacement dynamic should also include one-to-many, many-to-one and even one or many-to-none, and (3) the controls of these replacements have been vaguely discussed, they need to be strictly defined in terms of mechanisms and tolerances. This is assuming that plants already occupy spaces (i.e., it is not primary succession like found in landslides: Myster and Walker 1997), otherwise it would be a tree invasion and establishment without replacement.

Since plants are sessile organisms, and thus space is intimately related to how they grow and reproduce (e.g., Myster 2003; Myster and Pickett 1992b; Van Gardingen et al. 1997), I suggest that the best plant community models contain for each individual plant both a "neighborhood-space"—outside its plant body where it interacts with the environment through mechanisms and tolerances (sensu Turkington and Harper 1979)—and a "phyto-space" which is made up of both its necromass

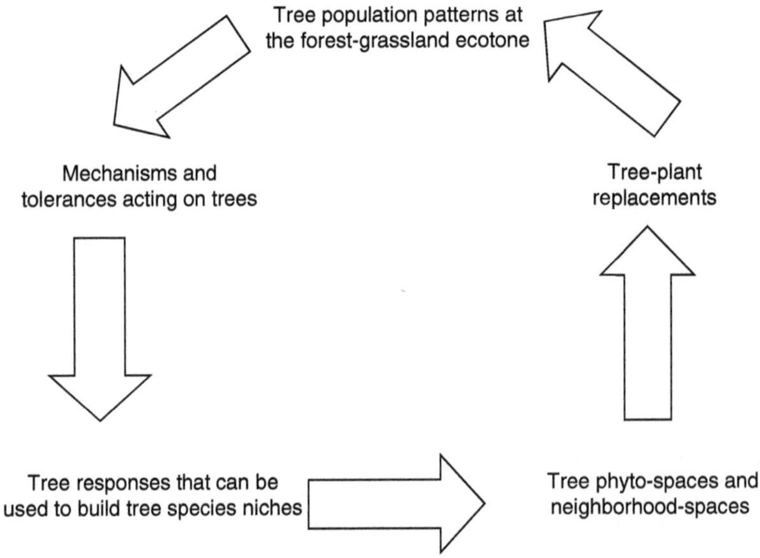

Fig. 1.1 A conceptual model of tree population dynamics at the ecotone between forest and grassland. Mechanisms and tolerances operating on individual trees lead to responses (e.g., survivorship, growth, allocation, architecture) which create spaces both within and without trees which change as trees are replaced by other plants (which can include other trees), creating tree population patterns. The new plants that compose those patterns are then acted on by mechanisms and tolerances as the cycle continues

and its biomass. While I agree with some past models that the fluid and continuous process of plant–plant replacement is at the core of plant community dynamics which generates all higher-level plant community patterns (Fig. 1.1: Busing 1996; Myster 2007b), I would focus mainly on replacements at the seed/seedling stage of the plant life cycle (Grubb 1977) where the neighborhood spaces and phyto-spaces of the seeds and seedlings "free-up" and are often reoccupied over time as replacements proceed.

I differ with past community models fundamentally, however, that conceptualize these spaces as equal among individual plants and, thus, present all replacements as one-to-one. In fact most replacements actually involve (1) a single individual plant being replaced by several other plants, (2) several plants being replaced by a single individual plant (common during succession: Myster 2007a), and/or (3) a plant or plants replaced by the growth of the still living plants which surround it (e.g., the thinning phase of forest stand development: Yoda et al. 1963). This complexity of the replacements can be seen by analysis of long-term permanent vegetation plots (see Myster 1993, 2007b and references therein).

I agree with those models that give replacements as caused by mechanisms and tolerances, but suggest that the niche (Gause 1934) of a plant species' does not pre-exist in communities "waiting to be filled" but instead is build from the responses of the individual plants of that species (Pickett and Bazzaz 1978; Parrish and Bazzaz

1982; Burton and Bazzaz 1991). Finally I suggest that an individual plant responds to a *mechanism* when it interacts directly with another living thing (e.g., dispersal, predation, pathogenic disease, herbivory) but *tolerates* a nonliving part of the environment by either responding to it directly (e.g., germination) or indirectly through another living thing (e.g., competition: Myster 2012b). As a plant fulfills its life cycles over time, these two spaces (phyto, neighborhood) change in response to the plant's genes and to the continuing reaction to mechanisms and tolerances. Eventually every plant will die. At which point, the neighborhood space will be available to other plants immediately, but for the phyto-space this will depend on local decomposition rates.

Taken together then I am suggesting that the best way to investigate forest-grassland ecotone patterns is to focus on these aspects of tree population dynamics (Mazia et al. 2001): (1) how mechanisms and tolerances affect tree seeds and seedlings producing responses, (2) how spaces both within and without these tree individuals change as a result, and (3) how those changes lead to trees replacing (or not replacing) grasses and forbs.

1.2 Case Study: The Cross Timbers

In the United States, a broad transitional boundary between the eastern deciduous forest and the tallgrass prairie of the southern Great Plains makes up a large and important ecotone called the Cross Timbers (Fig. 1.2: Dyksterhuis 1948; Engle et al. 1991; Johnson et al. 1972; Kroh and Nisbet 1983; Risser 1995). At one time, the Cross timbers may have covered nearly 8,000,000 ha (Kuchler 1964) from central Texas into eastern Kansas with most of it in Oklahoma (Duck and Flecher 1945). Humans have lived in the Cross Timbers for at least 7,000 years where Native Americans changed the Cross Timbers through hunting (including fire), gathering, fishing, and settlement (Francaviglia 2000). After white settlement much of the Cross Timbers was burned and then used for agriculture leaving today mostly secondary forest in those places where closed-canopy forest exists (Pogue and Schnell 2001), although small remnants of primary- or old-growth forest remain (Fig. 1.2: Therrell and Stahle 1998; Griffin et al. 2005).

Today the Cross Timbers ecotone is a mosaic characterized by patches of oak (*Quercus* sp.) dominated closed-canopy eastern deciduous forest, patches of tallgrass prairie, and patches containing large clones of shrubs (e.g., *Rhus copallina*) with a mix of small trees, grasses, and forbs (Arevalo 2002). Forest patches coalesce and increase in size as you move east across the Cross Timbers and/or into lower-lying, wetter areas. Likewise prairie patches coalesce and increase in size as you move west across the Cross Timbers and/or into higher, dryer areas (author, pers. obs). Shrubs and small trees are found in patches not dominated by forest or prairie species. Consequently, grass litter (Myster 2006), herbivory (Myster 2011), and competition (Weatherford and Myster 2011a, b; Myster 2012b) may be very important in controlling this ecotone.

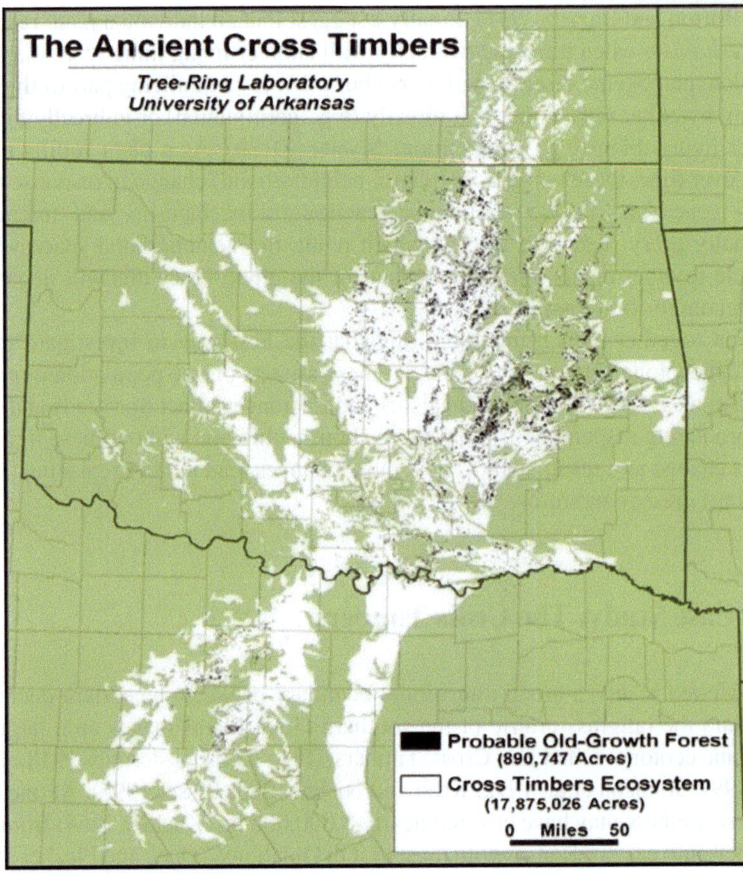

Fig. 1.2 Location of the Cross Timbers ecotone, between eastern deciduous broadleaf forest and tallgrass prairie in Oklahoma, Texas, Kansas, and Arkansas USA (reprinted with permission from D. W. Stahle, Tree-Ring laboratory, University of Arkansas)

The most dominant oaks are blackjack oak (*Quercus marilandica*) and post oak (*Quercus stellata:* Arevalo 2002), but Shumard oak (*Quercus shumardii*), bur oak (*Quercus macrocarpa*) and chinquapin oak (*Quercus muhlenbergii*) are also common (Petranka and McPherson 1979; Hoagland et al. 1999; Clark and Hallgren 2003). Other trees include red bud (*Cercis canadensis* L.), species of hickory (*Carya* sp. L.), slippery elm (*Ulmus rubra* Muhl.) and american elm (*Ulmus americana* L.). Understory trees, which may also associate with shrubs, include dogwood (*Cornus drummondii* C. A. Mey), red cedar (*Juniperus virginiana* L.), and mexican plum (*Prunus mexicana* S. Wats.). Dominant grasses include indiangrass (*Sorghastrum nutans*), little bluestem (*Schizachyrium scoparium*), switchgrass (*Panicum virgatum* L.) and big bluestem (*Andropogon gerardii*: Hoagland et al. 1999), and common forbs are milkweed (*Ascelepias* sp.), leadplant (*Amorpha* sp.), and various asters (Francaviglia 2000). Also within the Cross Timbers are streams where you find riparian trees such as willow (*Salix nigra*), cottonwood (*Populus deltoids*), hackberry (*Celtis occidentalis*), and pecan (*Carya illinoensis*).

Upland areas in the Cross Timbers have coarse textured soils derived from sandstone parent material and grassland areas tend to have finer textured soils (Johnson et al. 1972) but all soils have low fertility (Therrell and Stahle 1998). Furthermore, soils in the eastern part of the Cross Timbers have more sand and are more acidic, compared to the western areas (Francaviglia 2000). Temperatures range between 38 and −18 °C and annual precipitation ranges between 60 and 90 cm per year occurring mainly during the spring and summer (Hoagland et al. 1999; Francaviglia 2000).

I found at a Cross Timbers area near Lake Carl Blackwell (36′ 47″N, 96′ 25″W) that the average percentage of tree seed left in forest, prairie, and mixed patches after 9 days was 40% (Myster in press, similar to Haught and Myster 2008). Wind-dispersed white ash (*Fraxinus americana*) had more predation in the prairie patches compared to forest and shrub. American elm (*U. americana*) showed the opposite trend. For the bird-dispersed hackberry (*C. occidentalis*) and dogwood (*C. drummondii*), there was also more predation in the prairie and shrub patches compared to the forest. The bird-dispersed red cedar (*J. virginiana*) suffered little predation anywhere. Finally all three mammal-dispersed tree species, two Oaks (*Q. stellata*, *Q. marilandica*) and the pecan (*C. illinoensis*) showed the most predation in the forest and shrub patches compared to the prairie. In addition to these mechanistic results helping to fill out these tree niches and helping us understand tree replacement in this ecotone, seed predation may also alter fire intensity by removing biomass.

Also at Lake Carl Blackwell, 29% of tree seedlings set out in grass, red cedar, oak forest, and shrub microsites were alive after 1 year, mainly when protected from herbivores (Myster 2009b). Eastern redbud (*C. canadensis*), osage orange (*Maclura pomifera*), and lacebark elm (*Ulmus parrifolia*) all survived over 30%, rough-leaf dogwood (*C. drummondii*) at 25% and shumard oak (*Q. shumardii*) only at 5%. For those that survived, elm seedlings grew the fastest among all the test species but was hindered in the shrub patches with osage orange also growing well especially in grass patches (Myster 2009b). Another experiment at Lake Carl Blackwell found that Shumard Oak (*Q. shumardii*) seedlings without below-ground competition survived twice as much as those with it, seedlings in unburned areas survived up to four times more than those in burned areas, reduced below-ground competition lead to twice as much growth, and watering and placement under a shrub also lead to a twofold increase in growth (Myster 2009a).

In addition to the killing of oak and other tree seedlings, fire can facilitate grass growth by removing mats of dead grass biomass, thereby helping grass meristems get light and water. Advancing fires stop at the forest edge, however, because of the mineral and water content of the oak litter (Hoagland et al. 1999) and oaks may be able to allelopathically suppress other species by accumulation of their own tree leaf litter. Consequently global warming may be promoting grasslands by increasing fires, while managed fire suppression and introduction of large grazers should help trees invade.

The shrub effects I found conform well with other studies in the Cross Timbers that suggest that shrubs, after encroaching into the grassland asexually, can serve as "nuclei" for tree establishment, facilitating by shading and reducing herbaceous cover and vigor (Petranka and McPherson 1979), assuming that trees can avoid possible allopathic effects. This is a major similarity with tree invasion and establishment in eastern deciduous forests after agriculture, where shrub facilitation also included an increase in tree seed dispersal by birds (Myster 1993). Another similarity is the success of red cedar (*J. virginiana* L.) which may be a combination of its ease in dispersal, low seed predation and herbivory levels, and high resistance to moisture stress (Myster and Pickett 1992a, b, 1993; Myster 1994). Red cedar, however, cannot tolerate fire (Lassoie et al. 1983). Future experimentation in the Cross Timbers should include tree seed dispersal patterns (Myster and Pickett 1992b), litter effects on tree seed predation, germination, and seed losses to fungus (Myster and Pickett 1993; Myster 1994), and effects of herbivory, tree competition, and frost-heaving (Myster and McCarthy 1989; Myster and Pickett 1992a). Permanent plots are also needed to flesh out ecotone dynamics, spatial changes, and tree replacements.

Finally, due to the large number of severe wind events in Oklahoma, tornados are potentially important in structuring the Cross Timbers. I set up plots at Osage Hills State Park, seven miles west of Bartlesville, Oklahoma, USA (36′ 74″N, 98′ 17″W) in both an undisturbed and a Tornado touchdown area. I found that while clumping was observed in the undisturbed forest at the small spatial scale of 1–8 m and at the medium spatial scale of 30–36 m, after a Tornado clumping was only seen at small the spatial scales and then only for stems of medium size, for stems with branch damage and for stems that resprouted at or below 1 m height. In addition, the blowdown area showed domination by post oak (*Q. stellata*) with small stems (also seen by Shirakura et al. 2006), branch damage and stem resprouting above 1 m most common, had no trees that either lost their leaves without branch or stem damage or lived without some sort of resprouting, and showed a significant negative correlation between damage and resprouting (Myster and Malahy 2010). While a tornado may eliminate a forest's spatial heterogeneity and influence future growth and architecture, it can also increase tree persistence through resprouting. Because Tornados create large piles of woody debris, they can increase the risk and severity of future fires and fungal pathogen attack. These piles create refugia for small mammals, increasing their populations and consequent predation on seeds and seedlings.

In summary, Oak invasion into this ecotone is slowed down by mammalian seed dispersal (author, pers. obs.) as in old fields (Myster and Pickett 1992b) but can be helped by root sprouting if it as common in the ecotone as in the adjoining closed-canopy forest (Clark and Hallgren 2003). Likewise resprouting after a tornado may help trees to persist. If trees seeds can be dispersed into grassy patches, they will suffer low seed predation compared to shrub patches and underneath trees, and are likely to germinate (Myster 1994). After germination, those trees seedlings that can escape herbivory have a major advantage (Myster and McCarthy 1989). Oak seedlings growing under trees have a low mortality, but high mortality if growing under shrubs and in grass patches, perhaps due to intense blow-ground competition for water. Taken together then the slow oak invasion, and ecotonal change, that has been observed for decades (Dyksterhuis 1948; Engle et al. 1991; Johnson et al. 1972; Kroh and Nisbet 1983) may be due to poor oak dispersal and the lack of patches which have high survivorship and growth for both oak seeds and oak seedings. Suppression of fire may be the major reason behind any woody advance across this ecotone.

1.3 About This Book

The organizing theme in this book will be how forest-grassland ecotones have changed in the past, how they are changing today, and how they are likely to change in the future. Authors will investigate how the drivers for those changes (e.g., climate change, human disturbances, tree recruitment mechanisms, stress) differ now compared to the past where conceptual models will focus on the replacement, or barrier to replacement, of grasses/forbs by individual trees. This book has advantages over other ecotone books in that it focuses directly on forest-grassland ecotones, exploring the variation among forest and grassland types. The specific objectives are (1) to sample, plot and analyze this ecotone around the earth examining both its current spatial and temporal variation, and that of the past, (2) to relate the results of field experimentation along these ecotones, and (3) to use all available data to construct models of how these ecotones will change into the future.

The book is divided into three sections based on the kinds of grasslands that border each of three general forest types:

1. Ecotones that include these grasslands that border temperate forests: prairies, steppes, and pampas.
2. Ecotones that include these grasslands that border tropical forests: old fields, savannas, and llanos.
3. Ecotones that include these grasslands that border boreal and cloud forests: páramo, and other alpine grasslands.

After this introduction, we begin with exploration of changes in the Charpatian Basin during the Holocene and Weichsellan periods using pollen sequences within macrobotanical, charcoal, and phytolite remains. We then examine how synergistic relations between disturbance, climatic variability, and human impacts have dynamically shaped present forest-steppe ecotones in northern Patagonia, Argentina tree rings, repeat photography, and historical maps. We finish this first section with *Larix sibirica* water relations and photosynthetic performance in the steppe-forest ecotone of Northern Mongolia and the role of canopy disturbance, litter, and biotic interactions in limiting tree invasion in the pampa-forest ecotone in Argentina.

In the second section of the book, we start with an examination of a common and much studied old field ecotone in the Neotropics found in Puerto Rico as part of the National Science Foundation's Long-term Ecological Research (LTER) program. We then investigate how climate change interacts with fire frequency to affect woody plants of different functional types in the Brazilian savanna. We continue with a species replacement model which predicts changes to savanna and dry forest with conservation implications in Bolivia, and finish with using pollen data assigned to plant functional types and biomes along the llanos-forest ecotone in Colombia to prime a vegetative model that also uses climatic data to investigate the role of temperature and moisture.

In the third section of the book, we first use pollen analysis to document altitudinal migrations and the composition of the upper forest line of the páramo ecotone in Ecuador prior to deforestation and then use a stable carbon isotope analysis to explore how the competitive balance between C_3 and C_4 plants have changed in the ecotone over time. Then we map recent *Pinus uncinata* alpine tree line demographics in the Spanish Pyrenees in order to investigate recruitment trends and effects of human disturbance. Next we examine current demographic trends for *Pinus sylvestris* dynamics along the alpine treeline in Sweden and explore linkages to winter air and soil temperatures which include positive feedbacks and nonlinear responses. We conclude with how *Picea glauca* alpine tree lines change due to global climate change altered by wind exposure and proximity to the Atlantic coast.

We finish the book with a synthesis of specific results from each kind of ecotone first, followed by a view concerning forest-grassland ecotones in general, and finally a consideration of how ecotones and their research may change in the future, that is what lies ahead. Because ecotones will become more common in the future, and thus more important to the human race, we hope that this book can serve as a framework for that future research. Within the book chapters themselves, and the papers they reference, will be data that can be used by managers to more skillfully predict those changes but also to change the suggested social policies.

References

Allen CD, Breshears DD (1998) Drought-induced shift of a forest-woodland ecotone: rapid landscape response to climate variation. Proc Natl Acad Sci USA 95:14839–14842

Anschlag K, Broll G, Holtmeier F (2008) Mountain Birch seedlings in the treelineecotone, subartic Finland: variation in above- and below-ground growth depending on microtopography. Artic Antarctic Alpine Res 40:609–616

Arevalo JR (2002) Distribution of trees and saplings at the edge of Cross Timbers forests, Oklahoma USA. Nat Areas J 22:99–107

Bergengren JC, Thompson SL, Pollard D, Deconto RM (2001) Modeling global climate-vegetation interactions in a doubled CO^2 world. Clim Chang 50:31–75

Burton PJ, Bazzaz FA (1991) Tree seedling emergence on interactive temperature and moisture gradients and in patches of old-field vegetation. Am J Bot 78:131–149

Busing RT (1996) Estimation of tree replacement patterns in an Appalachian *Picea-Abies* forest. J Veg Sci 7:685–694

Butler DR, Malanson GP, Walsh SJ, ad Fagre DB (2009) The changing alpine treeline: the example of Glacier National Park, MT, USA. Developments in earth surface processes 12. Elsevier, Boston

Cadenasso ML, Pickett STA, Weathers KC, Bell SS, Benning TL, Carreiro MM, Dawson TE (2003) An interdisciplinary and synthetic approach to ecological boundaries. Bioscience 53:717–722

Cairns DM, Malanson GH (1998) Environmental variables influencing carbon balance at the alpine treeline ecotone: a modeling approach. J Veg Sci 9:679–692

Chave J, Muller-Landau HC, Levin SA (2002) Comparing classical community models: theoretical consequences for patterns of diversity. Am Nat 159:1–23

Clark SL, Hallgren SW (2003) Dynamics of oak (*Quercus marilandica* and *Q. stellata*) reproduction in an old-growth Cross Timbers forest. Southwest Nat 2:559–574

Clements FE (1904) Development and structure of vegetation. Report of the Botanical Survey of Nebraska. The Woodruff-Collins printing company. Lincoln, Nebraska vol 7. p 153

Czaran T, Bartha S (1992) Spatiotemporal dynamic models of plant populations and communities. Trends Ecol Evol 7:38–42

Duck LG, Flecher JB (1945) A survey of the game and fur bearing animals of Oklahoma. Bulletin no. 3, Oklahoma Game and Fish Commission, Norman

Dulamsuren C, Hauck M, Bader M, Osokhjargal D, Oyungerd S, Nyambayar S, Range M, Leuschner C (2008) Water relations and photosynthetic performance in *Larix sibirica* growing in the forest-steppe ecotone of northern Mongolia. Tree Physiol 29:99–110

Dyksterhuis EJ (1948) The vegetation of the western Cross Timbers. Ecol Monogr 18:276–325

Engle DM, Stritzke JF, McCollum FT (1991) Vegetation management in the Cross Timbers: response of understory vegetation to herbicides and burning. Weed Technol 5:406–410

Francaviglia R (2000) The Cast Iron forest: a natural and cultural history of the North American Cross Timbers. University of Texas Press, Austin

Gause GF (1934) The struggle for existence. Williams and Wilkins, Baltimore, p 163

Gosz JR (1993) Ecotone hierarchies. Ecol Appl 3:369–376

Griffin RD, Stahle DW, Therrell MD (2005) Repeat photography in the ancient Cross Timbers of Oklahoma, USA. Nat Areas J 25:176–182

Grubb PJ (1977) The maintenance of species richness in plant communities: the importance of the regeneration niche. Biol Rev Camb Philos Soc 52:107

Haught JA, Myster RW (2008) Effects of species, density, season and prairie-type on seed predation in Oklahoma. Am Midl Nat 159:482–488

Hoagland BW, Butler IH, Johnson FL, Glenn S (1999) The cross timbers. In: Anderson RC, Fralish JS, Baskin JM (eds) Savannas, barrens and rock outcrop plant communities of North America. Cambridge University Press, Cambridge, pp 231–245

Hoffmann WA, Orthen B, Franco AC (2004) Constraits to seedling success of savanna and forest trees across the savanna-forest boundary. Oecologia 140:252–260

Holland MM, Risser PG, Naiman RJ (1991) Ecotones: the role of landscape boundaries in the management and restoration of changing environments. Chapman and Hall, New York

Holtmeier F (2009) Mountain timberlines: ecology, patchiness and dynamics. Springer, New York

Hufkens K, Scheunders P, Ceulemans R (2009) Ecotones in vegetation ecology: methodologies and definitions revisited. Ecol Res 24:977–986

Jia GJ, Epstein HE, Walker DA (2006) Spatial heterogeneity of tundra vegetation response to recent temperature changes. Glob Chang Biol 12:42–55

Johnson KS, Branson CC, Curtis NM, Ham WH, Marcher MV, Roberts JF (1972) Geology and earth resources of Oklahoma. Educational publication I: Oklahoma Geologic survey. Steck-Warlick, Dallas, Texas

Knapp AK, Briggs JM, Hartnett DC, Collins SL (1998) Grassland dynamics: long -term ecological research in Tallgrass prairie. Oxford University Press, Oxford

Kroh GC, Nisbet J (1983) Some structural aspects of a western Cross Timbers forest in North Central Texas. Tex J Sci 35:42–45

Kuchler AW (1964) Potential natural vegetation of the conterminous United States. Special Publication 36. American Geographical Society, Washington

Kupfer JA, Cairns DM (1996) The suitability of montane ecotones as indicators of global climatic change. Prog Phys Geogr 20:253–272

Lassoie JP, Dougherty PM, Reich PB, Hinckley TM, Metcalf CM, Dina SJ (1983) Ecophysiological investigations of understory eastern red cedar in central Missouri. Ecology 64:1355–1366

Lortie CJ, Brooker RW, Choler P, Kikvidze Z, Michalet R, Pugnaire FI, Callaway RM (2004) Rethinking plant community theory. Oikos 107:433–438

Marchant R, Berrio JC, Behling H, Boom A, Hooghiemstra H (2006) Colombian dry moist forest transitions in the Llanos Orientalas—a comparison of model and pollen-based biome reconstructions. Palaeogr Palaeoclimatol Palaeoecol 234:44–88
Mazia CN, Chaneton EJ, Ghersa CM, Leon RJC (2001) Limits to tree species invasion in pampean grassland and forest plant communities. Oecologia 128:594–602
McCook LJ (1994) Understanding ecological community succession: causal models and theories: a review. Plant Ecol 110:115–147
Myster RW (1993) Tree invasion and establishment in old fields at Hutcheson Memorial Forest. Bot Rev 59:251–272
Myster RW (1994) Contrasting litter effects on old field tree germination and emergence. Vegetatio 114:169–174
Myster RW (2003) Seed regeneration mechanisms over fine spatial scales on recovering Coffee plantation and pasture in Puerto Rico. Plant Ecol 166:199–205
Myster RW (2006) Species-specific effects of grass litter mass and type on emergence of three tall-grass prairie species. Ecoscience 13:95–99
Myster RW (2007a) Post-agricultural succession in the Neotropics. Springer, New York
Myster RW (2007b) Introduction. In: Myster RW (ed) Post-agricultural succession in the Neotropics. Springer, New York, pp 2–32
Myster RW (in press) The role of seed predation in the maintenance of the Cross Timbers ecotone of Oklahoma, USA. Journal of plant interaction
Myster RW (2009a) Controls on Shumard Oak (*Quercus shumardii*) establishment into the Cross Timbers ecotone of Oklahoma: implications for restoration. Restor Ecol 17:893–899
Myster RW (2009b) Tree survivorship, growth, and allocation in the Cross Timbers ecotone of Oklahoma. Plant Ecol 205:193–199
Myster RW (2010) A comparison of tree replacement models in oldfields at Hutchenson Memorial Forest. J Torrey Bot Soc 137:113–119
Myster RW (2011) Above-ground vs. below-ground interactive effects of mammalian herbivory on tallgrass prairie plant and soil parameters. J Plant Interact 8:1–8
Myster RW (2012a) Plants replacing plants: the future of community modeling and research. Bot Rev 78:2–9
Myster RW (2012b) How does the precipitation gradient interact with common disturbances to structure prairies? J Plant Interactions 7:10–18
Myster RW, Malahy MP (2010) Spatial heterogeneity of Tornado damage and resprouting in the Cross timbers ecotone of Oklahoma. J Plant Ecol 3:157–163
Myster RW, McCarthy BC (1989) Effects of herbivory and competition on survival of *Carya tomentosa* (Juglandaceae) seedlings. Oikos 56:145–148
Myster RW, Pickett STA (1992a) Dynamics of associations between plants in ten old fields through 31 years of succession. J Ecol 80:291–302
Myster RW, Pickett STA (1992b) Effects of palatability and dispersal mode on spatial patterns of tree seedlings in old fields. Bull Torrey Bot Club 119:145–151
Myster RW, Pickett STA (1993) Effects of litter, distance, density and vegetation patch type on postdispersal tree seed predation in old fields. Oikos 66:381–388
Myster RW, Walker LR (1997) Plant successional pathways on Puerto Rican landslides. J Trop Ecol 13:165–173
Nelson DM, Hu FS (2008) Patterns and drivers of Holocene vegetational change near the prairie-forest ecotone in Minnesota: revisting McAndrews transect. New Phytol 179:449–459
Parrish JAD, Bazzaz FA (1982) Niche responses of early and late successional tree seedlings on three resource gradients. Bull Torrey Bot Club 109:451–456
Peltzer DA (2001) Plant responses to competition and soil origin across a prairie-forest boundary. J Ecol 89:176–185
Peters DPC, Gosz JR, Pockman WT, Small EE, Parmenter RP, Collins SL, Muldavin E (2006) Integrating patch and boundary dynamics to understand and predict biotic transitions at multiple scales. Landsc Ecol 21:19–33

Petranka JW, McPherson JK (1979) The role of *Rhus copallina* in the dynamics of the forest-prairie ecotone in North-central Oklahoma. Ecology 60:956–965

Pickett STA, Bazzaz FA (1978) Organization of an assemblage of early successional species on a soil moisture gradient. Ecology 59:1248–1255

Pogue DW, Schnell GD (2001) Effects of agriculture on habitat complexity in a prairie-forest ecotone in the Southern Great Plains of North America. Agric Ecosyst Environ 87:287–298

Risser PG (1995) The status of the science examing ecotones. Bioscience 45:318–325

Scholes RJ, Archer SR (1997) Tree-grass interactions in Savannas. Annu Rev Ecol Syst 28:517–544

Shirakura F, Sasaki K, Arevalo JR, Palmer MW (2006) Tornado damage of *Quercus stellata* and *Quercus marilandica* in the Cross Timbers, Oklahoma, USA. J Veg Sci 17:347–352

Simonetta AM (1990) On the significance of ecotones in evolution and on the significance of morphologic differences in primitive taxa: reflections and hypotheses. Ital J Zool 57:325–330

Smith TB, Wayne RK, Girman DJ, Bruford MW (1997) A role for ecotones in generating rainforest biodiversity. Science 276:1855–1857

Stevens GC, Fox JF (1991) The causes of treeline. Annu Rev Ecol Syst 22:177–191

Therrell MD, Stahle DW (1998) A predictive model to locate ancient forests in the Cross Timbers of Osage county, Oklahoma. J Biogeogr 25:847–854

Turkington R, Harper JL (1979) The growth, distribution and neighbour relationships of *Trifolium repens* in a permanent pasture. I. Ordination, pattern and contact. J Ecol 67:201–218

van Gardingen PR, Russell G, Foody GM, Curran PJ (1997) Science of scaling: a perspective on future challenges. In: van Gardingen PR, Foody GM, Curran PJ (eds) Scaling-up: from cell to landscape. Cambridge University Press, Cambridge, pp 371–378

Walker S, Wilson JB, Steel JB, Rapson GL, Smith B, King WM, Cottam YH (2003) Properties of ecotones: evidence from five ecotones objectively determined from a coastal vegetation gradient. J Veg Sci 14:579–590

Walter H (1979) Vegetation of the earth and the ecological systems of the Geo-biosphere. Springer, New York

Weatherford JA, Myster RW (2011a) Effects of species, nitrogen, and water on grass competition in Oklahoma. Prairie Nat 43:45–51

Weatherford JA, Myster RW (2011b) Interactive effects of simulated grazing, species and below-ground resources water on competitive outcome among three prairie grasses. J Torrey Bot Soc 138:107–119

Yoda K, Kira T, Ogawa H, Hozumi K (1963) Self-thinning in overcrowded pure stands under cultivated and natural conditions. J Biol Osaka City Univ 14:107–129

Part I
Temperate Forest-Grassland Ecotones: Prairies, Steppes, and Pampas

Chapter 2
Woodland-Grassland Ecotonal Shifts in Environmental Mosaics: Lessons Learnt from the Environmental History of the Carpathian Basin (Central Europe) During the Holocene and the Last Ice Age Based on Investigation of Paleobotanical and Mollusk Remains

Pál Sümegi, Gergő Persaits, and Sándor Gulyás

2.1 Introduction

A treeline is a boundary used for marking the edge of the habitat at which trees are capable of growing. In ecology an upper and lower treeline is generally highlighted (Ødum 1979) with an additional transitional zone (ecotone) found between the referred boundary and the adjacent open vegetation areas. Temperature besides precipitation is the major factor that controls the growth and sustainment of trees in an area. Nevertheless, several local ecological factors including such parameters as soil type, local vegetation, snow cover, elevation, geomorphology, rainshadow, gravity-induced mass movement, lightning, volcanic eruptions, wildfires caused by meteorite impact or wind shear can alter or prevent the sustainment of an arboreal vegetation in an area locally or regionally.

In areas of alpine or arctic, also recorded as upper treeline low temperatures generally reduce biomass production. This is strikingly disadvantageous for trees as their ecological potential is based on the storage of large biomass, rendering them ideal in ousting other competitive plants. The emergence of transitionary zones between woodlands and grasslands (ecotone) is generally controlled by the availability of water/humidity as a limiting factor. This boundary is mostly referred to as lower or dry treeline (Stevens and Fox 1991) in contrast to the alpine and arctic treelines. The area of the Carpathian Basin is highly unique from the point that both treeline zones are present.

Within the belt of the Alps, Dinarides and the Capathians, embracing the basin itself, an ecotone related to the upper or alpine treeline developed roughly at an elevation of 1,700–2,300 m due to mainly temperature constraints. Conversely, climatic and geomorphological endowments within the heart of the basin favored the

P. Sümegi (✉) • G. Persaits • S. Gulyás
Department of Geology and Paleontology, University of Szeged,
6722 Szeged, Egyetem u.2-6, Hungary
e-mail: sumegi@geo.u-szeged.hu; enric.batllori@ub.edu

R.W. Myster (ed.), *Ecotones Between Forest and Grassland*,
DOI 10.1007/978-1-4614-3797-0_2,

emergence of a forest-steppe ecotone along a dry or lower treeline. A treeline does not necessarily indicate that beyond the referred boundary no arboreal elements are present at all (Arno and Hammerly 1984). Dwarf bushes are characteristic elements of a tundra or alpine landscape. These plants however remain dwarf even among better conditions as well, say at lower altitudes.

Thus a timberline would rather mark the zone of closed woodland beyond which scattered tree stands may occur. However, these are generally small and often do not reproduce due to the unfavorable conditions. Nevertheless, this clearly illustrates the problem that the exact determination of the timberline is fundamentally based on how the concept of tree is understood. Based on size measures trees are generally considered to fall between 2–8 m in height. This approach is often used in distinguishing tree line and the actual woodland boundary as trees higher than 8 m are generally confined to the woodland areas below the upper and above the lower treeline. According to another definition, the criterion of woodland is based on the presence of a closed canopy with a closure rate of 30–40 % and a soil intervowen by roots of arboreal elements.

There is a drastic fall in total biomass in the transition zone between the actual woodland and the tree line from ca. 20 to 0.6 kg/m^2 due to the replacement of trees by smaller bushes and non-arboreal elements. The average annual biomass production is also significantly reduced; in the Alps this value is from ca. 1,000 to <200 g/m^2/year. The actual height of the tree line also varies from area to area in accordance with latitude, rainfall, and exposition. The referred alpine ecotone is relatively narrow corresponding to an altitudinal difference of ca.100–200 m. Anthropogenic influences however may enhance the widening of this ecotone (Myster 2007). Thus this zone can even reach a width of 300–400 m as well, as a result of extensive pastoralism in alpine pasturelands. The zone of the ecotone connected to either the arctic or the dry lower tree line can have extremely large spatial extent due to the emergence of a parkland type landscape studded by widely spaced stands of trees.

The area of the Great Hungarian Plains hosts a lower or dry treeline with an unusually wide ecotone, where the actual steppe zone is not uniform (Molnár et al. 2007a, b) but rather constitutes a mosaic of more or less isolated grassland patches (Sümegi 2011). Nevertheless, it must be kept in mind that the modern landscape is highly transformed due to intensive human activities present during the past 8,000 years. So the original vegetation must have been only partially preserved. One must turn to various environmental historical archives including pollen, charcoal, phytolith, and plant macrofossil data, as well as terrestrial mollusks if he or she wants to reconstruct shifts in the lower and upper treeline for the late ice age and the Holocene.

Investigations aimed at reconstructing fluctuations of the timberline are generally focusing on the analysis of lacustrine and marshland catchment basins and their paleoecological records. In case of an upper treeline, the analysis of cave or rock-shelter sediments is likewise informative. All these sequences yield rich pollen material in good preservation in general. On the other hand, wind-borne transportation of pollen grains to large distances largely hampers the identification of small-scale regional fluctuations in the tree line. Analysis of plant and animal macrofossils including plant tissue, seed and fruit remains, as well as charcoal and mollusks, not to mention plant opalites (phytoliths), which are preserved locally without transportation offer a better tool in such investigations. The spatial resolution and reliability of information gained from these archives is thus a couple of meters instead of kilometers in contrast pollen grains. This makes them ideal in small-scale spatial vegetation reconstructions. Not to mention the higher reliability of taxonomic identification of macro remains to the lowest level (species or even subspecies) in contrast to pollen grains.

In order to understand the development of *forest-steppe* ecotone at the boundary of woodland, grassland biotopes nestled in the heart of the Carpathian Basin, one must understand the present and past characteristics, as well as behavior of ecological factors responsible for the emergence and sustainment of the referred ecotone. To accurately capture past fluctuations of these factors we turned to archives of loess/paleosol sequences and catchment basins recording environmental fluctuations for the terminal part of the last ice age. For the Holocene the records preserved in smaller lacustrine and marshland basins, as well as rock-shelters were utilized. For archives we relied on data from pollen and plant opalites, as well as charcoal and macrobotanical remains. As an independent control, we turned to terrestrial mollusks, which on the basis of the ecological needs of taxa identified indirectly reflect the composition of the vegetation, as well as other nonliving (abiotic) ecological parameters such as humidity and temperature (Ant 1963; Horsak et al. 2010; Meng and Hoffmann 2009). Furthermore, the taxa used in our study are highly sensitive to vegetational shifts and that of the referred abiotic parameters.

2.2 Modern Woodland-Grassland Ecotone in the Carpathian Basin and Controversies Around Definitions

Identification of woodland-grassland ecotone connected to a lower or dry tree line in the heart of the Carpathian Basin, known and termed also as the *Pannonian forest-steppe*, is in a certain sense the so-called "holy grail" of Hungarian botanical research. Controversies between representatives of various Hungarian botanical schools and researchers hampered the correct definition of the concept and is a constant subject of scientific debates. Issues debated include aspects, age, and trajectories of vegetation development in general. The majority of the hard data comes from modern phytogeographic and botanical surveys and resulted in such a diversity of scientific views between schools that even the actual definition of a forest-steppe has not been uniformly accepted yet (Varga et al. 2000: p. 7).

These debates stem from the overwhelming pressure from a group of Hungarian botanists, who tried to decide the question on the basis of postulations from modern survey data alone regardless of the initial results of paleobotanical studies (Kerner 1863; Borbás 1900; Boros 1929; Rapaics 1918; Soó 1926, 1929; Tuzson 1915). Consequently, these early hypotheses must be regarded as working hypotheses rather than accepted scientific theories. It's highly unfortunate though that the uniform prevalence of these early postulations in the scientific mind is still distorting

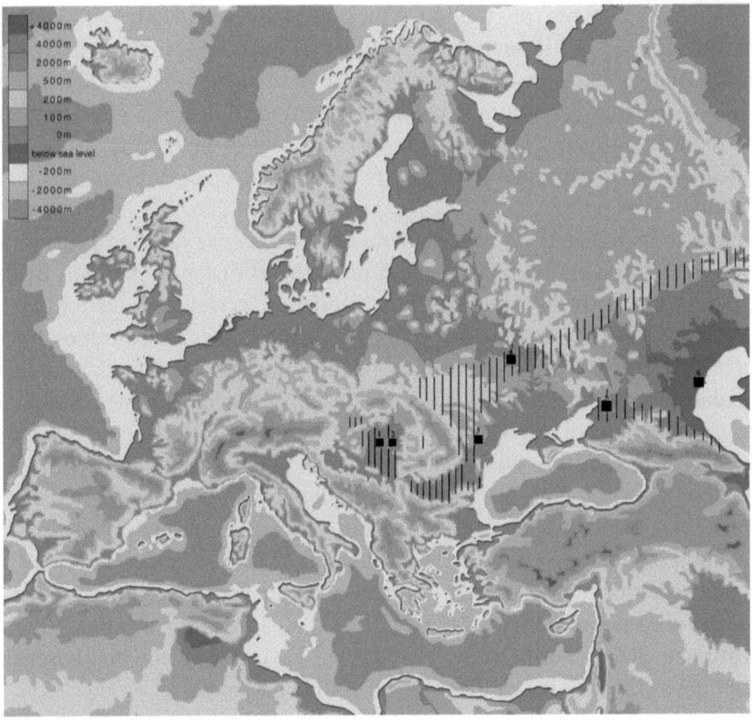

Fig. 2.1 The geographical setting of the Carpathian Basin in the heart of Europe and the extension of the European forest-steppe belt and the Pannonian forest-steppe with the location of meteorological stations referred (after C.S. Hammond's World Atlas, 1936)

the evaluation and acknowledgement of results of recent paleobotanical studies. Nevertheless, the true cause must be linked to the unique geographical setting of the Carpathian Basin found in the heart of Europe (Fig. 2.1). As a result of its location, in contrast to the rest of the European continent, a unique climatic interface developed here seen in the overlap of four major (Sümegi 1995, 1996, 2004a, b, 2005) climatic (Oceanic or Atlantic, Continental, Submediterranean and Alpine) influences (Fig. 2.2) (Réthly 1948; Borhidi 1961; Bacsó 1959; Zólyomi et al. 1992).

Besides the referred versatility in the climate, a similarly large-scale variance is observed in the geology (bedrock, soils) of the area (Trunkó and Földvári 1996) hosting a likewise highly heterogenous vegetation. As a consequence, oak and beech woodlands characteristic of Central and Western Europe are complemented by woodlands containing floral elements of the West and East Balkans as well. Not to mention the zone of the woodland-grassland ecotone subject of this paper covering an area of ca. 100,000 km^2 nestled in the heart of the basin (Fig. 2.3).

According to our present scientific knowledge, woodland-grassland ecotones generally emerge at the continental interface of large closed woodlands and adjacent grasslands, such as the case of the North American oak prairie. A major question to be

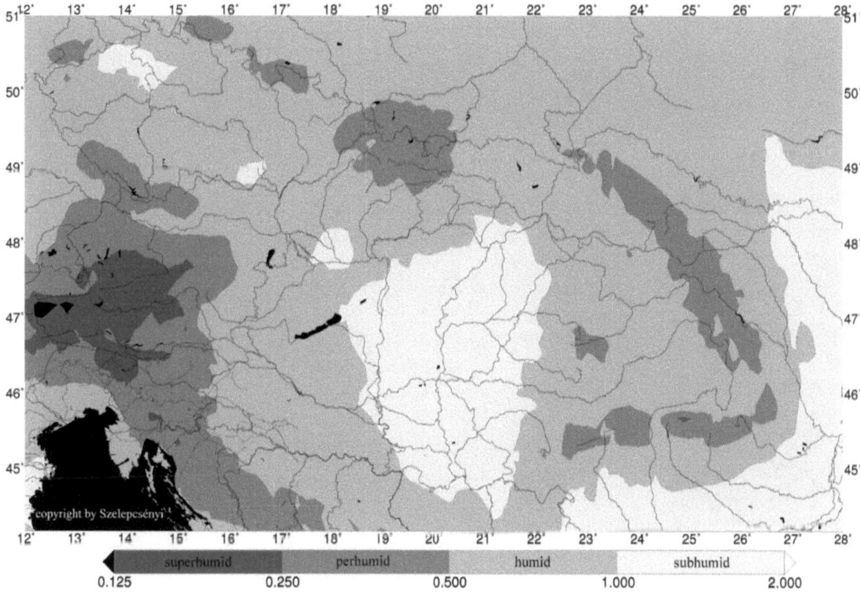

Fig. 2.2 The climate of the area of the Carpathian Basin (numbers correspond to POE values of Fig. 2.3) (Szelepcsényi et al. 2009)

decided in our case, whether or not the same holds true for the area of the Carpathian Basin. If yes, then what factors and how did they influence the emergence and sustainment of a woodland-grassland ecotone. In general three prevailing theories are available in the literature for the development of the so-called Pannonian forest-steppe.

According to the first, the forest-steppe found in the heart of the Carpathian Basin is an interim continuation of the Eastern European forest-steppe belt, which emerged as a result of the extreme drought literally exterminating arboreal elements in the area (Kerner 1863). This concept, held for over 150 years with only slight modifications (Soó 1926; Borhidi 1961, 2003; Varga 1989), considers the modern Pannonian forest-steppe as an independent westernmost island-like fragment of the European continental oak forest-steppe, which emerged at the transitionary climatic zone of closed woodlands and grasslands separated from the main belt by the ranges of the Carpathian Mts (Lavrenko 1980; Lavrenko and Karamyseva 1991). This concept regards the woodland-grassland ecotone in the basin as local manifestation of the climate-zonal vegetation belt stretching roughly 8,000 km from the heart of Europe to the Far East (Varga et al. 2000).

The second one considers so-called edaphic factors (soil, geomorphology) being responsible primarily for the emergence of forest-steppe ecotone in the basin (Zólyomi 1942, 1958, 1987). According to this concept, the heart of the Carpathian Basin is considered to be a part of the woodland belt from the point of climate-zonal classification. Thus the opening of closed woodland and the appearance of parkland and grassland patches must be attributed to local abiotic ecological factors. Some of

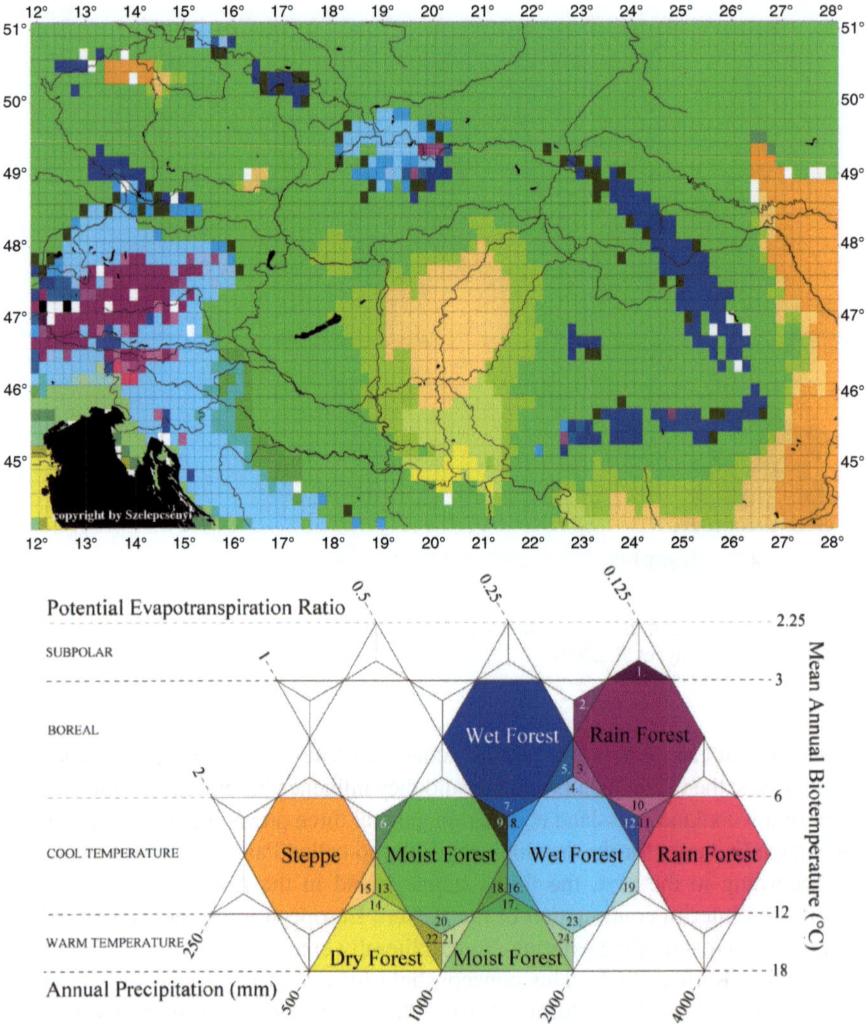

Fig. 2.3 The Holdrige type classification of the vegetation of the Carpathian Basin (Szelepcsényi et al. 2009)

these factors might have direct influence on vegetation development, such as the interaction of soils and groundwater in relation to local geomorphology (shallow soils, alkaline and calcareous sandy soils and low groundwater table).

Edaphic forest-steppes generally emerge along the northern margin of the steppe belt, where favorable soil, morphology and hydrological conditions enable the emergence of woodlands patches. The opposite may also be true when unfavorable local hydrology, soil, and morphology brings about the opening up of closed woodland creating isolated grassland patches. Unfortunately, these edaphic factors cannot explain the collective presence of phytogeographically, climatically exclusive

Continental and Submediterranean forest-steppe elements in the Carpathian Basin (Zólyomi 1952, 1957; Zólyomi and Fekete 1994; Borhidi 1956, 1966).

The third explanation claims human activities being responsible for the opening of the original woodland vegetation and the emergence of a Pannonian forest-steppe (Bernátsky 1914; Rapaics 1918). The first such disturbances are linked to the first farming cultures settling in the basin. Initial clearings gradually expanded as human activities intensified parallel with cyclical population growth. These activities thus contributed to the sustainment of a highly variegated, mosaic-like forest-steppe vegetation in the Great Hungarian Plains as early as prehistoric times. This concept is still widely acknowledged despite the contrasting results of recent paleoecological investigations which pointed to the emergence of natural steppe-forest-steppe communities here, covering several 100 km^2, as early as the terminal part of the last ice age and the opening of the Holocene (Sümegi 1989, 1995, 1996, 2005).

Conversely, these paleoecological data have corroborated the importance of first farming groups in enhancing the expansion of forest-steppe areas within an originally woodland environment and the creation of the modern forest-steppe in the basin.

2.3 Profiles Selected and Methods Applied in Modeling Woodland-Grassland Ecotone Shifts in the Carpathian Basin

Unfortunately, methods applied so far for studying the Quaternary deposits of the Carpathian Basin are extremely heterogenous starting from sampling through processing and evaluation, hampering correct correlation and comparison of findings and often yielding contrasting results (Sümegi 2001a). For the sake of accurate modeling, only those results and profiles were included in our work where certain common prerequisites of methodological congruence were present. Among these criteria were undisturbed sampling and reliable chronology based on radiocarbon and/or OSL, IRSL or uranium-series dating.

Pollen data using the spore tablet approach (Stockmarr 1971) was accepted alone. Similarly, results on plant macrofossil remains were included only if they had been evaluated using the QLCMA method developed in Southampton (Barber et al. 1994), or a modified version of it (Jakab and Sümegi 2004). During the phytolith extraction of samples deriving from loess/paleosol sequences, infills of archeological features, soils, marshlands, lakes, burial mounds and tells a method developed on the basis of personal experienced was systematically adopted (Persaits et al. 2008; Persaits 2010; Persaits and Sümegi 2011), which was primarily based on the methodologies postulated by Dolores R. Piperno and Marco Madella and colleagues (Piperno 2006; Madella et al. 1998). For mollusk studies, an aliquot of 5.7 kg of sediment sample was generally screen-washed to retrieve the shells. However, in case of archeological sites and where material was available and the investigation aims made it necessary sometimes several tons of sediment samples were processed using a sieve of 0.5 mm mesh size (Sümegi 2005).

2.3.1 The Climate-Zonal Hypothesis Put to the Test

One of the most significant pitfalls of this hypothesis is that macro-climate conditions generally characteristic of the steppe belt (Lavrenko and Karamyseva 1991: p. 254) have never fully developed and stabilized in the Carpathian Basin during the Quaternary (Réthly 1933; Bacsó 1959; Dobosi and Felméry 1977). Winters are not cold and summers are not dry enough in the referred study area to enable the long-term macroclimatic sustainment of such vegetation (Fig. 2.4), as it is clearly seen in the map of the western, European margin of the Eurasian steppe belt as well (Figs. 2.4 and 2.5). Thus proving the macroclimatic nature of the ecotone in the area of the Carpathian Basin between the belt of European woodland and Eurasian steppe is quite ambiguous.

Climatic influences characteristics of the Köppen type steppe zone (BS) (Köppen 1923) are traceable in the heart of the Carpathian Basin alone and appear with a frequency of two to three times during a decade. Furthermore, their development is by no means regular and cyclical but rather sporadic (Dobosi and Felméry 1977). So by using the Köppen type climate classification system, the area of the Carpathian Basin falls mainly into the category Cfb and subordinately into Cfa. The marginal hilly and montane areas belong to the category D according to this system (Fig. 2.6) (Szelepcsényi et al. 2009).

Conversely, in order to better highlight the climatic background of the woodland-grassland ecotone system, which is present in the heart of the Carpathian Basin, the vegetation classification system of Holdridge (Holdridge 1947, 1967) is better suited than the climate classification system of Köppen (Szelepcsényi et al. 2009). According to the Holdridge classification (Fig. 2.3), the major part of the basin is put to the transitional category found between those of cold, temperate grasslands, cold, temperate humid woodlands and warm, temperate dry woodlands, where the first (steppe) and the last categories (dry woodland) also turn up climatically in the form of scattered patches. This grassy area forming an ecotone between the actual grasslands and dry and humid woodlands corresponds to the Pannonian forest-steppe vegetation of the Great Hungarian Plains.

From the point of geographical setting, the area of the Carpathian Basin is found between the latitudes 45–47°N. The affinity of the study area with the western part of the Eurasian forest-steppe belt is even more apparent when the climate diagrams presented on Fig. 2.1 are compared with those of Eastern Europe along the latitude 46°N (Chisinau, Kiev, Rostov) (Figs. 2.4 and 2.5). Nevertheless, it is also clearly visible that there are major differences regarding the frequency of strong cold winter episodes and annual average temperature fluctuations (continentality), which are considered as the most important control of arboreal vegetation development. The strong continentality observable in Eastern Europe occurred with a frequency of only 30 % during the past century in the Carpathian Basin. Despite these climatic differences, the heart of the Carpathian Basin and the Great Hungarian Plains itself has strong affinities with the forest-steppe zone of Eastern Europe and thus must be regarded as an isolated fragment of it.

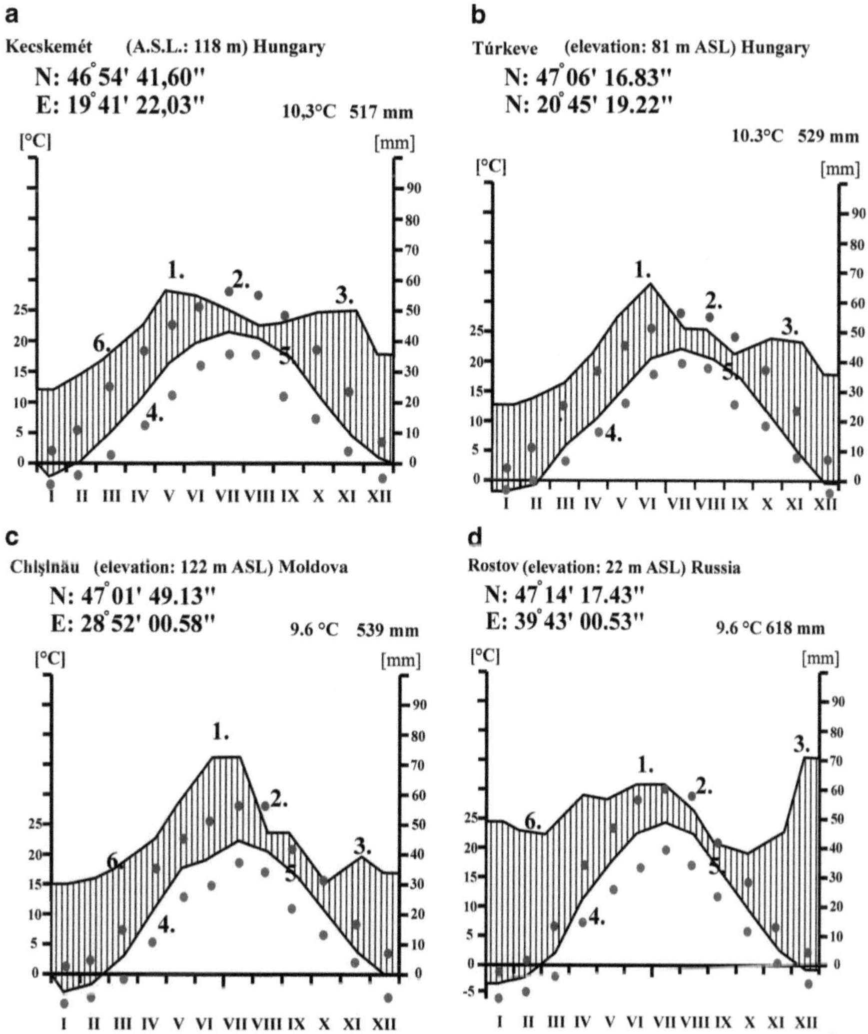

Fig. 2.4 Walter-Lieth diagrams of selected stations located between the latitudes 46–47°N (**a**) Túrkeve, (**b**) Kecskemét, (**c**) Chisinau, (**d**) Rostov. 1. First rainfall maximum in early summer, 2. Average high temperature, 3. Second rainfall maximum in fall, 4. Average low temperature, 5. Daily mean temperature, 6. Precipitation (mm)

As similarly to the presented climatic parameters, an unambiguous floral relationship exists between the areas of Eastern Europe and the Carpathian Basin (Berg 1958; Borhidi 1966; Fekete 2000; Jakucs et al. 1959; Molnár et al. 2007a, b, 2008; Soó 1958; Zólyomi 1957).

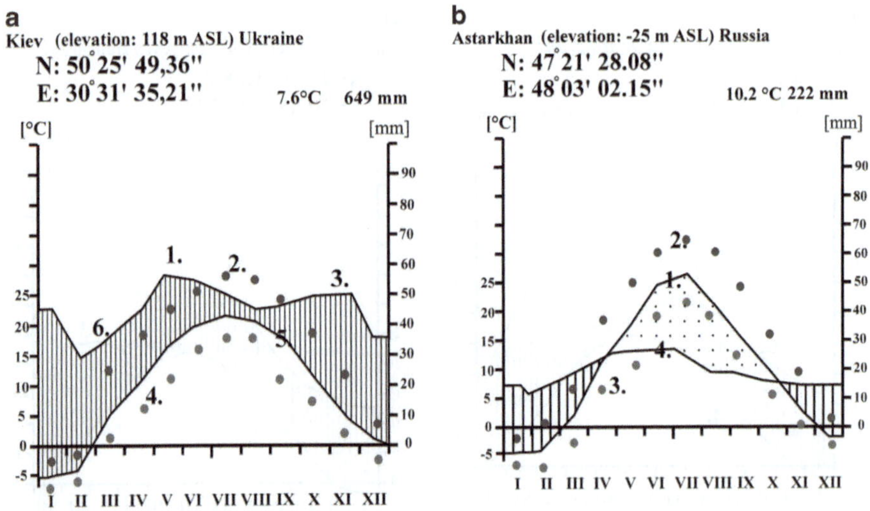

Fig. 2.5 Walter-Lieth diagrams of selected stations of the Eurasian steppe belt between the latitudes of 46–50°N (**a**) Kiev, (**b**) Astrakhan. 1. Daily mean temperature, 2. Average high temperature, 3. Average low temperature, 4. Precipitation (mm)

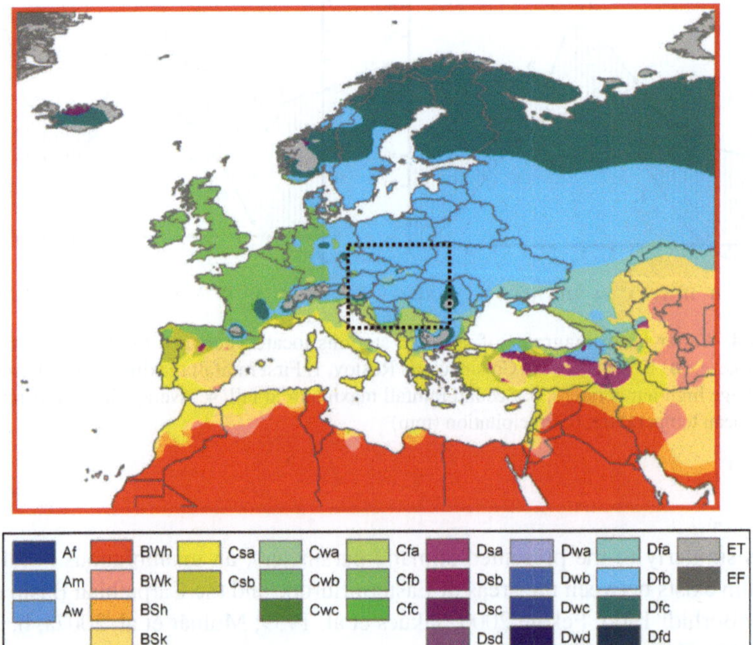

Fig. 2.6 Köppen type climate zonation of the study area of the Carpathian Basin

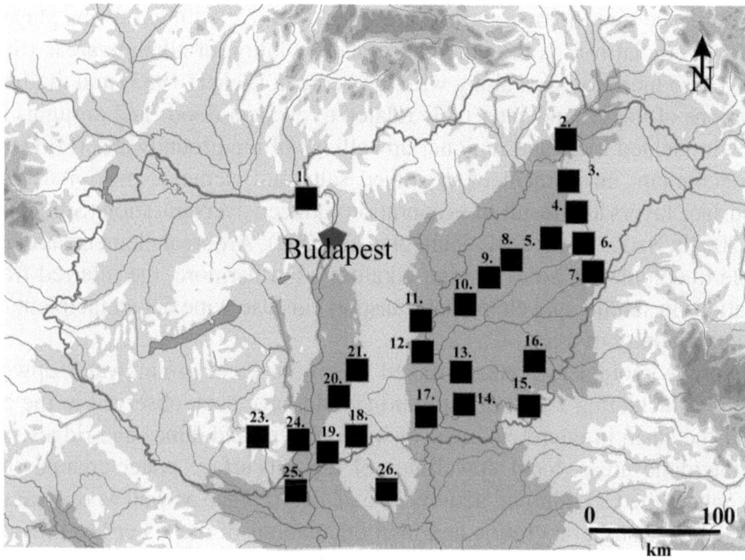

Fig. 2.7 The location of loess/paleosol profiles used for the modeling (after Sümegi 2005, and new data). 1. Basaharc, 2. Tokaj, 3. Hajdúnánás, 4. Hajdúböszörmény, 5. Hajdúszoboszló, 6. Debrecen, 7. Pocsaj, 8. Püspökladány, 9. Karcag, 10. Kisújszállás, 11. Lakitelek, 12. Tiszaalpar, 13. Szentes, 14. Hódmezővásárhely, 15. Makó, 16. Békéscsaba, 17. Szeged, 18. Madaras (Hupuczi and Sümegi 2010), 19. Katymár, 20. Császártöltés, 21. Kecel, 22. Kiskunfélegyháza, 23. Szászvár, 24. Dunaszekcső, 25. Zmajevac, 26. Crvenka (Zech et al. 2010; Stevens et al. 2011)

2.3.2 *Testing the Model of Edaphic Ecological Factors*

The unique salt transport observed in connection with numerous local upwellings of the groundwater in several parts of the Great Hungarian Plains (Sümegi 1989; 1997; Szöőr et al. 1991) and the related alkalization of these areas (Sümegi 1989; Sümegi et al. 1999), as well as the precipitation of freshwater carbonates in some of these alkaline lacustrine systems (Miháltz 1954; Molnár 1961; Mucsi 1963; Sümegi et al. 2005a) are the best evidences available to justify the importance of so-called edaphic factors in the development of the unique forest-steppe vegetation. A relative high-stand of the groundwater and concomitant enrichment of the root zone and near-surface layers in dissolved salts, as well as the presence of large inundated areas related to the elevated zone of saturation in certain parts of the Great Hungarian Plains largely limited the sustainment and spreading of arboreal vegetation giving way to non-arboreal elements.

In other parts mostly characterized by a complex elevated geomorphology, like those of loess plateaus, loess-covered island-like lag-surfaces on the floodplain (Sümegi 1989, 2003, 2004a, 2005; Sümegi et al. 2005b), the extremely low level of groundwater, often below 5 m preceding the period of modern canalizations and river regulations (Rónai 1961), in times of extensive drought hampered the colonization of arboreal elements and thus contributed to the spread of non-arboreal vegetation. The morphology and the presence of so-called microforms played and still play a crucial role in the emergence and evolution of alkaline areas (Strömpl 1931). These morphologies along with the bedrock (Scherf 1935) and the frequent fluctuations of the groundwater table related to cyclical floods and droughts on the floodplain yielded local and regional upwellings of the groundwater (Fórizs 2003) and concomitant enrichment of dissolved salts and elements precipitated in the near-surface layers as a result of evaporation-induced oversaturation (Sümegi 1989, 1997, 2004b; Szöőr et al. 1991; Sümegi et al. 2000). Arboreal elements are usually not able to colonize in areas of high alkalinity. Furthermore, the referred frequent fluctuations in salinity and elemental transport can also wipe out an incipient, rather fragile community of trees.

Numerous geological features linked to similar local groundwater upwellings and highly fluctuating zones of saturation could have been observed in the central areas of the basin in interdune areas of wind-blown sand dunes and ridges. Within these interdune areas, local upwellings of the subsurface water created shallow ponds, where the CO_2 binding activity of calcareous algae as well as oversaturation connected to strong evaporation and high fluctuations of the water level resulted in the precipitation of freshwater carbonates in basinal and near-shore areas (Mucsi 1963; Molnár et al. 1981).

The resulting carbonate beds also limited the areas suitable for the colonization of trees in bedrock of wind-blown sand. According to our recent paleoecological findings presented in the following chapters, alkalization was continuously present in the area for the past 40 kys and thus must have been an important edaphic factor contributing to the development of steppe, forest-steppe areas of the Great Hungarian Plains. The formation of freshwater carbonates could have been attested from the opening of the Holocene. Although constrained to certain areas, the responsible ecological components must have been present as well continuously since the terminal part of the ice age and must have affected the development of the vegetation besides alkalization.

Based on our findings (Sümegi 1989, 2004b), the geochemistry of the bedrock, the depth and fluctuations of the groundwater table, and the local geomorphology must have been the most influential ecological factors controlling and being responsible for the modern pattern of forest-steppe development in the area of the Carpathian Basin as we shall see in the following chapters. The mutual interaction of these factors must have contributed to the opening of the woodland vegetation locally and regionally by either hampering colonization of the arboreal elements or wiping them out. However, these factors alone cannot be blamed for the development of a unique forest-steppe covering several 100 km^2 in the study area.

2.3.3 Testing the Idea of Human-Induced Ecotone Development

The hypothesis of human-induced forest-steppe development was first postulated at the opening of the twentieth century in relation to modern paleobotanical and phytogeographical surveys implemented in the area of the Carpathian Basin. Archeology and the identification of potential influences of archeological cultures were highly incipient at this stage. Although the oldest archeological excavations date back as far as 1860 in the Great Hungarian Plains, the question of influences of settling cultures on their environment, including the vegetation as was rather subordinate at this time. Similarly, the application of paleoecological methods including archeobotany and archeozoology was also negligible. The lack of reliable, quantitative data regarding the number of sites, the size of the population further hampered the correct answering of such hypothetical questions.

Nevertheless, it is important to note, that it was this period when the first major war (WW1) restricted wide-scale scientific studies on the one hand. On the other hand, the first negative outcomes of the nineteenth century river regulations also began to turn up during this period fueling notions of devastating human activities on local ecosystems. The general lack of archeological data on local cultures hampered the deduction of correct postulations regarding the influence of humans on the environment. Thus often the findings of similar scope studies related to distant high cultures of Egypt, Mesopotamia, and the Near East were erroneously adapted to the area of the Carpathian Basin as well. These initial postulations were so influential that they still influence the acknowledgement of the results of new archeological, paleobotanical, and archeozoological studies using correctly obtained hard data in Hungarian environmental historical research.

The emergence of new archeological theory and methodology in Hungary (Kosse 1979; Sherratt 1982, 1983; Jerem et al. 1985, 1986; Bökönyi 1992) was accompanied by the introduction of complex paleoecological investigations of catchment basins serving as environmental background areas local cultures (Sümegi 2001a, b, 2003, 2004b, 2005, 2007, 2010, Sümegi et al. 2005a, 2009, 2011). A combination of the correctly dated archeological and paleoecological results enabled the clarification of long-term spatial and temporal interaction of man and environment in the basin. Despite the promising outcome of these first efforts, findings of archeobotanical studies from archeological sites are still not fully disseminated in modern Hungarian botanical studies aimed to elucidate the development of the forest-steppe vegetation of the Great Hungarian Plains.

2.4 The Vegetation History of the Great Hungarian Plains as Inferred from the Evaluation of Quaternary Paleoecological and Environmental Historical Data

The original ecosystem of the Great Hungarian Plains hosting the Pannonian forest-steppes was fundamentally altered during the middle of the nineteenth century as a result of human intervention into probably the most important ecological component of the landscape: the fluvial system of the Tisza and its watershed. As a result of the extensive river regulation and flood protection measures, a 4,000 km-long dike system and a canal system of 22,000 km was established. Thanks to these activities in addition to the artificial cut of selected meanders the complete hydrological system of the Great Hungarian Plain was transformed seriously affecting not only the landscape and the vegetation but also the trajectory of regional and local climate as well.

The active floodplain originally covering an area of ca. 30,000 km^2 was reduced to a mere 3,000 km^2. The dried up floodplain areas were transformed into arable and pasturelands. The major part of the gallery forests fringing the riverbed was logged down to reduce costs of the regulation measures. Not only the majority of the original floodplain vegetation was destroyed, but by modifying the natural hydrological link between the river and the floodplain the hydrological system transporting the excess waters of the surrounding mountains coming from the precipitation was fundamentally altered as well during the past 160 years.

As only a single cartographic map sequence, prepared on a nongeometric basis, is available for the landscape of the Great Hungarian Plains preceding the nineteenth century river regulations (Map of the first Austrian Military Survey, 1782: Timár et al. 2008), the only way to track the vegetation history including those of forest-steppes is to turn to environmental historical and paleoecological records.

In order to highlight the vegetation history of the Carpathian Basin for the period from the terminal part of the last ice age, results of extensive paleoecological investigations implemented on 20 selected loess/paleosol sequences, including charcoal analysis (Rudner and Sümegi 2001), mollusk analysis (Sümegi 2005; Sümegi and Krolopp 2002; Hupuczi and Sümegi 2010; Molnár et al. 2010), and phytolith analysis (Persaits and Sümegi 2011) has been adopted (Fig. 2.7). In addition, findings of similar type of studies (palynology, mollusk and plant macrofossil analyses) deriving from radiocarbon-dated, undisturbed core sequences of 20 catchment basins have been utilized (Fig. 2.8). These data have been complemented by records of mollusk, pollen and phytolith analysis implemented on nearby archeological sites (Fig. 2.8). A combination of data enabled us to track not only temporal but also spatial differences in the trajectory of vegetation evolution.

2.4.1 Vegetation Development During Last Ice Age

Based on our findings, two major paleoenvironmental regions could have been identified within the area of the Carpathian Basin, which were characterized by fundamentally different endowments and developmental histories during the course of cyclical coolings characterized by loess formation and intervening warm-ups resulting in the temporary cessation of dust accumulation (Sümegi 1996, 2005).

The resolution of our records enabled us to capture the environmental history of the northern region for the past 70 kys (Sümegi and Hertelendi 1998). The warm-up periods were all characterized by the general spread of spruce. Within the emerging open parkland type spruce woodland ecosystems (Willis et al. 2000), sheltered spots hosted stands of Scots pine and hornbeam (Rudner and Sümegi 2001), while the

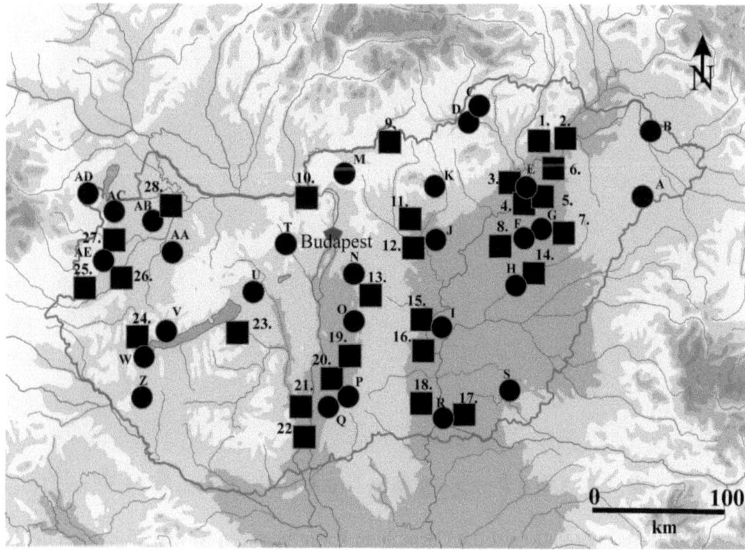

Fig. 2.8 The location of catchment basins and archeological sites used for the modeling. A = Bátorliget, marshland, B = Csaroda, Nyíres—lake, C = Kelemér, Kis-Mohos, D = Kelemér, Nagy-Mohos, E = Polgár, F = Hortobágy—Halasfenék, G = Nagyiván—Kunkápolnás, H = Ecsegfalva—Kiritó, I = Tiszapüspöki—Karcsúér, J = Jászberény, Megyes-forest, K = Sirok, Nyírjes peatbog, M = Nagybárkány, Nádas Lake, N = Ocsa—Selyemrét, O = Izsák, Lake Kolon, S = Kardoskút, Fehér (*White*) Lake, T = Sóskút, Benta creek, U = Sárkeszi—Sárrét, V = Balatonederics—Lake Balaton, W = Fenékpuszta—Little-Balaton, Z = Kaszó, Lake Baláta, AA = Mezőlak—Szélmező, AB = Osli—peatland, AC = Hidegség, Lake Fertő, AD = Sopron—Kismalomtó pond, AE = Zanat, Surányi creek. 1. Tarcal, Citrom mine, 2. Bodrogkeresztúr, Henyetető, 3. Polgár, Csőszhalom, 4. Polgár Kenderföld, 5. Polgár, Ferencihát, 6. Tedej, Lyukashalom (kurgan), 7. Hortobágy, Szálkahalom (kurgan), 8. Nagyiván, Ecsehalom (kurgan), 9. Salgótarján, Tesco parkoló, 10. Pilismarót, Pálrét, 11. Jászfelsőszentgyörgy, Szúnyogos, 12. Jászberény, Káplártanyag, 13. Dunatetétlen, Oltóhalom (kurgan), 14. Ecsegfalva, Kiritó, 15. Tiszapüspöki, karancspart, 16. Felgyő, road excavation, 17. Maroslele, Pana, 18. Kiskundorozsma, Vágóhíd, 19. Dusnok, road excavation, 20. Fajsz, Garadomb, 21. Alsónyék, 10b site, 22. Bátaszék, Mérnöktelep, 23. Zamárdi, Kútvölgyi-dűlő, 24. Keszthely, Fenékpuszta, 25. Szombathely, Oladi-plató, 26. Szombathely, Zanat, 27. Vát, 86-os road excavation, 28. Tóköz, middle age trenches

higher areas were populated by stands of Swiss pine and larch. Water bank areas hosted populations of willow, hairy birch, and green alder (Fig. 2.9).

Moving towards the centre of the Great Hungarian Plains arboreal elements become less frequent components of the vegetation yielding a special tree-steppe ecosystem characterized by the dominance of non-arboreal elements even during the warm-ups as well (Fig. 2.9). At the same time, the inner, drier areas, characterized by a prevalence of grassland-type vegetation also hosted scattered stands of Scots pine, juniper, alder, birch, and willow (Sümegi et al. 2005c), restricted mainly to the banks of watercourses. One of the driest areas of the Great Hungarian Plains, the Hortobágy, which is extensive alkaline grassland today, was hosting some alkaline elements (*Artemisia maritima*, *Sueda*) even during this early period as well.

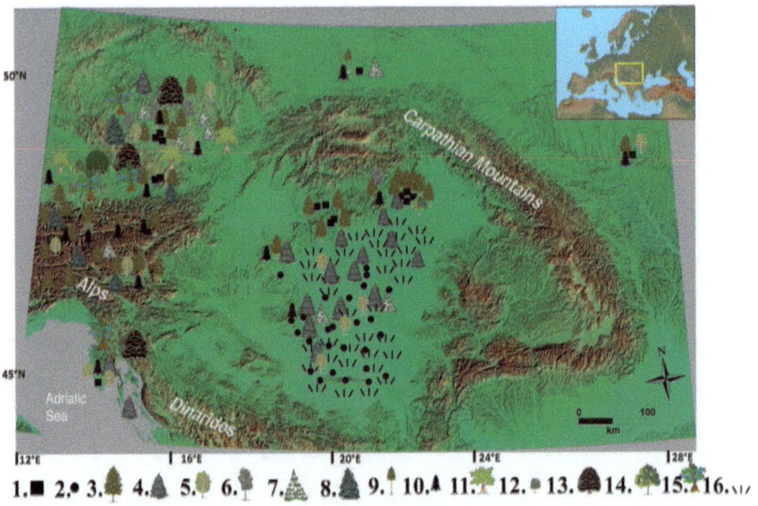

Fig. 2.9 The inferred vegetation of the Carpathian Basin between 32 and 25 kys cal BP. 1. Upper Palaeolithic site, 2. Warmth-loving, open forest-steppe—steppe dweller snails, *Granaria frumentum* site, 3. Spruce (*Picea*) remains, 4. Scotch pine (*Pinus sylvestris*) remains, 5. Birch, willow, alder (*Betula, Salix, Alnus*) remains 6. Swiss pine (*Pinus cembra*) remains, 7. Fir (*Abies*) remains, 8. Juniper (*Juniperus*) remains, 9. Larch (*Larix*) remains, 10. Pine (Pinacae) remains, 11. Oak (*Quercus*) remains, 12. Hazel (*Corylus*) remains, 13. Elm (*Ulmus*) remains, 14. Hornbeam (*Carpinus*) remains, 15. Beech (Fagus), 16. Herbs (Non-arboreal plants)

However, the banks of minor watercourses also hosted a loose canopy gallery forest composed of Scots pine, alder, and birch here as well.

The same warm-up period was characterized by similar tree-steppe vegetation in the region south of the centre of the Great Hungarian Plain. However, this similarity was true for the structure of the vegetation alone. The taxonomic composition of the flora and the fauna was fundamentally different (Fig. 2.9). The dry loess plateaus of the southern part of the Carpathian Basin hosted a mix of arboreal elements (Scots pine, birch, and fir) during the referred warm-ups (Rudner and Sümegi 2001; Sümegi 2005). However, as shown by data from charcoal and phytolith analysis, the presence of arboreal elements was spatially highly dissected and highly subordinate compared to that of non-arboreal forms. Conversely, areas with an ideal hydrology and higher groundwater table, like along the banks of rivers and watercourses, lush mixed gallery forests emerged characterized by a 70–80 % closed canopy. Thus the dry tree-steppe vegetation of the Great Hungarian Plains studded by scattered stands of trees was highly dissected at the regional or meso level by closed-canopy gallery forests along the riverbanks during the ice age.

Furthermore, in accordance with the local morphological, geological, and hydrological endowments the emergence of full hydro series from the riverbed up to the elevated dry plateaus of the floodplain could have been attested. This hydro series started off with a belt of bulrush (*Typha*) and reed (*Phragmites*) right at the littoral zone and was exchanged by tall herb meadows (*Thalictrum*, *Sanguisorba*, *Angelica*,

Campanula, *Filipendula*) and arctic fens (*Armeria maritima*, *Selaginella selaginoides*, *Pinguicaula*) moving inwards the floodplain besides stands of willow still hosted by the littoral zone. The interfluvial ridges were populated by representatives of the taxa *Hippophaë rhamnoides*, *Ephedra*, and cold continental steppe communities (Poaceae, *Artemisia*, heliophyte compositae). The presence of some alkaline plants such as *A. maritima* refers to the emergence of scattered spots of alkaline soil on the floodplain at the end of the Pleistocene.

The pollen spectrum of the catchment basin of White Lake at Kardoskút (Sümegi et al. 1999) shows high relative frequencies of arboreal pollen types (70 %) with pine, spruce, and birch being the most abundant in this interstadial phase. This indicates the presence of boreal forest-steppe vegetation around the referred lake. Forested areas around the lake likely included *Pinus sylvestris*, *Pinus mugo*, *Pinus cembra*, and *Larix*. In addition, *Pinus*, *Picea* with frequencies around 10 % in this zone suggest the local occurrence of habitats with more humid conditions, likely in the floodplain zone. The sporadic appearance of broad-leaved deciduous tree pollen grains like *Betula*, *Quercus*, *Carpinus betulus*, *Acer*, *Populus*, *Ulmus* is also characteristic for this zone and indicates scattered presence of these trees in the lowland pine woodlands. Away from the riverbank the proportion of arboreal elements was significantly reduced parallel with a decrease of the groundwater table. In the higher elevated areas, their presence is highly subordinate. An increase in dry grassland areas was accompanied by the spread of xerophylous, grassland mollusk taxa. It was this zone, where the representatives of the character species of modern Pannonian steppe/forest-steppe areas *Granaria frumentum* first turn up in the southern parts of the basin. The accessory fauna is very similar in composition to that of the modern Pannonian forest-steppe areas.

Accordingly, the warmer periods of the terminal Pleistocene must have created ideal conditions for the spread of this type of mollusk fauna in the southern parts of the Carpathian Basin. On the basis of the paleodistribution of the mollusk species *G. frumentum*, the area of the Pannonian steppe/forest-steppe belt must have expanded as far north as the heart of the basin during the interstadials (Sümegi and Krolopp 2002). However, this zone managed to conquer the foothills of the Carpathian Mts. during the last interglacial (Sümegi and Krolopp 2006; Sümegi et al. 2002).

Nevertheless, despite the structural similarities, the general composition of the vegetation during the interstadials of the terminal ice age was utterly different from the modern one. Deciduous elements (elm, hornbeam, maple, oak, birch, poplar) were only subordinately present under a dominance of Scots pine complemented by spruce, fir, larch, and Swiss pine. The contradiction between a cold climate vegetation and a warmth-loving mollusk fauna is only seemingly important. During the interstadials, a relative increase in the continentality is discernible in the area of the Carpathian Basin; i.e., the emergence of cold winter spells besides mild, dry summers and an increase in the average annual temperature fluctuations. As the winter season is a period of hibernation for mollusks this enables them to survive unfavorable conditions as well. On the other hand, a major part of plants, including the temperate deciduous arboreal elements are very sensitive to negative shifts observed

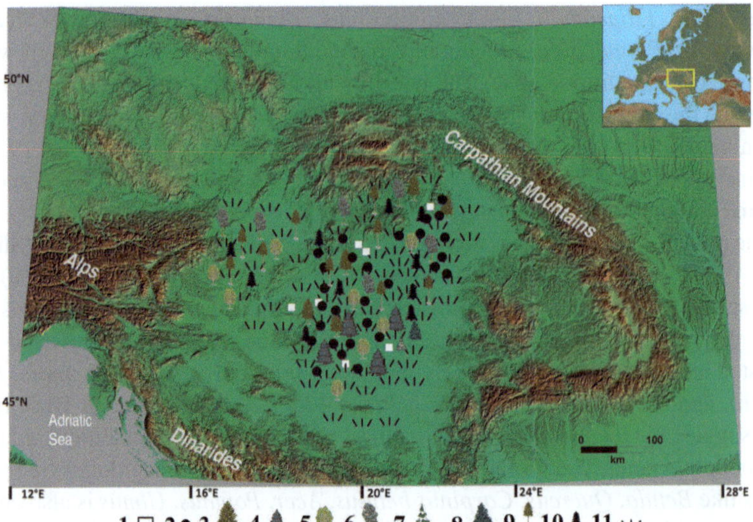

Fig. 2.10 The inferred vegetation of the Carpathian Basin during the interstadial between 25 and 18 kys cal BP. 1. Upper Palaeolithic site, 2. Cryophilous, recent Central Asian xeromontane snails, *Vallonia tenuilabris* site 3. Spruce (*Picea*) remains, 4. Scotch pine (*P. sylvestris*) remains, 5. Birch, willow, alder (*Betula, Salix, Alnus*) remains 6. Swiss pine (*P. cembra*) remains, 7. Fir (*Abies*) remains, 8. Juniper (*Juniperus*) remains, 9. Larch (*Larix*) remains, 10. Pine (Pinacae) remains, 11. Herbs (non-arboreal plants)

in the length of the growth period, the onset of early winters and high annual temperature fluctuations. Coniferous elements to a certain extent are much more resistant to these negative influences.

The temporal resolution of our records enabled us to continuously capture vegetation changes in the southern parts of the Carpathian Basin for the past 140 kys; i.e., from ca. the last interglacial (Zech et al. 2010). The presence of a steppe/forest-steppe vegetation hosting the marker taxon *G. frumentum* could have been attested in the area of the Great Hungarian Plains and the marginal part of the Carpathians as early as 110–130 kys (Fig. 2.9). In contrast, the foothills hosted a closed woodland vegetation (Krolopp 1983; Lozek 1964). Nevertheless, besides the warm-ups during the terminal part of the ice age referred earlier, significant coolings could have also inferred cause a drastic reduction in the temperature of the summer months, the length of the growth season resulting in a transformation of the flora and the fauna of the basin as well (Fig. 2.10). Taxa characteristic of alpine cold steppes and the tundra turn up and become dominant in the northern part of the basin bringing about a retreat of open parklands hosting spruce, Scots pine, and scattered deciduous elements locally and regionally.

The dominant taxa of the arboreal vegetation are Swiss pine (*P. cembra*), juniper (*Juniperus*), and larch (*Larix*). However, the proportion of arboreal elements and woodlands was highly reduced surviving only in sheltered refugees of the basin (Willis et al. 2000). The dominant elements of the flora during these times were heliophyllic non-arboreal plants including artemisia (*Artemisia*), grasses (*Gramineae*), and members of the orpine family (*Crassulaceae*). At the same time, several cold-loving tundra elements like stone-breakers (*Saxifraga*) and other perennial elements (*Androsace*) also turn up. A dominance of a treeless steppe vegetation could be postulated mixed with perennial elements and scattered trees. Thus despite the emergence of a dominantly loess steppe environment in the northern areas of the Carpathian Basin, the mosaic development of environmental parameters enabled the sporadic survival of a forest-steppe vegetation with an extremely low proportion of arboreal elements as well.

During the LGM, vegetation development in the northern part of the Carpathian Basin followed a clear and characteristic path. Namely, the prevalence of a treeless steppe was interrupted by rapid phases of forest expansion lasting for several centuries twice. This is seen in the trajectories of open parkland boreal pine forest expansion and the accompanying woodland element of the Carpathian spindle snail (*Vestiga turgida*). These forest expansion periods were first identified on the basis of shifts observable in the composition of the mollusk fauna (Sümegi and Krolopp 2002; Sümegi 2005), and later corroborated by pollen (Magyari et al. 1999; Willis et al. 2000), charcoal (Rudner and Sümegi 2001) and plant alkane data (Schatz et al. 2011) (Fig. 2.11). The inferred LGM mosaic of cold steppe studded by spots of tundra and woodland patches could have been correlated with the modern landscape of the Altai Mts and its foothill areas on the basis of the composition of the accessory mollusk fauna (Sümegi 1996, 2005; Horsak et al. 2010; Meng and Hoffmann 2009). Thus, the modern alpine boreal forest-steppe areas of Southern Siberia serve as an ideal analogy for the ice age vegetation of the Carpathian Basin (Sümegi 1996, 2005; Sümegi et al. 1999).

During the most prominent coolings, the cold-resistant and cold-loving elements turned up in the southern parts of the Carpathian Basin parallel with a gradual retreat of warmth-loving Southeastern European mollusk elements, including the character elements of the Pannonian forest-steppe. The analysis of mollusk faunas dated to the LGM in profiles from the Southern Great Hungarian Plains yielded surprising results. Here the expansion of cold-loving turndraic elements (*Columella columella*, *Vallonia tenuilabris*) was accompanied by the spread of Central European woodland elements (*Clausilia dubia*, *Clausilia pumila*, *Cochlodina laminata*, *Macrogastra ventricosa*, *Aegopinella ressmanni*, *Discus ruderatus*, *Orcula dolium*) (Sümegi 2005; Hupuczi and Sümegi 2010). On the basis of these findings, we may postulate only a slight decrease in the temperature for areas located at lower latitudes during the LGM. At the same time this decrease in the temperature resulted in a significant rise in relative humidity favoring the spread of arboreal elements (trees, bushes). The general composition of the mollusk fauna indicates the emergence of a forest-steppe vegetation in the southern parts of the Carpathian Basin during the LGM, extremely rich in species and with a dominance of cold-resistant and eurytopic species (Fig. 2.12).

On the basis of pollen data (Sümegi et al. 1999, 2011), the dominant arboreal elements were those of Scots pine, spruce, birches, and subordinately deciduous elements (oak, elm, alder, hazel, and willow). This pollen spectrum displays good affinity with the dry intamontane forest-steppes of the Altai and Sajan Mts (Pelánková

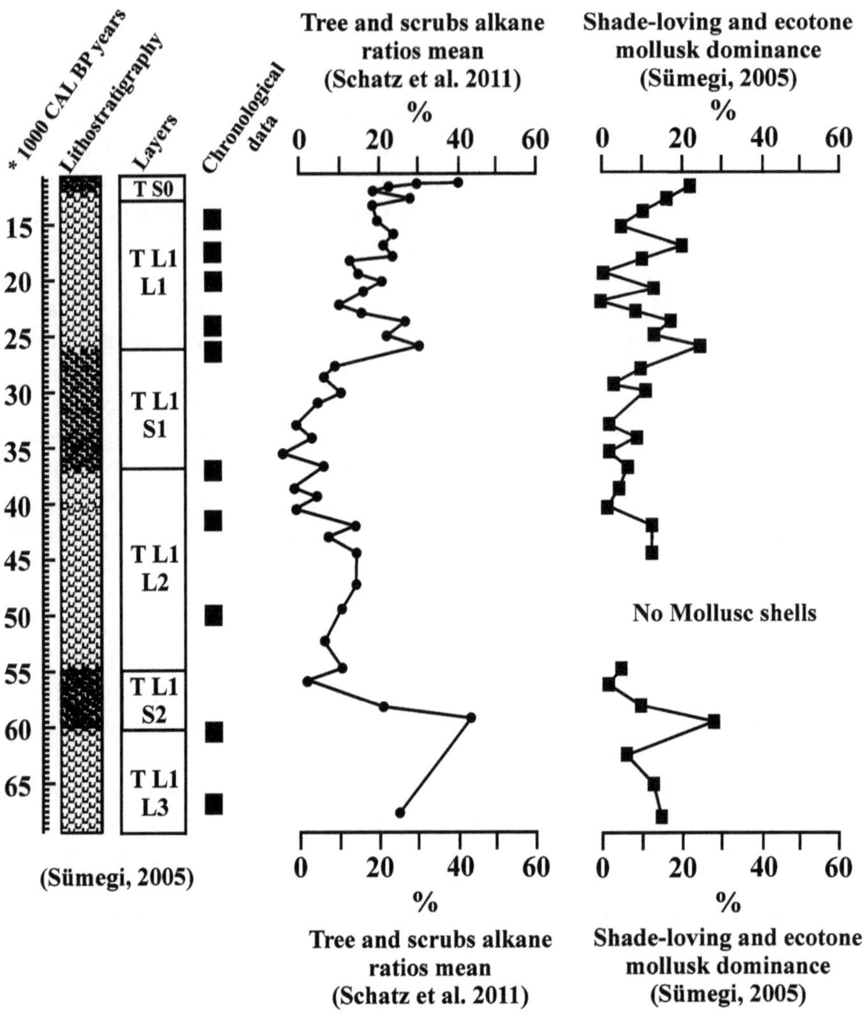

Fig. 2.11 Changes observed in the composition of plant alkanes and the terrestrial mollusk fauna in the area of the Northern Great Hungarian Plains for the past 70 kys. TS0 = Tokaj, modern soil, TL1L1 = late Pleniglacial loess layer, TL1S1 = middle Pleniglacial paleosol layer, TL1L2 = middle Pleniglacial loess layer, T1L1S2 = early Pleniglacial paleosol layer, TL1L3 = early Pleniglacial loess layer

and Chytry 2009). These results of mollusk and pollen analysis were clearly corroborated by those of the analysis of plant alkanes (Zech et al. 2010) (Fig. 2.12). Based on the collective evaluation of these findings, it seems to us that the inferred higher temperatures of the warmings must have favored the sustainment of a temperate steppe/forest-steppe in the southern parts of the basin by decreasing humidity. Conversely, the coolings resulted only in a minor drop of the temperatures but a more prominent increase in relative humidity favoring the emergence of a boreal open parkland-type forest-steppe present in Southern Siberia today.

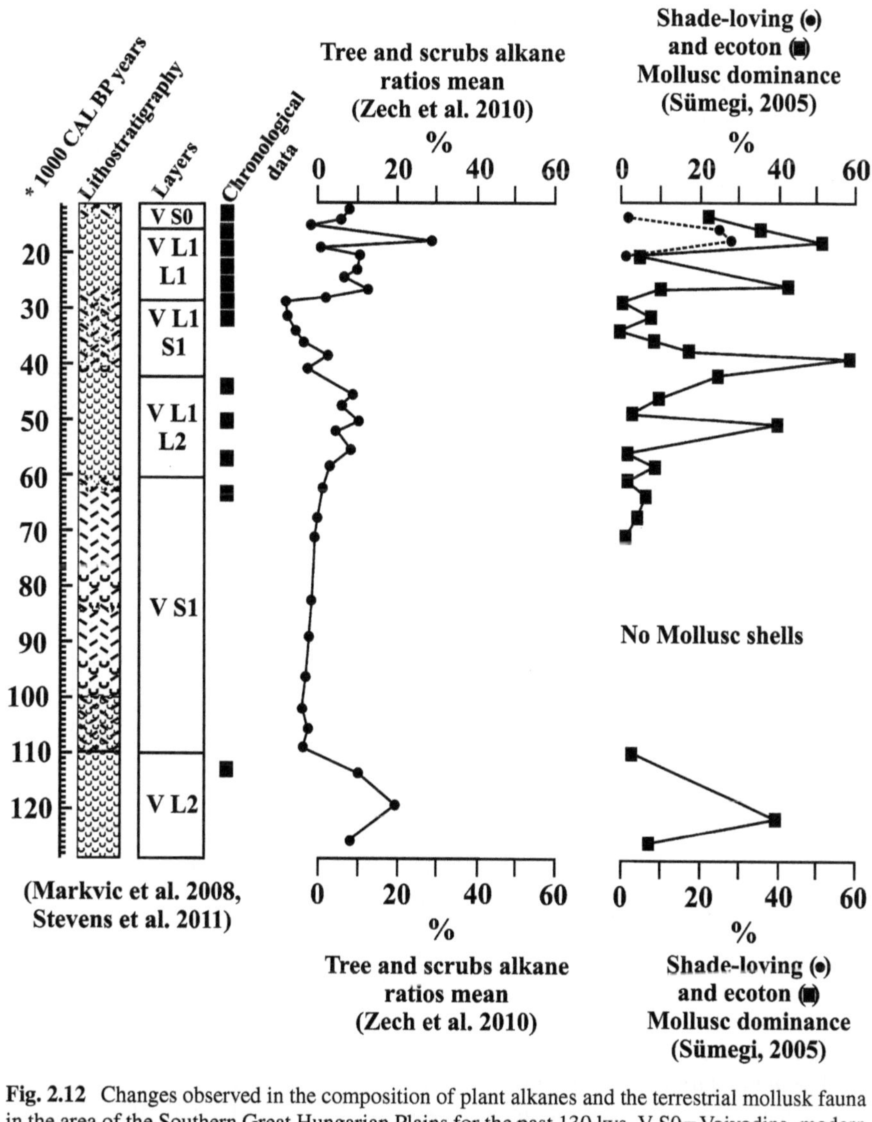

Fig. 2.12 Changes observed in the composition of plant alkanes and the terrestrial mollusk fauna in the area of the Southern Great Hungarian Plains for the past 130 kys. V S0 = Vojvodina, modern soil, V L1L1 = late Pleniglacial loess layer, V L1S1 = middle Pleniglacial paleosol layer, V L1L2 = early Pleniglacial paleosol layer, V S1 = last interglacial and early Weichselian paleosol layers, V L2 = Saalian loess layer

To sum up, based on our results, a relatively stable woodland-grassland ecotone with fluctuating proportion of arboreal elements and taxa was the dominant vegetation type of the Carpathian Basin between ca. 140 and 16 kys. This ecotone, as shown by the paleoecological record was by no means uniform, but a multiple mosaic patterning is presumed observable on local, regional, and basinwide scale as well. The emergence of macro or basin-scale mosaic patterning was attributable to the interplay of climatic influences, including the rainshadow effect of the basin and the actual distance from the continental and alpine ice cover. As a result, the presence of two distinct paleoenvironmental, paleobiogeographical units covering an area of ca. 50–100,000 km^2 could be postulated for the ice age of the Carpathian Basin. One of these entailed the northern areas, while the other the southern areas of the basin. Regional and local differences in the bedrock geology, soils, geomorphology, and hydrology further dissected these two major units to smaller mosaic components at a scale of some 100 or some km^2.

This multiple mosaic patterning of the paleoenvironment attributable to climatic, orographical, geological, and hydrological endowments was the most important of feature of the Carpathian Basin enabling the collective presence and survival of floral and faunal elements sometimes with contrasting ecological needs. It was this mosaic patterning of the paleoenvironment which enabled the long-term sustainment of a woodland-grassland ecotone in the basin within the course of fluctuating climatic evolution of the ice age (Sümegi 1996). Furthermore, the presence of multiple ecological habitats also ensured the survival of cold-resistant taxa during the warmings and warmths-loving taxa during the coolings in specially sheltered habitats. This phenomenon is known as the dual refugee effect in the literature (Willis et al. 1995). The general structure of a woodland-grassland ecotone was preserved even during the largest coolings, during phases of a treeless steppe. However, locally there must have been some homogenization of the vegetation in the northern parts and plant mosaics must have been preserved in shelters of favorable microclimate, along the watercourses or on slopes, walls, valleys with a southern exposure (Willis et al. 2000).

2.4.2 *Vegetation Development During the Terminal Part of the Last Ice Age*

The first emergence of the Pannonian forest-steppe, corresponding to the westernmost isolated part of the Eastern European forest-steppe belt, is generally dated to the opening of the Holocene (between ca. 10 and 8.5 kys cal BP); i.e., the Boreal in the literature (Járai-Komlódi 1987, 2000; Zólyomi and Fekete 1994). According to this concept, the emergence of an extremely dry climate during the referred period must have triggered the opening of the mixed taiga hosted by the area of the Great Hungarian Plains due to specific edaphic factors enhancing the invasion of heliophyl steppe and Pontic elements into the initial clearings and the retreat of pine to the higher, cooler areas and cold-spots of the basin. This hypothesis considers the opening of a mixed taiga as the main trigger in the formation of a temperate forest-steppe during the initial part of the Holocene.

Conversely, as it was stated in the previous chapter, the emergence of a stable forest-steppe could have been attested even during the ice age as well on the basis of paleoecological data. Furthermore, representatives of the Southeast European warmth-loving mollusk *G. frumentum*, which is also a character element of the

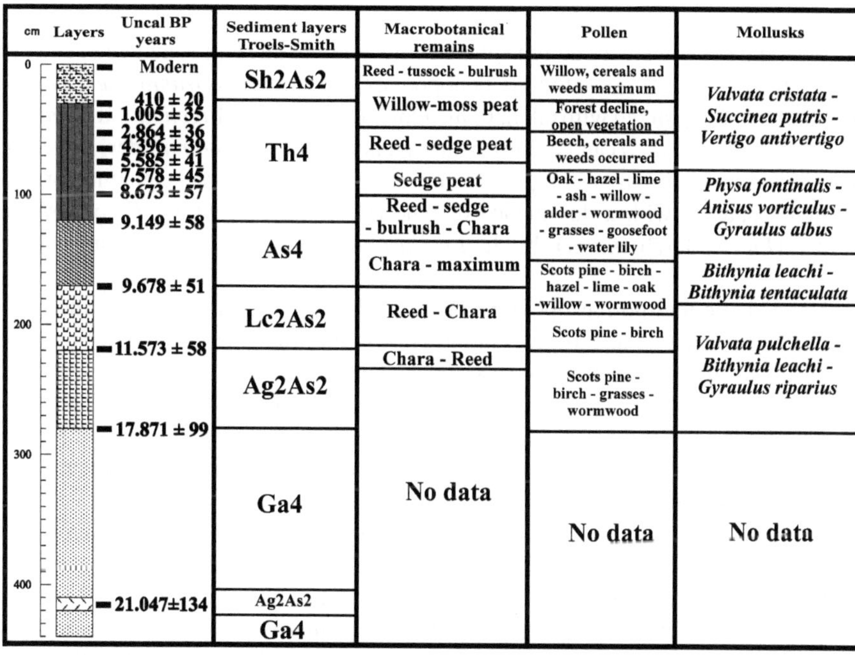

Fig. 2.13 The findings of paleoecological investigations for the catchment of Lake Kolon

modern Pannonian forest-steppe were present during this period in the southern parts of the basin and managed to survive even the largest coolings (Sümegi et al. 1998; Sümegi and Krolopp 2002). So on the basis of our findings, the most important elements of the forest-steppe within a mosaic setting were present preceding the Holocene as well (Sümegi 1996). In order to test this idea, a radiocarbon-dated profile was chosen, where environmental changes could have been continuously traced back as far as 24 ky cal BP (LGM). The core of our analysis was given by the profile of the catchment basin of Lake Kolon located in the centre of the Pannonian forest-steppe (N: 46° 46′ 23.82″, E: 19° 20′ 42.49″) where 17 radiocarbon dates (Sümegi et al. 2011) enabled us to capture environmental changes within a section of 2.8 m for the past 24 kys (Fig. 2.13). In order to highlight regional differences and capture a spatial pattern of vegetation development, the results of this profile was correlated with those of 32 additional radiocarbon-dated undisturbed core sequences along N–S and W–E transects (Fig. 2.13).

Between 24 and 16 kys cal BP the deposition of silt-rich, minerorganic lacustrine clays with minimum organic content was inferred for the catchment of the Kolon Lake. The proportion of plant macrofossils was negligible. Nevertheless, the composition of the mollusk fauna with a dominance of cold-loving elements (*Valvata pulchella*, *Bithynia leachi*, *Gyraulus riparius*) refers to the emergence of a cold, clear lake with a water depth of ca. 3 m during this period within an interdune basin. The general pollen composition (ratio of AP:NAP between 40:60 and 60:40 %)

implies a forest-steppe vegetation fringing the lacustrine basin. The dominant plant types of the LGM were arctic fens plants, cold continental steppe elements, such as sedge (Cyperaceae), goosefoot (Chenopodiaceae), grasses (*Gramineae*), wormwood (*Artemisia*) mixed with tall herb meadows (*Thalictrum*, *Angelica*, *Campanula*, *Centaurea*, *Filipendula*) and cold-resistant plants such as *H. rhamnoides*, *Ephedra*, *S. selaginoides*, heliophyte compositae, and *Pinguicaula*. Arboreal elements formed scattered patches with a dominance of pine (*P. mugo*, *P. cembra*, *P. sylvestris*), spruce (*Picea*), larch (*Larix*), birch (*Betula nana*, *Betula pubescens*) and willow (*Salix*).

Not a single thermo-mesophylous taxon could have been identified in the horizon dated between 24 and 16 kys cal BP. This implies the emergence of a Southern Siberian, Mongolian boreal forest-steppe, characterized by extremely scattered stands of arboreal elements in the study area during the LGM. A similar vegetation type was inferred for the central and northern parts of the Carpathian Basin for the same period based on the evaluation of other coeval pollen profiles (Sümegi et al. 1999, 2011; Magyari et al. 1999).

There is a marked transformation in the pollen composition of the starting profile of Lake Kolon between 16 and 11.5 kys cal BP. It is characterized by a general retreat and complete disappearance of cold-resistant and cold-loving plants (*P. mugo*, *B. nana*) accompanied by an increase in pine, primarily *P. sylvestris* sometimes exceeding 60 %. Thus, Scots pine must have been a dominant element of the local arboreal vegetation during the referred period. Besides Scots pine, the proportion of birch, willow, and alder was outstanding. The total ratio of AP exceeded 70 % in this horizon. The presence of several warmth-loving elements could have also been attested like oak, elm, lime, hornbeam, and hazelnut.

The total ratio of thermo-mesophylous APs was around 5–10 % here. The ratio of NAPs was around 30 % with a dominance of reed, bulrush and grasses (*Gramineae*), wormwood (*Artemisia*), goosefoot (Chenopodiaceae), and heliophyl compositae. Conversely, the scattered presence of *Achillea*, *Aster*, *Centaurea*, *Filipendula*, *Thalictrum* pollen grains indicate the preservation of large open areas within a mixed pine woodland, characterized by a dominance of birch and pine. Natural grasslands hosting warm steppe elements intervened the referred woodlands. This implies the emergence of a mesophylous forest-steppe in the heart of the Carpathian Basin during the terminal part of the ice age.

The appearance of warm steppe, forest-steppe elements in the pollen spectrum could have been correlated with the intrusion of warmth-loving, steppe/forest-steppe dweller mollusks in the loess profiles of the Southern Great Hungarian Plains dated between 16 and 15 kys cal BP and including the character species of the Pannonian forest-steppe *G. frumentum*. Based on the paleorecords, a woodland-grassland ecotone emerged in areas of higher and lower groundwater table of the plains. The dominant elements were those of temperate steppe in areas of lower groundwater table with a subordinate presence of trees. Conversely, areas with more favorable hydrologies hosted a mixed taiga studded by grassland spots of Artemisia steppe. So according to these records, the immigration of warmth-loving floral and faunal elements responsible for the creation and sustainment of a modern temperate forest-steppe in the basin must have initiated as early as 16/15 kys cal BP in the southern parts of the Carpathian Basin.

These changes observed in the terrestrial vegetation were equally attested in the aquatic habitats of Lake Kolon as well, characterized by a rapid increase in the organic content and the proportion of *Chara* remains in the lacustrine deposits dated to this period. Plant macrofossils rich in these deposits talk about the same story (*Phragmites communis*, *Typha*, *Juncus*, *Carex elata*, *Carex rostata*). A clear alteration in the aquatic habitat is attested by an increase in water lily remains as well (*Nymphae*). The mollusk fauna was likewise characterized by the presence of warmth-loving elements (*Bithynia tentaculata*, *Vertigo antivertigo*, *Vertigo angustior*, *Helicopsis striata*) and forest-steppe dwellers (*Vallonia costata*, *Vitrea crystallina*).

2.4.3 Vegetation Development During the Pleistocene/Holocene Transition

There is a marked transformation in the aquatic habitats of Lake Kolon, our starting profile dated between 11.5 and 9.5 kys cal. BP. Here, a general increase in the organic content of the deposit was observed parallely with a rise in the proportion of reed, bulrush, and sedge macro remains. The proportion of plant remains marking a relatively high water level was also significant here. The appearance of mosses and some character elements of marshy reeds (*Senecio paludosus*) as well as ostracod shells and *Chara* remains in increasing proportions mark an incipient paludification of the lacustrine basin. The aquatic mollusk fauna also refers to a marshy lake with the general warming of the water (*Anisus vorticulus*, *Physa fontinalis*, *Gyraulus albus*). Water lily remains also have a peak here. The proportion of pine pollen grains goes below 10 % here giving way to the advent of thermo-mesophylous trees and bushes including oak, lime, elm, hornbeam, and alder. The high concentration of flue ash within this horizon marks the outbreak of natural fires in elevated temperatures. No wonder the ratio of AP:NAP was reduced to 50 % in this horizon. The dominant APs were wormwood and grasses besides *Achillea*, *Aster*, *Centaurea*, and compositae. All this indicates the development of a Subcontinental-Submediterranean oak forest-steppe in the area of our central profile. For the same time period, the areas of hills and foothills of Transdanubia and northern Hungary as well as the main watercourses witnessed a rapid expansion of temperate woodlands composed of mainly oak, lime, elm, hornbeam, and hazelnut (Willis 1997, 2007, Willis et al. 1995, 1997, 2000; Sümegi 1999, 2004b, 2005; Magyari et al. 1999, 2002). In contrast, the heart of the basin witnessed changes similar to the one observed in the profile of Lake Kolon leading to the emergence of a species-rich temperate forest-steppe studded by alkaline steppes (Sümegi 1989, 2001a, b, 2004b; Sümegi and Törőcsik 2007; Sümegi et al. 2000, 2004, 2006).

Thus according to the paleorecord, the environmental changes dated to the Pleistocene/Holocene transition favored the spread of woodlands (mixed taiga exchanged by deciduous woodlands) in the marginal foothills of the Carpathians as well as the hilly and mid-mountain regions of the basin. At the same time, the heart of the basin witnessed the emergence of a temperate forest-steppe. According to the data available, the intrusion of thermo-mesophylous trees and steppe elements occurred within a mesophylous mixed forest-steppe vegetation after the LGM between 16 and 15 kys cal BP. Then increasing temperatures and a rise in natural fires contributed to the disappearance of cold-resistant elements and the emergence of a Subcontinental-Submediterranean oak forest-steppe. Based on investigation results from the profile of Lake Kolon, a marked local and regional decrease in relative humidity could have been inferred during the opening of the Holocene parallel with increasing temperatures. This might have favored the general reduction of woodlands, expansion of grasslands and thus the sustainment of temperate forest-steppes during the Holocene in the area of the Great Hungarian Plains.

All in all, paleoenvironmental changes preceding the Holocene transition must have favored the intrusion and expansion of thermo-mesophylous elements from their refugees found in the southern parts of the Carpathian Basin into a mosaic of boreal type forest-steppe, which had emerged in the terminal ice age (Sümegi et al. 1998). By ousting the previously dominant cold-resistant and mesophylous taxa, they must have contributed to the gradual emergence of a temperate oak forest-steppe in the Carpathian Basin even preceding the actual boundary of Pleistocene/Holocene. Natural fires set in a woodland of easily flammable pines must have significantly accelerated the development of forest-steppes. A very similar vegetation change leading to the emergence of a Subcontinental-Submediterranean oak forest-steppe was modeled for the area of the Eastern European Plains based on paleobotanical records (Novenko et al. 2011). However, as shown by the chronology, this type of transformation must have occurred much earlier in the area of the Carpathian Basin, probably as a result of heightened drop of humidity attributable to the basin effect related to increasing temperatures, plus the relative vicinity of refugees of the warmth-loving, steppe elements aiding quick colonization (Willis et al. 1995, 2000; Sümegi et al. 1998).

To sum up, the development of an end-Pleistocene forest-steppe, characterized by comparable structural makeup but different composing taxa to its Holoocene counterpart, was mainly controlled by global climatic changes. However, the local and regional climate, morphology, hydrology, and geology have significantly influenced the trajectory of evolution. Nevertheless, one must also bear in mind the potential influences of ice age megaherbovires of the so-called mammoth steppes in the sustainment of a woodland-grassland ecotone (Guthrie 2001; Zazula et al. 2002, 2003, 2007). The presence of mammoth, wild horses, reindeer, moose, and buffalos could have been univocally attested in the study area for the referred period (Kretzoi 1977; Jánossy 1979; Vörös 1987; Kordos 1987; Pazonyi 2004). The general composition of this Upper Würmian mammal fauna also indicates a species-rich forest-steppe vegetation in the area of the Carpathian Basin (Sümegi 2005) as numerous woodland elements (moose, reindeer) were collectively present with typical steppe dwellers (Guthrie 2001).

Furthermore, these herds of megaherbivores must have significantly contributed to the reduction of arboreal vegetation in the area. The notion of a forest-steppe is corroborated by data on rodents (Kordos 1977; Kordos and Rinder 1991) and birds, where typical steppe or tundra elements like the great bustard (*Otis tarda*), or ptarmigan (*Lagopus lagopus*, *Lagopus mutus*) are complemented by typical woodland elements like various taxa of woodpeckers (*Caprimulgus europaeus*, *Dendrocopos major*, *D. medius*, *D. leucotos*, *Picus canus*, *Jynx torquilla*) in sites dated to the terminal part of the last ice age (Jánossy 1979).

A characteristic feature of the Carpathian Basin is that the extant megaherbivores of the ice age were replaced woodland elements like roe-deer, deer, auroch, wild boar. At the same time in sites of the Great Hungarian Plains remains of typical steppe elements like wild horse (*Equus ferus*) and onager (*Equus hemionus*) turn up in large numbers (Vörös 1987). The presence of wild horse could have been attested from 62, while that of onager from 21 Holocene sites in Hungary (Vörös 1987). Remains of steppe dweller wild ass (*Asinus hydruntinus*) have also been retrieved from various archeological sites of the Great Hungarian Plains dated between 7,000 and 11,000 years (Bökönyi 1984, 1992). This latter taxon went extinct ca. 6–7 kys ago, but the presence of wild horse and onager could have been traced as far as the Bronze Age (3000–3500 BC).

Domesticates replacing these wild forms and other grazers (caprines, cattle, horses, pigs) have also ensured the sustainment of a forest-steppe, steppe environment in the basin. The collective presence of new woodland elements and steppe/forest-steppe elements of the wild fauna is another independent proof of the mosaic patterning of the Holocene forest-steppe vegetation of the Great Hungarian Plains well before the emergence of first agricultural groups. A similar duality is seen in the record of early Holocene rodent remains. Woodland elements are complemented by typical steppe/forest-steppe elements like birch mouse (*Sicista*), mole rat (*Spalax*), gopher (*Citellus citellus*), hamster (*Cricetus cricetus*) (Cserkész et al. 2009; Kordos 1978). The early Holocene avifauna of the Great Hungarian Plains is likewise dominated by steppe/forest-steppe elements including the Great bastard (*O. tarda*), the Blacktail Godwilt (*Limosa limosa*) in same proportions as other woodland elements (Gál 2007).

Thus besides the referred edaphic and climatic factors, the new steppe/forest-steppe grazers and rodents must have contributed to the sustainment of a Subcontinental-Submediterranean oak forest-steppe of the Early Holocene of the Carpathian Basin.

2.4.4 *Vegetation History of the Carpathian Basin from the Settlement of the First Farmers*

The first farming groups, connected to the early Neolithic Körös culture turn up in the Holocene Subcontinental-Submediterranean oak forest-steppe of the Great

Hungarian Plains roughly 8,000 years ago (Kutzián 1944; Kalicz 1970; Kosse 1979; Sherratt 1982; Bökönyi 1992; Whittle 2007). Many researchers blamed these first farming groups for the initial human-induced vegetation changes. However, recent paleoecological and paleoenvironmental studies of Mesolithic and Neolithic cultures of the Carpathian Basin and the Great Hungarian Plains yielded surprising results. In the horizon of woodland profiles dated to the Late Mesolithic a clear, iterative signal of vegetation changes could have been identified seen in periodic increases in flue ash, the appearance of open area taxa and the cyclical reduction of elm and hazelnut pollen concentrations (Sümegi 1999, 2004b, 2005, 2008). These changes were interpreted as the side-effects of human activities related to the creation of seasonal hunting camps and paths, intensive foraging and the use of twigs and leaves as fodder (Sümegi 1999) favoring the expansion of open areas in closed woodlands and the spread of heliophyl marginal vegetation.

Conversely, no such changes could have been identified in the horizons of catchments of small lacustrine basins and marshlands from the heart of the basin (Great Hungarian Plains) found next to Early Neolithic sites and dated to the Late Mesolithic and Early Neolithic (Willis 1997, 2007; Sümegi 2004b). Furthermore, besides an obvious increase in cereal pollen grains, no other signs of vegetation disturbance related to agricultural activities could have been attested (Sümegi 2004b). Despite their extensive settlement of more than 3,000 sites identified, no clear sign of fire-induced deforestation by members of the Early Neolithic Körös culture could have been proven in the heart of the basin (Willis 2007). The same holds true for our primary profile of Lake Kolon, where no significant transformation in the formerly emerged oak forest-steppe could have been observed in the horizon dated to the Early Neolithic either. Conversely, results of phytolith and mollusk analysis implemented on material from the Early Neolithic sites themselves (Madella 2007; Windland 2007; Persaits and Sümegi 2011; Sümegi 2007; Gulyás and Sümegi 2011) have corroborated the presence of extensive grassy spots and arable lands in the direct vicinity of the sites.

Representatives of the Körös culture chose to settle on the floodplain of the Tisza River and its tributaries. These settlements however were not confined to the actual riverbed but were placed on top of the flood-free, loess-covered island-like lag-surfaces at an interface of multiple ecotones as shown by the findings of detailed geoarcheological investigations implemented in the vicinity of Körös sites in Hungary (Fig. 2.14) (Sümegi 2003, 2004b, 2007, 2011; Sümegi and Molnár 2007). These island-like lag-surfaces were covered by chernozem soils, which developed on the loessy bedrock, hosting a steppe/forest-steppe vegetation. The sites thus were located at the interface of the higher floodplain hosting a steppe/forest-steppe and the adjacent low floodplain hosting a hardwood gallery forest. This mosaic of multiple ecotones ensured the engagement of multifocal subsistence on the one hand. On the other hand, thanks to the relative openness of the natural highs hosting the settlement there was no need of deforestation for the creation of living space and arable/or pasturelands.

This is a highly intriguing example of how the mosaic-like multiple ecotone of woodland-grassland areas influenced settlement and subsistence strategies of first

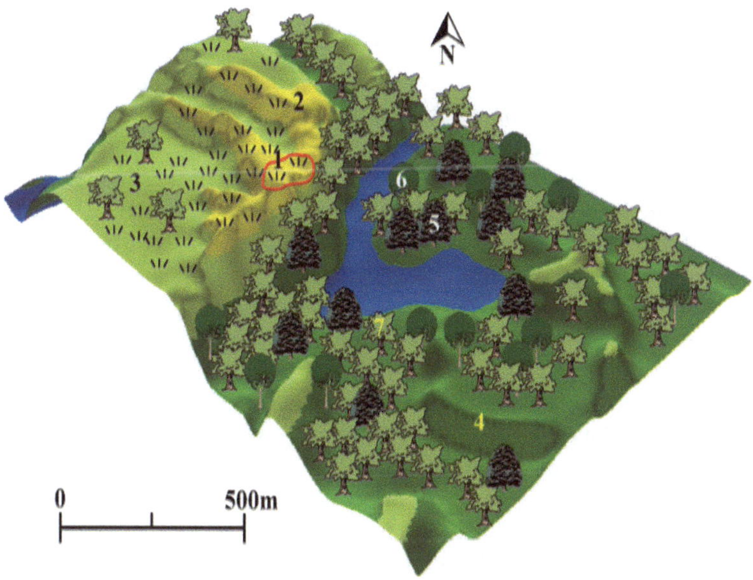

Fig. 2.14 Geomorphology and inferred vegetation of the Early Neolithic Körös site of Nagykörü Orchard. 1. Körös settlement, 2. Steppe, 3. Forest-steppe, 4. Backswamp, 5. Elm of modern hardwood gallery forest, 6. Lime of modern hardwood gallery forest, 7. Oak of modern hardwood gallery forest

farming groups of the Great Hungarian Plains. It is interesting to note though that numerous even more extensive steppe-forest-steppe areas are found several kilometers away from the rivers like those of the Hortobágy and the Hajdúság (Sümegi 1989, 1997, 2004b, 2005; Sümegi et al. 2000, 2005b) yet they were colonized during the second half of the Neolithic only. All in all based on our data, the continuous presence of a forest-steppe nestled in the heart of the Carpathian Basin made human disturbances obsolete for ensuring economic activities of the first farming groups. Sings of human disturbances could have been attested in the marginal areas of the basin, where the natural development of the landscape favored the sustainment of closed woodland and did not result in the emergence of forest-steppe areas (Willis et al. 1995; Juhász 2007).

Vegetation changes related to conscious human intervention in the landscape like burning, the expansion of weeds, and the reduction of the arboreal vegetation in a forest-steppe environment are traceable from the Late Neolithic and Early Copper Age (Willis et al. 1995; Willis 2007). This pattern is visible in the profile of Lake Kolon as well (Fig. 2.13). Here the observed increase in cereal pollen concentrations were concomitant with a rise in weed pollen grains marking treading and signaling the creation of arable lands and pasturelands. Furthermore, a significant rise in NAPs marks the opening of the vegetation. These human influences are gradually intensified in the Great Hungarian Plains from the Copper Age. However, their spatial and temporal appearance is by no ways uniform but changes from area to area and period to period (Sümegi and Törőcsik 2007). This must be attributed to the differential demography and settlement, subsistence strategies of the various cultures settling in the area (Sherratt 1982, 1983). Nevertheless, certain well-traceable stages of human-induced vegetation transformation can be clearly identified. Five such stages were differentiated from the Late Neolithic, Early Copper Age.

Stage one is hallmarked by the settlement of the Late Copper Age Badenian and the Pit Grave Cultures. Settlement was so extensive that human influences were attested in alkaline areas of the Great Hungarian Plains as well. Stage two is put to the Middle Bronze Age and the emergence of multilayered (tell) settlements hallmarking the height of preurban societal evolution in the area. The vicinity of the tell settlements was intensively exploited hallmarked by such activities as the establishment of floodplain orchards (walnut production), a complete transformation of hardwood gallery forests etc. (Sümegi 1999, 2010; Sümegi and Bodor 2000; Willis et al. 1998). The extremely focused exploitation of the landscape during the establishment of the tell settlements brought about a complete disappearance of the boundaries between closed woodlands and adjacent forest-steppe areas contributing to the expansion of the ecotonal elements to the former areas of gallery forests and the closed woodlands of the hills and foothills as well as mid-mountains of the basin. Signs of deforestation are most pronounced in fortified tell sites, where the creation of palisades required large amounts of timber (Sümegi et al. 1998).

The continuation of human disturbances characteristic of the Middle Bronze Age is postulated for the Late Bronze Age and the Early Iron Age as well. However, the foci of human activities were displaced from the area of the original, natural forest-steppes to those of natural woodlands in Transdanubia and the Subcarpathian region (Sümegi and Bodor 2005; Sümegi et al. 2008). Around the newly established fortified military centers, located in a woodland setting a full transformation of the landscape and the vegetation is discernible yielding the emergence of wide open areas. These transformations favored the intrusion of non-arboreal elements indigenous of adjacent natural forest-steppes as well as that of weeds. The resulting woodland-grassland ecotone in a natural woodland setting was the clear outcome of human activities here dated to the Late Bronze and Early Iron Ages.

Signs of extensive deforestation during the referred periods are recorded in the catchment basins, hallmarking intensive soil erosion and inwash and contributing to the paludification and silting up of minor creek beds, ponds under the Early Iron Age rainfall maximum. The littoral region of larger lakes and the banks of larger rivers were silted up to such an extent, that a clear advent of reed, bulrush, and sedge was traceable there (Sümegi and Bodor 2005).

The next stage is connected to the invasion of Celtic tribes conquering the entire basin and in possession of high-tech iron tools. Human-induced forest-steppe development could have been attested along numerous settlement points (Jerem et al. 1986). Based on the analysis of plant remains retrieved from Celtic features, a clear transformation of the agricultural activities can be postulated. A drastic drop in the areas under woodland cover is likewise discernible parallel with the expansion of arable lands, orchards, and pasturelands becoming the focus of economic activities during the Late Iron and the Imperial Ages. The array of plants cultivated during the

Early Iron Age (wheat, oat, barley, rye, and grapes) was complemented by new elements including cherry, walnut, pear, chestnut, poppy, cucumber, rowan, horse bean, and cress. An increase in weeds such as brome-grass, corn-poppy, bearbind, mercury, and dandelion marks a clear expansion of pastures and arable lands. Thus these Celtic tribes were engaged in extensive plant cultivation, high-class gardening, and animal husbandry. The strong presence of wine grapes and chestnut in the array of plants cultivated by the Celts is outstanding in our understanding of the emergence of Submediterranean agriculture of the Imperial Age as it clearly shows the local roots of such activities.

The highly intertwined agricultural activities of the Late Iron Age and Imperial Age reach such heights in the area of the Carpathian Basin, that they are even recorded in the diaries of imperial travelers as well dated to the third and fourth centuries AD. Other hallmarks of intensive human activities include the well-constructed and maintained network of roads and the defense line of the limes. The transformation of the floodplain of the Danube valley during the reign of the Roman emperors reached such heights that it brought about a complete disappearance of the natural woodlands on the side of Pannonia province. When counter-fortresses were constructed on the other side of the Danube, it also resulted in the disappearance of the natural woodland there as well.

An excellent canary in the coal mine of the period hallmarking the extent of human disturbances in the landscape is the dry, rock-dweller and steppe-dweller Northern Balkanic *Helicella obvia*. This mollusk species was able to invade even the westernmost Alpine foothills along the Danube, where the original climatic and vegetation conditions would clearly hamper its settlement (cool, humid, woodland). The creation of artificial steppes accompanying the construction of roads and defense systems favored the colonization of the species to these areas as well.

It must be noted here that although human-induced transformations of the landscape are largely recorded in the Transdanubian side of the Pannonia province for the referred period, changes were likewise large-scale on the other side of the basin populated by barbarian herds. Sarmatian tribes characterized by extensive animal husbandry and high population numbers and later adopting Roman type economic activities had similarly large impact on the landscape of the Great Hungarian Plains. Incipient wind-blown sand movements in areas under an original forest-steppe cover are perhaps the best markers of intensified landscape exploitation (Kiss et al. 2006). Overgrazing of sandy steppes, forest-steppes by members of a group characterized by nomadic pastoralism was an important trigger of anthropogenic sand movement.

These changes however by no means were unique. Similar negative outcome could have been attested to tribes settling the area during the Migration Age, including the nomadic Avars and the Hungarians as well. Cumanians settling the area before and after the Mongolian invasion caused similar landscape changes in the wind-blown sandy areas of the Great Hungarian Plains (Sümegi 2001b; Kiss et al. 2006). The pastoral practices of these nomadic tribes, when complemented by dry spells of the climate resulted in the creation of semi-desert environment in certain parts. These were counterbalanced by periods of higher rainfall and the transformation of agricultural practices to avoid the full degradation of the landscape yielding a forest-steppe again in the long run. Nevertheless, these highly degraded landscapes have also pointed to the high vulnerability of this unique ecosystem. Although these transformations clearly highlighted periods, when the limit of sustainability was reached no truly irreversible transformations can be inferred in the landscape until the nineteenth century.

The period of the Middle Ages, although characterized by intensive human activities, favored the preservation of mosaic patterning present in the vegetation thanks to the array of productive techniques adopted ensuring the collective presence of meadows, pasturelands, cultivated arable lands, fallows, remnant natural, and artificial woodlands. Thus the original structure of the forest-steppe ecotone and the most important composing taxa was more or less preserved. The appearance of Turkish rule and the desertification of certain urban areas of the Great Hungarian Plains further enhanced the natural regeneration of the Pannonian forest-steppe and these conditions prevailed up to the second half of the seventeenth century. It was the time when maize (Turkish wheat in original name) was first introduced via Turkish means to the area of the Great Hungarian Plains. This crop altered the way people thought about animal husbandry and resulted in the introduction of intensive stall-feeding where corn gave the main fodder.

The introduction of this new line of thinking fundamentally transformed the way landscape was treated. The traditional system of meadows-pasturelands and arable lands was abandoned and there was a huge hunger for the exploitation of new lands to provide higher yields. This could have been achieved by the drainage of waterlogged areas, peatlands, and the plowing of meadows and pasturelands alone. The lack of male labor force as an outcome of the Napoleon wars resulting in extremely high wheat prices was a further push factor to achieve these goals in hope of a greater profit putting the owners of the agrarian system to the side of river regulations and the acquisition of new land for production. The complete destruction of the original structure and composition of the Pannonian forest-steppe is attributable to the nineteenth century flood protection and river regulation measures seen in the creation of vast dike and canal systems and the drainage of the landscape to acquire new land suitable for intensive cereal production (Fig. 2.15). It was this time when the ecological foundation of the area was overthrown: the hydrology.

As the natural supply of climate-induced loss of water of the landscape via floods ceased, the ecological systems whose functioning is based on the availability of water collapsed. Industrial agricultural production resulted a complete homogenization of the vegetation creating a so-called cultural landscape or better to say *cultural desert*. This process is still active and the extreme droughts and increasing temperatures of the past 50 years favored the emergence of a steppe-like environment in the heart of the basin (Pálfai 1989). A major deal is whether or not this process will halt at a certain point or will result in a complete degradation of the landscape. It's quite clear though that the natural system of multiple ecotones was seriously damaged and could be restored by special measures of experts working collectively for the restoration and conservation of the landscape.

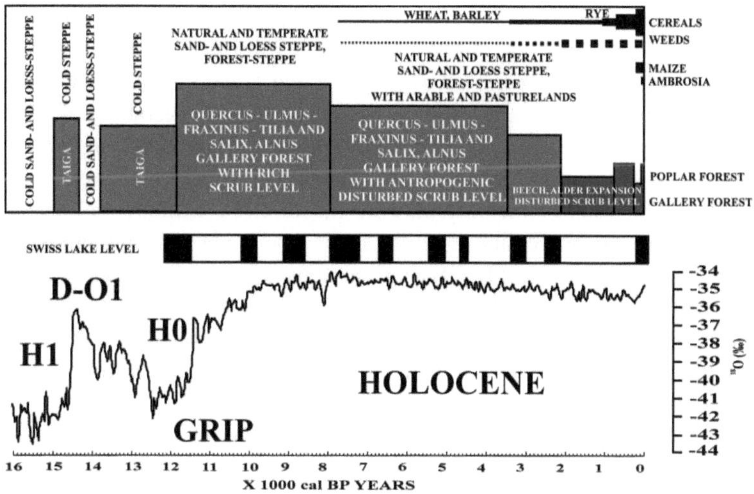

Fig. 2.15 The trajectory of Pannonian forest-steppe evolution within the framework of global climate change (d18O GRIP) for the past 16 kys

2.5 Summary

As a result of its unique natural endowments a boreal type of forest-steppe vegetation emerged in the heart of the Carpathian Basin during the LGM, which was invaded by numerous thermo-mesophylous elements initiating around 16 kys cal BP. This species-rich forest-steppe characterized by the dominance of coniferous elements was transformed into a Subcontinental-Submediterranean oak forest-steppe by the beginning of the Holocene. These vegetation changes display strong affinity with the processes observed in the Eastern European Plains at a later period.

However, as a result of its geographic location, the area of the Carpathian Basin was characterized by an overlap of strong Atlantic, Submediterranean, and Subcarpathian climatic influences not present in the Eastern European Plains. As a result of the fluctuating interface of these climatic influences, numerous elements with Pontian, Balkanic, Atlantic, and Subcarpathian affinity are present in the continental forest-steppe yielding an interface and overlap of vegetation zones in the basin similar to the system of climatic influences.

The most important climatic parameters favoring the development and sustainment of a forest-steppe are those of the average temperature of the summer months, rainfall and evaporation as well as evapotranspiraton. The spatial distribution and patterning of these parameters is clearly congruent with the observed boundaries of the forest-steppe vegetation of the Great Hungarian Plains. The forest-steppes nestled in the heart of the basin are fringed by closed woodlands in the higher, marginal hilly and alpine areas enjoying higher rainfall. According to the Holdridge classification system, the major part of the basin is put to the transitional category found between those of cold, temperate grasslands, cold, temperate humid woodlands and warm, temperate dry woodlands, where the first (steppe) and the last categories (dry woodland) also turn up climatically in the form of scattered patches. This grassy area forming an ecotone between the actual grasslands and dry and humid woodlands corresponds to the Pannonian forest-steppe vegetation of the Great Hungarian Plains. The trajectory of evolution of the Pannonian temperate forest-steppe was fundamentally controlled by local edaphic factors (orography, geology, hydrology, soils) besides the interplay of the referred climatic influences. The influences of these edaphic factors on the landscape are best visible in the alkaline areas, which emerged as a result of the unique subsurface hydrological cycle and the rock–soil–water interaction of the area discernible from the last ice age. Natural fires induced by increasing temperatures as well as the invasion of the megaherbivores and rodents enhanced the sustainment of the original vegetation. The highly complex mosaic of woodlands, grasslands, and intervening transitional ecotones hosted the first foraging Mesolithic and farming Neolithic groups of the basin.

As the settlements of the first farmers were confined mainly to the open vegetation areas, these communities caused no significant alteration of the original forest-steppe, although their activities have contributed to its sustainment. New elements of the vegetation were also introduced as a side effect of human activities. At the same time, the marginal vegetations of the ecotone corresponding to the zone of cool, fresh woodlands was significantly transformed by representatives of the same communities favoring the expansion of heliophyl elements. Based on the available data, besides the macroclimatic and regional edaphic environmental factors human activities from the Mesolithic have significantly contributed to the sustainment of the original woodland-grassland ecotone.

A general increase in the animal stock as well as the population characterizes the period from the Late Neolithic and the Early Copper Age favoring the expansion of open areas and weeds and reflecting intensifying human influences on the landscape. Another major change occurred in the wild fauna around 1500 BC hallmarked by a complete disappearance of large-bodied wild herbivores enhancing the preservation of the mosaic nature of the vegetation. This change is attributable to either overhunting of the natural population or the increase in domestication. At the same time, the new domesticates substituted the previous wild grazers in the sustainment of the original structure of the vegetation. As a result of population growth, higher social hierarchy and preurban settlement development during the Bronze Age characterized by intensive forest burning and landscape exploitation the trajectory of natural evolution was distorted resulting in the emergence of artificial steppes and woodlands. The proportion of disturbed vegetation areas continued to rise with the emergence of the metal cultures and human influences on the vegetation were extended to new areas formerly hosting closed woodlands.

This brought about the emergence of artificial forest-steppe ecotones. These transformations further intensified during the Late Iron and Imperial Ages turning almost the entire basin into an artificial forest-steppe. The nomadic pastoral groups invading the area of the basin from the East European Plain fundamentally altered the evolution of this transformed landscape. Overgrazing of steppes complemented by extreme dry spells has initiated the emergence of semi-deserts and wind-blown sand movement. These wounds were quickly healed as a result of altered agricultural techniques and landscape exploitation complemented by recurrent periods of heightened rainfall. Agricultural techniques during the Middle Ages and the initial part of the New Age have contributed to the enhancement of species-richness and the sustainment of the structure of the Pannonian forest-steppe. Unfortunately the introduction of in-stall animal husbandry and the acquisition of new farmlands to ensure higher crops and profits initiated a full transformation of the landscape via river regulation measures dated to the second part of the nineteenth century.

By intervening in the basis of the ecological system, the hydrology, these measures managed to put the Pannonian forest-steppe to its margin of existence. Only the last minute conservation measures of the twentieth century could save the witness of old times from complete extirpation via the revitalization of overdeveloped agricultural areas, reintroduction of forest-steppe floral elements and wild fauna (wild horses) ensuring the sustainment of a natural forest-steppe ecotone. Nevertheless, the fate of the Pannonian forest-steppe, a unique woodland-grassland ecotone is still not settled partly because global warming favors the conquest of steppe vegetation by increasing aridity. Furthermore, as the area of modern forest-steppes enjoying protection is highly constrained and fragmented, this ecosystem is highly vulnerable to both natural and artificial environmental changes.

References

Ant H (1963) Faunistische, ökologische und tiergeographische Untersuchungen zur Verbreitung der Landschnecken in Nordwestdeutschland. Abhandlungen des Landesmuseums für Naturkunde Münster 25:1–125

Arno SF, Hammerly RP (1984) Timberline. Mountain and arctic forest frontiers. The Mountaineers, Seattle

Bacsó N (1959) Magyarország éghajlata. Akadémiai Kiadó, Budapest

Barber KE, Chambers FM, Maddy D, Brew J (1994) A sensitive high resolution record of the Holocene climatic change from a raised bog in northern England. Holocene 4:198–205

Berg LS (1958) Die geographischen Zonen der Sowjetunion. Teubner, Leipzig

Bernátsky J (1914) A magyar Alföld fás növényzete. Erd Kís 16:129–180

Bökönyi S (1984) Die neolithische Wirbeltierfauna von Battonya-Gödrösök. In: Gy G (ed) BattonyaGödrösök, eine neolithische Siedlung in Südost-Ungarn. Békés megyei Tanács VB Tudományos-Koordinációs Szakbizottság, Békéscsaba, pp 119–169

Bökönyi S (ed) (1992) Cultural and landscape changes in South-east Hungary. I. Reports on the Gyomaendrőd Project. Budapest

Borbás V (1900) A Balaton Flórája (A Balaton tavának és partmellékének növényföldrajza és edényes növényzete). In: Lóczy L (ed) A Balaton Tudományos Tanulmányozásainak Eredményei. Franklin Társulat, Budapest, pp 1–431

Borhidi A (1956) Die Steppen und Wiesen im Sandgebiet der Kleinen Ungarischen Tiefebene. Acta Bot Acad Sci Hungar 2:241–274

Borhidi A (1961) Klimadiagramme und Klimazonale Karte Ungarns. Annal Univ Sci Budapestensis Sectio Biologica 4:21–50

Borhidi A (1966) Erdőtanulmányok a Szovjetunió erdőssztyepp-övében. Waldunter-suchungen in der Waldsteppen-Zone der Sowietunion. Bot Közlem 53:85–188

Borhidi A (2003) Magyarország növénytársulásai. Akadémiai Kiadó, Budapest

Boros Á (1929) A Nyírség flórája és növényföldrajza. Math Term Tud Ért 46:48–59
Cserkész T, Kitowski I, Czochra K, Rusin M (2009) Distribution of the Southern birch mouse (*Sicista subtilis*) in East-Poland. Morphometric variations in discrete European populations of superspecies *S. subtilis*. Mammalia 73:221–229
Dobosi Z, Felméry L (1977) Klimatológia. Tankönyvkiadó, Budapest
Fekete G (2000) A középső orosz erdőssztyepp. In: ZS Molnár, Kun A (eds) Alföld erdőssztyepp maradványok Magyarországon. WWF Publishing, Budapest, p 33
Fórizs I (2003) Isotopes as natural tracers in the water cycle: examples from the Carpathian Basin. Studia Universitatis Babeş-Bolyai, Phys Spec Issue 48:69–77
Gál E (2007) Fowling in lowlands. Neolithic and Copper Age bird bone remains from the Great Hungarian Plain and South-East Romania. Archaeolingua (Series Minor), Budapest
Gulyás S, Sümegi P (2011) Climate-induced transformation of alluvial environments and subsistence crisis in first farming groups of the Carpathian Basin during the 6th millenium BC. Geophys Res Abstr 13:EGU2011–EGU12105
Guthrie RD (2001) Origin and causes of the mammoth steppe: a story of cloud cover, woolly mammal tooth pits, bucklets, and inside-out Beringia. Quat Sci Rev 20:549–574
Holdridge LR (1947) Determination of world plant formations from simple climatic data. Science 105:367–368
Holdridge LR (1967) Life zone ecology. Tropical Science Center, San Jose, Costa Rica
Horsák M, Chytrý M, Pokryszko BM, Danihelka J, Ermakov N, Hájek M, Hájková P, Kintrová K, Kocí M, Kubesová S, Lustyk P, Otýpková Z, Pelánková B, Valachovic M (2010) Habitats of relict terrestrial snails in southern Siberia: lessons for the reconstruction of palaeoenvironments of full-glacial Europe. J Biogeogr 37:1450–1462
Hupuczi J, Sümegi P (2010) The Late Pleistocene paleoenvironment and paleoclimate of the Madaras section (South Hungary), based on preliminary records from mollusks. Cen Eur J Geosci 2:64–70
Jakab G, Sümegi P (2004) A lágyszárú növények tőzegben található maradványainak határozója mikroszkópikus bélyegek alapján. Kitaibelia 9:93–129
Jakucs P, Fekete G, Gergely J (1959) Angaben zur Vegetation der Moldau und der Dobrudscha. Annls Hist-Nat Mus Natn Hung 51:211–225
Jánossy D (1979) A magyarországi pleisztocén tagolása gerinces faunák alapján. Akadémiai Kiadó, Budapest
Járai-Komlódi M (1987) Postglacial climate and vegetation history in Hungary. In: Pécsi M, Kordos L (eds) Holocene environment in Hungary. Geographical Research Institute, Hungarian Academy of Sciences, Budapest, pp 37–48
Járai-Komlódi M (2000) A Kárpát-medence növényzetének kialakulása. Tilia 9:5–59
Jerem E, Facsar G, Kordos L, Krolopp E, Vörös I (1985) A Sopron-Krautackeren feltárt vaskori telep régészeti és környezetrekonstrukciós vizsgálata. I. Archeológiai Értesítő 111:141–169
Jerem E, Facsar G, Kordos L, Krolopp E, Vörös I (1986) A Sopron-Krautackern feltárt vaskori telep régészeti és környezetrekonstrukciós vizsgálata. II. Archeológiai Értesítő 112:3–24
Juhász I (2007) The pollen sequence from Alsópáhok. In: Zatykó CS, Juhász I, Sümegi P (eds) Environmental archaeology in Transdanubia, Varia Archaeologica Hungarica XXI, MTA Régészeti Intézet, Budapest, pp 29–35
Kalicz N (1970) Agyag istenek. A neolitikum és a rézkor emlékei Magyarországon. Hereditas sorozat. Corvina Kiadó, Budapest
Kerner A (1863) Das Pflanzenleben der Donauländer. Wagner, Innsbruck
Kiss T, Nyári D, Sipos GY (2006) Homokmozgások vizsgálata a történelmi időkben Csengele területén. In: Kiss A, Mezősi G, Sümegi Z (eds) Táj, környezet és társadalom. Ünnepi tanulmányok Keveiné Bárány Ilona professzor asszony tiszteletére. Szegedi Tudományegyetem, Szeged, pp 373–383
Kordos L (1977) Changes in the Holocene climate of Hungary reflected by the "vole-thermometer" method. Földrajzi Közlemények 25:144–160
Kordos L (1978) Historico-zoogeographical and ecological investigations of the subfossil vertebrate fauna of Aggtelek Karst. Vertebrata Hungarica 18:85–100

Kordos L (1987) Climatostratigraphy of Upper Pleistocene Vertebrates and the condition of loess formation in Hungary. GeoJournal 15:163–166

Kordos L, Ringer Á (1991) A magyarországi felső-pleisztocén Arvicolidae-sztratigráfiájának klimato- és archeosztratigráfiai korrelációja. MÁFI Jelentése 1989 évről:523–533

Kosse K (1979) Settlement ecology of early and middle Neolithic Körös and Linear Pottery cultures in Hungary. BAR International Series, 64. British Archaeological Reports, Oxford

Köppen W (1923) Die Klimate der Erde — Grundriß der Klimakunde, Verlag Walter de Gruyter & Co., Berlin und Leipzig

Kretzoi M (1977) Ecological conditions of the "loess period" in Hungary as revealed by vertebrate fauna. Földrajzi Közlemények 25:75–89

Krolopp E (1983) Biostratigraphic division of Hungarian Pleistocene Formations according to their Mollusc fauna. Acta Geol Hung 26:62–89

Kutzián I (1944) Körös-kultúra. Dissertationes Pannonicae, Series II. 23. Péter Pázmány University, Budapest

Lavrenko EM (1980) Steppi (The Steppes). In: Gribova SA, Isachenko TI, Lavrenko M (eds) Rastitelnost Evropeiskoi chasti SSSR (vegetation of the European SSR). Nauka Publisher, Leningrad, pp 203–273

Lavrenko EM, Karamyseva ZV (1991) Steppes of the former Soviet Union and Mongolia. In: Coupland RT (ed) Natural Grassand: eastern hemisphere and résumé. Ecosystems of the World, 8b. Elsevier, London, pp 3–60

Ložek V (1964) Quartärmollusken der Tschechoslowakei. Rozpravy Ústredniho ústavu geologického 31:1–374

Madella M (2007) The silica skeletons from the anthropic deposits. In: Whittle A (ed) The early neolithic on the Great Hungarian Plain: investigations of the Körös culture site of Ecsegfalva 23, Co. Békés, Varia Archaeologica Hungarica XXI, MTA Régészeti Intézet, Budapest, pp 447–460

Madella M, Powers-Jones AH, Jones MK (1998) A simple method of extraction of opal phytoliths from sediments using a non-toxic heavy liquid. J Archaeol Sci 25:801–803

Magyari E, Jakab G, Rudner E, Sümegi P (1999) Palynological and plant macrofossil data on Late Pleistocen short term climatic osscilations in North-east Hungary. Acta Palaeobotanica suppl 2:491–502

Magyari E, Sümegi P, Braun M, Jakab G (2002) Retarded hydrosere: anthropogenic and climatic signals in a Holocene raised bog profile from the NE Carpathian Basin. J Ecol 89:1019–1032

Meng S, Hoffmann M (2009) Pupilla loessica LOŽEK 1954 (Gastropoda: Pulmonata: Pupillidae)—"A Living Fossil" in Central Asia? Quat Sci J 58:55–69

Miháltz I (1954) A Duna-Tisza köze déli részének földtani felvétele. MÁFI Jelentése 1951-ről:61–68

Molnár B (1961) A Duna-Tisza közi eolikus rétegek felszíni és felszín alatti kiterjedése. Földtani Közlöny 91:303–315

Molnár B, Szónoky M, Kovács S (1981) Recens hiperszalin dolomitok diagenetikus és litifikációs folyamatai a Duna-Tisza közén. Földtani Közlöny 111:119–144

Molnár C, Bölöni J, Pál R, Türke JI, Jakab G, Kállayné Szerényi J (2007a) Növényföldrajz és flóra. In: Ilyés E, Bölöni J (eds) Löszsztyeppek, lejtősszteppek és erdőssztyepp rétek Magyarországon. Private Edition, Budapest, pp 170–177

Molnár C, Türke IJ, Csathó AI (2007b) Botanikai megfigyelések Dél-Bukovina térségében. [összehasonlító erdőssztyepp-tanulmányok I.]. Kanitzia 15:19–34

Molnár C, Türke IJ, Kelemen A, Korompai T, Schmidt J (2008) Botanikai tanulmányút Moldovába. [összehasonlító erdőssztyepp-tanulmányok II.]. Botanikai Közlemények 95:127–154

Molnár D, Hupuczi J, Galovic L, Sümegi P (2010) Preliminary malacological investigation on the loess profile at Zmajevac, Croatia. Cen Eur J Geosci 2:52–56

Mucsi M (1963) Finomrétegtani vizsgálatok a kiskunsági édesvízi karbonát-képződményeken. Földtani Közlöny 93:373–386

Myster RW (2007) Introduction. In: Myster RW (ed) Post-agricultural succession in the neotropics. Springer, New York, pp 3–21

Novenko E, Yu, Voskresenskaya KV, Suganova IS (2011) Late Pleistocene paleoenvironments in the centre of European Russia (reconstruction by pollen and plant macrofossil data from loess and alluvial sediments). In: Jary Z, Mroczek P, Owzarek P, Przybylski B, Zöller L (eds) Closing the gap—North Carpatian loess traverse int he Eurasian loess belt. International Workshop, 6th Loess Seminar, Wroclaw, Poland, pp 15–16

Ødum S (1979) Actual and potential tree line in the North Atlantic region, especially in Greenland and the Faroes. Holarct Ecol 2:222–227

Pálfai I (1989) Az Alföld aszályossága. Alföldi Tanulmányok 13:7–25

Pazonyi P (2004) Mammalian ecosystem dynamics in the Carpathian Basin during the last 27,000 years. Palaeogeogr Palaeoclimatol Palaeoecol 212:295–314

Pelánková B, Chytrý M (2009) Surface pollen-vegetation relationships in the forest-steppe, taiga and tundra landscapes of the Russian Altai Mountains. Rev Palaeobot Palynol 157:253–265

Persaits G (2010) A fitolitok szerepe a geoarcheológiai minták értékelésében. PhD dissertation. SZTE TTIK Földtani és őslénytani Tanszék, Szeged

Persaits G, Sümegi P (2011) A fitolitok szerepe a régészeti geológiai és környezettörténeti minták értékelésében. In: Unger J, Pál-Molnár E (eds) Geoszférák 2010. GeoLitera, Szeged, pp 307–354

Persaits G, Gulyás S, Sümegi P, Imre M (2008) Phytolith analysis: environmental reconstruction derived from a Sarmatian kiln used for firing pottery. In: Szabó P, Hédl R (eds) Human nature: studies in historical ecology and environmental history. Institute of Botany of the Czech Academy of Sciences, Pruhonice, pp 116–122

Piperno DR (2006) Phytoliths. A comprehensive guide for archaeologists and paleoecologists. Altamira Press, Oxford

Rapaics R (1918) Az Alföld növényföldrajzi jelleme I-II. Erd Kís 20(1–2):1–97, 20(3–4): 183–247

Réthly A (1933) Kísérlet Magyarország klímatérképének szerkesztésére a Köppen-féle klímabeosztás értelmében. Időjárás 37:105–115

Réthly A (1948) Magyarország éghajlata. In: Réthly A, Aujenszky L (eds) Agrometeorológia. Quick Kiadó, Budapest, pp 95–109

Rónai A (1961) Az Alföld talajvíztérképe. MÁFI Kiadványa, Budapest

Rudner E, Sümegi P (2001) Recurring taiga forest steppe habitats in the Carpathian Basin in the Upper Weichselian. Quat Int 76(77):177–189

Schatz AK, Zech M, Buggle B, Gulyás S, Hambach U, Sümegi P, Marković S, Scholten T (2011) The late Quaternary loess record of Tokaj, Hungary—reconstructing palaeoenvironment, -vegetation and -climate using stable C and N isotopes and biomarkers. Quat Int 240:52–61

Scherf E (1935) Alföldünk pleisztocén és holocén rétegeinek geológiai és morfológiai viszonyai és ezeknek összefüggése a talajkutatással, különösen a sziktalajképződéssel. Földtani Intézet Évi Jelentése 1925–1928-ról:1–37

Sherratt A (1982) The development of Neolithic and Copper Age settlement in the Great Hungarian Plain. Part 1: the regional setting. Oxf J Archaeol 1:287–316

Sherratt A (1983) The development of Neolithic and Copper Age settlement in the Great Hungarian Plain. Part 2: site survey and settlement dynamics. Oxf J Archaeol 2:13–40

Soó R (1926) Die Entstehung der ungarischen Puszta. Ungarische Jahrbücher 6:258–276

Soó R (1929) Die vegetation und die Entstehung der ungarischen Puszta. J Ecol 17:329–350

Soó R (1958) Összehasonlító vegetációtanulmányok a Szovjetunió erdőssztyepp övéből. MTA Biológiai Csoport Közleményei 1:209–222

Stevens GC, Fox JF (1991) The causes of treeline. Annu Rev Ecol Syst 22:177–191

Stevens T, Markovic SB, Zech M, Hambach U, Sümegi P (2011) Dust deposition and climate in the Carpathian basin over an independently dated last glaciale/interglacial cycle. Quat Sci Rew 30:662–681

Stockmarr J (1971) Tablets with spores used in absolute pollen analysis. Pollen Et Spores 13:615–621

Strömpl G (1931) A szik geomorfológiája. Földrajzi Közlemények 4–5:62–74

Sümegi P (1989) Hajdúság felső-pleisztocén fejlődéstörténete finomrétegtani (üledékföldtani, őslénytani, geokémiai) vizsgálatok alapján. unpublished DrUniv thesis, Debrecen

Sümegi P. 1989. Hajdúság felső-pleisztocén fejlődéstörténete finomrétegtani (üledékföldtani, őslénytani, geokémiai) vizsgálatok alapján. (Evaluation of the Upper Pleistocene of the Hajdúság region based on fine-stratigraphical (sedimentological, paleontological, geochemical) analyses). dr.univ. dissertation, p. 96. Debrecen (Hungary)

Sümegi P (1995) Az utolsó 30.000 év változásainak rekonstrukciója őslénytani adatok alapján a Kárpát-medence centrális részén. In: "Berényi Dénes professzor születésének 95. Évfordulója" tiszteletére rendezett tudományos emlékülés előadásai. MTA Debreceni Területi Bizottsága, Meteorológiai Munkabizottság és KLTE Meteorológiai Tanszék Kiadvány, Debrecen, pp 244–258

Sümegi P (1996) Az ÉK-magyarországi löszterületek összehasonlító őskörnyezeti és sztratigráfiai értékelése. PhD thesis, Debrecen

Sümegi P (1997) The Process of Sodification on Hortobágy in Space and Time according to geopedological investigation. Proceedings and Field Trip of Hydro-Petro-Geology and Hungary, Budapest, pp 237–242

Sümegi P (1999) Reconstruction of flora, soil and landscape evolution, and human impact on the Bereg Plain from late-glacial up to the present, based on palaeoecological analysis. In: Hamar J, Sárkány-Kiss A (eds) The Upper Tisa Valley. Tiscia Monograph Series, Szeged, pp 173–204

Sümegi P (2001a) Környezetrégészet problémái Magyarországon. MUMOSZ, I. Fiatal őskoros Kutatók I. Összejövetelének konferenciakötete, Debrecen, pp 17–49

Sümegi P (2001b) A Kiskunság a középkorban—geológus szemmel. In: Horváth F (ed) A Csengelei kunok ura és népe. Archaeolingua Kiadó, Budapest, pp 313–317

Sümegi P (2003) Early Neolithic man and riparian environment in the Carpathian Basin. In: Jerem E, Raczky P (eds) Morgenrot der Kulturen. Archaeoligua Press, Budapest, pp 53–60

Sümegi P (2004a) Findings of geoarcheological and environmental historical investigations at the Körös site of Tiszapüspöki-Karancspart Háromága. Anthaeus 27:307–342

Sümegi P (2004b) The results of paleoenvironmental reconstruction and comparative geoarcheological analysis for the examined area. In: Sümegi P, Gulyás S (eds) The geohistory of Bátorliget Marshland. Archaeolingua Press, Budapest, pp 301–348

Sümegi P (2005) Loess and Upper Paleolithic environment in Hungary. Aurea Press, Nagykovácsi

Sümegi P (2007) Description of the sampling locations in the Balaton Uplands. In: Zatykó C, Juhász I, Sümegi P (eds) Environmental archaeology in Transdanubia (Hungary). Varia Archaeologica Hungarica XX. MTA Régészeti Intézet, Budapest, pp 112–113

Sümegi P (2008) Palaeogeographical background of the Mesolithic and Early Neolithic settlement in the Carpathian Basin. In: Proceedings of the XV. World Congress UISPP. Archeopress, London, pp 45–53

Sümegi, P.-Törőcsik, T. 2007. Hazánk növényzete az éghajlatváltozások tükrében (The fate of the flora of Hungary in course of climate changes). Természet Világa, 138: 292–295.

Sümegi P (2010) Az Északi középhegység negyedidőszak végi őstörténete. Ember és környezet kapcsolata a szubkárpáti (felföldi) régióban. In: Guba S, Tankó K (eds) "Régről kell kezdenünk"...Studia Archaeologica in honoreum Pauli Patay. Szécsényi Múzeum, Szécsény, pp 295–326

Sümegi P (2011) Az Alföld élővilágának fejlődése a jégkor végétől napjainkig. In: Rakonczai J (ed) Környezeti változások és az Alföld. A Nagyalföld Alapítvány kötetei 7. Nagyalföld Alapítvány, Békéscsaba, pp 35–44

Sümegi P, Bodor E (2000) Sedimentological, pollen and geoarcheological analysis of core sequence at Tököl. In: Poroszlai I, Vicze M (eds) Százhalombatta archaeological expedition. Archeolingua Press, Budapest, pp 83–96

Sümegi P, Bodor E (2005) Geoarcheological and archaeobotanical investigations in the valley of the Benta (Békás) creek. In: Poroszlai I, Vicze M (eds) SAX Reports 2—Field Season 2000–2003. Matrica Múzeum, Százhalombatta, pp 209–236

Sümegi P, Hertelendi E (1998) Reconstruction of microenvironmental changes in Kopasz Hill loess area at Tokaj (Hungary) between 15.000-70.000 BP years. Radiocarbon 40:855–863

Sümegi P, Krolopp E (2002) Quartermalacological analyses for modeling of the Upper Weichselian palaeoenvironmental changes in the Carpathian Basin. Quat Int 91:53–63

Sümegi P, Krolopp E (2006) A basaharci téglagyári löszfeltárás Mollusca-faunája. Malakológiai Tájékoztató 24:15–30

Sümegi P, Molnár S (2007) The Kiritó meander: sediments and the question of flooding. In: Whittle A (ed) The Ecsegfalva Project. Varia Archaeologica Hungarica sorozat, XXI. MTA Régészeti Intézet, Budapest, pp 67–82

Sümegi P, Törőcsik T (2007) Hazánk növényzete az éghajlatváltozások tükrében. Természet Világa 138:292–295

Sümegi P, Krolopp E, Hertelendi E (1998) A Ságvár-Lascaux interstadiális őskörnyezeti rekonstrukciója. Acta Geographica Geologica Et Meteorologica Debrecina 34:181–202

Sümegi P, Magyari E, Daniel P, Hertelendi E, Rudner E (1999) A kardoskúti Fehér-tó negyedidőszaki fejlődéstörténetének rekonstrukciója. Földtani Közlöny 129:479–519

Sümegi P, Molnár A, Szilágyi G (2000) Szikesedés a Hortobágyon. Természet Világa 131:213–216

Sümegi P, Krolopp E, Rudner E (2002) Negyedidőszak végi őskörnyezeti változások térben és időben a Kárpát-medencében. Földtani Közlöny 132:5–22

Sümegi P, Mucsi M, Fényes J, Gulyás S (2005a) First radiocarbon dates from the freshwater carbonates of the Danube Tisza Interfluve. In: Hum L, Gulyás S, Sümegi P (eds) Environmental historical studies from the late tertiary and quaternary of Hungary. University of Szeged, Szeged, pp 103–117

Sümegi P, Csökmei B, Persaits G (2005b) The evolution of Polgár Island, a loess covered lag surface and its influences on the subsistence of settling human cultural groups. In: Hum L, Gulyás S, Sümegi P (eds) Environmental historical studies from the late tertiary and quaternary of Hungary. University of Szeged, Szeged, pp 141–163

Sümegi P, Bodor E, Törőcsik T (2005c) The origins of sodification in the Hortobágy region in the light of the palaeoenvironmental studies at Zám–Halasfenék. In: Gál E, Juhász I, Sümegi P (eds) Environmental archaeology in North-Eastern Hungary. Varia Archaeologica Hungarica XIX. MTA Régészeti Intézet, Budapest, pp 115–126

Sümegi P, Bodor E, Törőcsik T (2006) A hortobágyi szikesedés eredete. In: Kiss A, Mezősi G, Sümeghy Z (eds) Táj, környezet és társadalom. Ünnepi tanulmányok Keveiné Bárány Ilona professzor asszony tiszteletére. SZTE Kiadványa, Szeged, pp 633–641

Sümegi P, Törőcsik T, Jakab G, Gulyás S, Pomázi P, Majkut P, Páll GD, Persaits G, Bodor E (2008) The environmental history of Fenékpuszta with a special attention to the climate and precipitation of the last 2000 years. J Environ Geogr 2:5–14

Sümegi P, Ilon G, Jakab G, Páll DG, Törőcsik T (2009) Neolit és rézkori régészeti kultúrák és környezeti hátterük az Alpokaljáról. In: Bende L, Lőrinczy G. Szentes (eds) Medinától Etéig. Régészeti Tanulmányok Csalog József születésének 100. évfordulójára. Koszta József Múzeum, pp 189–195

Sümegi P, Molnár M, Jakab G, Persaits G, Majkut P, Páll DG, Gulyás S, Jull AJT, Törőcsik T (2011) Radiocarbon-dated paleoenvironmental changes on a lake and peat sediment sequence from the central part of the Great Hungarian Plains (Central Europe) during the last 25.000 years. Radiocarbon 52:85–97

Szelepcsényi Z, Breuer H, Ács F, Kozma I (2009) Biofizikai klímaklasszifikációk (1. rész: a módszerek bemutatása). Légkör 54:21–26

Szöőr G, Sümegi P, Balázs É (1991) Sedimentological and geochemical facies analysis Upper Pleistocene fossil soil zones discovered in the Hajdúság region, NE Hungary. In: Pécsi M, Schweitzer F (eds) Quaternary environment in Hungary. Studies in geography in Hungary 26. Akadémiai Kiadó, Budapest, pp 47–59

Timár G, Székely B, Molnár G, Ferencz C, Kern A, Galambos C, Gercsák G, Zentai L (2008) Combination of historical maps and satellite images of the Banat region—re-appearance of an old wetland area. Glob Planet Chang 62:29–38

Trunkó G, Földvári GZ (1996) Geology of Hungary. Borntraeger Press, Berlin

Tuzson J (1915) A magyar Alföld növényföldrajzi tagolódása. Mathem Term Ért 33:170–176

Varga Z (1989) Die Waldsteppen des pannonischen Raumes aus biogeographiser Sicht. Düsseldorfer Geobotanische Kolloquien 6:35–50

Varga Z, Borhidi A, Fekete G, Debreczy Z, Bartha D, Bölöni J, Molnár A, Kun A, Molnár Z, Lendvai G, Szodfrid I, Rédei T, Facsar G, Sümegi P, Kósa G, Király G (2000) Az erdőssztyepp

fogalma, típusai és jellemzésük. In: Molnár Z, Kun A (eds) Alföld erdőssztyepp maradványok Magyarországon. WWF, Budapest, pp 7–19
Vörös I (1987) Large mammalian faunal changes during the Late Upper Pleistocene and Early Holocene times in the Carpathian Basin. In: Pécsi M (ed) Pleistocene environment in Hungary. Geographical Research Institute, Hungarian Academy of Sciences, Budapest, pp 81–101
Whittle A (ed) (2007) The Early Neolithic on the Great Hungarian Plain: investigations of the Körös culture site of Ecsegfalva 23, Co. Békés. Varia Archaeologica Hungarica sorozat, XXI. MTA Régészeti Intézet, Budapest
Willis KJ (1997) The impact of early agriculture upon the Hungarian landscape. In: Chapman J, Dolukhanov PP (eds) Landscapes in flux: central and Eastern Europe in antiquity. Oxbow Books Ltd, Oxford, pp 193–209
Willis KJ (2007) Impact of the early Neolithic Körös culture on the landscape: evidence from palaeoecological investigations of Kiri-tó. In: Whittle A (ed) The Early Neolithic on the Great Hungarian Plain: investigations of the Körös culture site of Ecsegfalva 23, Co. Békés. Varia Archaeologica Hungarica XXI. MTA Régészeti Intézet, Budapest, pp 83–99
Willis KJ, Sümegi P, Braun M, Tóth A (1995) The late quaternary environmental history of Bátorliget, N.E. Hungary. Palaeogeogr Palaeoclimatol Palaeoecol 118:25–47
Willis KJ, Braun M, Sümegi P, Tóth A (1997) Does soil change cause vegetation change or vice-versa? A temporal perspective from Hungary. Ecology 78:740–750
Willis KJ, Sümegi P, Braun M, Bennett KD, Tóth A (1998) Prehistoric land degradation in Hungary: who, how and why? Antiquity 72:101–113
Willis KJ, Rudner E, Sümegi P (2000) The full-glacial forests of central and southeastern Europe: evidence from Hungarian palaeoecological records. Quat Res 53:203–213
Windland P (2007) Phytoliths of the Kiri-tó. In: Whittle A (ed) The Early Neolithic on the Great Hungarian Plain: investigations of the Körös culture site of Ecsegfalva 23, Co. Békés. Varia Archaeologica Hungarica XXI. MTA Régészeti Intézet, Budapest, pp 99–107
Zazula GD, Froese DG, Telka AM, Mathewes RW, Westgate JA (2002) Plants, bugs, and a giant mammoth tusk: Paleoecology of Last Chance Creek, Yukon Territory. In: Emond DS, Lewis LL (eds) Yukon exploration and geology. Exploration and Geological Services Division, Yukon Region, Indian and Northern Affairs Canada, pp 251–258
Zazula GD, Froese DG, Schweger CE, Mathewes RW, Beaudoin AB, Telka AM, Harington CR, Westgate JA (2003) Ice-age steppe vegetation in east Beringia. Nature 423:603
Zazula GD, Froese DG, Elias SA, Kuzmina S, Mathewes RW (2007) Arctic ground squirrels of the mammoth-steppe: paleoecology of Late Pleistocene middens (24 000–29 450 14C yr BP), Yukon Territory, Canada. Quat Sci Rev 26:979–1003
Zech M, Buggle B, Leiber K, Marković S, Glaser B, Hambach U, Huwe B, Stevens T, Sümegi P, Wiesenberg G, Zöller L (2010) Reconstructing Quaternary vegetation history in the Carpathian Basin, SE Europe, using n-alkane biomarkers as molecular fossils: problems and possible solutions, potential and limitations. Quat Sci J 85:150–157
Zólyomi B (1942) A középdunai flóraválasztó és a dolomitjelenség. Bot Közlem 39:183–193
Zólyomi B (1952) Magyarország növénytakarójának fejlődéstörténete az utolsó jégkorszaktól. MTA Biol Oszt Közl 1:491–544
Zólyomi B (1957) Der Tatarenahorn-Eichen-Lösswald der zonalen Waldsteppe. Acta Bot Hung 3:401–424
Zólyomi B (1958) Budapest és környékének természetes növénytakarója. In: Pécsi M, Marosi S, Szilárd J (eds) Budapest Természeti Képe. Akadémiai Kiadó, Budapest, pp 509–642
Zólyomi B (1987) Coenotone, ecotone and their role in preserving relic species. Acta Bot Hung 33:3–18
Zólyomi B, Fekete G (1994) The Pannonian loess steppe: differentation in space and time. Abstr Bot 18:29–41
Zólyomi B, Kéri M, Horváth F (1992) A szubmediterrán éghajlati hatások jelentősége a Kárpát-medence klímazonális növénytársuásainak összetételére. In: Hegyfoky Kabos klimatológus születésének 145. évfordulája alkalmából rendezett tudományos emlékülés előadásai. Debrecen-Túrkeve: MTA Debreceni Területi Bizottsága

Chapter 3
Ecotones as Complex Arenas of Disturbance, Climate, and Human Impacts: The Trans-Andean Forest-Steppe Ecotone of Northern Patagonia

Thomas Kitzberger

3.1 Introduction

Terrestrial ecotones are possibly the most dynamic regions of the world where global change impacts will earlier become most evident. It has been therefore suggested that ecotones could be considered indicators of global changes where monitoring efforts should be primarily directed (di Castri et al. 1988; Nielson 1991). A basic assumption for this is that small changes in some (limiting) condition, bottom up resource or top down process will, when approaching some threshold, produce rapid and abrupt responses such as shifts in the distribution of dominant species and associated communities or patches.

This rather simplistic view of ecotone dynamics has been challenged because environmental changes are complex usually involving interacting conditions, resources, and processes (Kupfer and Cairns 1996), so that changing one driver (e.g., temperature) may amplify the effect of another driver (e.g., fire) thus producing complex ecotone responses. Another complicating issue is that responses are scale dependent so that large shifts in dominant species may produce large physiognomic changes without necessarily producing biome boundary shifts (Nielson 1991). Furthermore, changes in ecotones are more likely to occur in response to changes in variability and frequency of extreme events rather than to changes in mean conditions.

Finally, ecotones (particularly low elevation forest grassland transitions) are often regions that have been heavily used by humans and where land use and land use changes have largely influenced its dynamics. A better understanding of the nature of ecotones and the complexity of interrelated natural and anthropogenic controls that regulate their internal dynamic is necessary before adequate assessments

T. Kitzberger (✉)
Laboratorio Ecotono, Universidad Nacional del Comahue, INIBIOMA-CONICET,
Quintral 1250, 8400 Bariloche, Argentina
e-mail: kitzberger@gmail.com

R.W. Myster (ed.), *Ecotones Between Forest and Grassland*,
DOI 10.1007/978-1-4614-3797-0_3, © Springer Science+Business Media New York 2012

of their usefulness as early warning indicators of global changes can be made. In this chapter I will attempt to illustrate some of these complexities by synthesizing extensive research performed along the trans-Andean forest-steppe ecotone of northern Patagonia during the last 3 decades.

3.2 Physical and Biological Setting of Forest-Steppe Ecotone of Northern Patagonia

3.2.1 Abiotic Transition

Biomes location and boundaries in southern South America are controlled by large-scale atmospheric circulation patterns modified by the topographic effects of the Andes (Hueck 1978). In particular at mid-latitudes, south of c. 37°S, within the realm of the dominant westerly circulation, this continuous N–S stretching mountain range forces the orographic uplift of moist Pacific air producing annual rainfalls on the windward slopes of 3,000–5,000 mm. By the time air masses have reached the Patagonian plains, normally about 100 km eastwards of the main continental divide, annual rainfall has declined an order of magnitude (Veblen 2007).

Concomitant with these strong rainfall gradients are also steep W–E gradients in continentality. Western slopes of the Andes and the Chilean central valley lowlands are strongly buffered by the Pacific Ocean from extreme temperatures, while the Patagonian plateau, usually drier and cloud free, displays larger diurnal and inter annual temperature amplitudes with regular frosts periods and warmer summer temperatures (Burgos 1985; Kleinebecker et al. 2007). Because of the seasonal influence of subtropical high pressure building up during the summer, Mediterranean-type precipitation seasonality develops over mid-latitudes producing stronger water deficits to plants during the summer warmer on the drier eastern slopes and Patagonian plains.

An additional effect of the Andean Cordillera is the existence of chains of active volcanoes, mostly located in Chilean territory (c. 50 between 36°S and 44°S; Besoain 1985) which, in combination with the dominant westerly wind direction, largely control soil parent material. Therefore western soils that lie within the domain of volcanic plumes are Andesitic soils of distance decreasing granulomery from the ejection sources, whereas eastern soils are more relatively more dominated by gaciofluvial deposits. While physical fertility is higher in western volcanic-derived soils, chemical fertility is higher in eastern soils derived from glaciofluvial deposits (Gaitán and Lopez 2007; Morales et al. 2010). Volcanic soils are rich in allophanes (amorphous clays), have low bulk density, high water retention capacity and contain more organic carbon, therefore providing high macroporosity, aeration, and drainage capacity. However, on the other hand, amorphous Al associated to allophanes and humic complexes effectively retain available P. In high precipitation areas where volcanic deposits are coarser, pumicite base lixiviation is higher. Therefore base cations, cation exchange capacity, and P availability all decrease from east to west thus decreasing chemical fertility (Morales et al. 2010). Low physical fertility (soil compaction due to high clay contents with depth) has been proposed as an impediment for root development in eastern soils so that toward the eastern extreme of the rainfall gradient plants remain shallow-rooted and thus unable to compensate with rooting depth for lower water availability (Schulze et al. 1996).

3.2.2 Ecosystem Properties Across the Transition

As in many other rainfall-controlled ecotones total plant biomass sharply decreases with decreasing rainfall (and physical soil fertility despite higher chemical fertility). Mean annual precipitation explains more than 60% of the variance in NDVI, an estimator of aboveground net primary production which varies from c. 2,500 kg ha^{-1} in the wettest portion (i.e., forests) to c. 500 kg ha^{-1} in the driest extreme (i.e., steppe) of the biogeographical transition (Paruelo et al. 1993, 1998). However, belowground biomass decreases at a lower rate, thus increasing root: shoot ratios. Irrespective of the position along the rainfall gradient most roots are located in shallow layers despite availability of deeper water at least during parts of the year (Schulze et al. 1996). Therefore water balance alone cannot fully explain vegetation along this ecotone and poses the question of why native trees do not invade the grassland and vice versa. One explanation is that zonation could be reflecting more establishment conditions than adult growth conditions. So that trees may not be able to successfully establish in the grasslands where they could potentially grow as adult individuals and grassland plants could not establish in wetter habitats due to light competition despite having abundant water availability for potential growth as adults (Schulze et al. 1996). This may be one reason why trees planted as juveniles or adults (e.g., exotic plantations) are successful in grassland areas.

3.2.3 Plant Communities and Plant Diversity Across the Transition

Broad-scale native vegetation types mirror the sharp W–E climatic gradients (Fig. 3.1). At mid-latitudes of Southern South America (37–43°S) temperate Valdivian rainforests dominate low-mid elevations of the western windward slopes of the Andes with some outskirts of rainforest occurring on the leeward westernmost areas of Argentinean Andes where low passes allow produce high precipitation and mild temperature microclimates.

The Valdivian rainforest represent the forests with maximum biomass and arboreal species richness (Arroyo et al. 1995) due to regional coexistence of disturbance-dependent emergent shade intolerant trees (*Nothofagus dombeyi, Eucryphia cordifolia, Weinmannia trichosperma, Fitzroya cupressoides*) with shade-tolerant

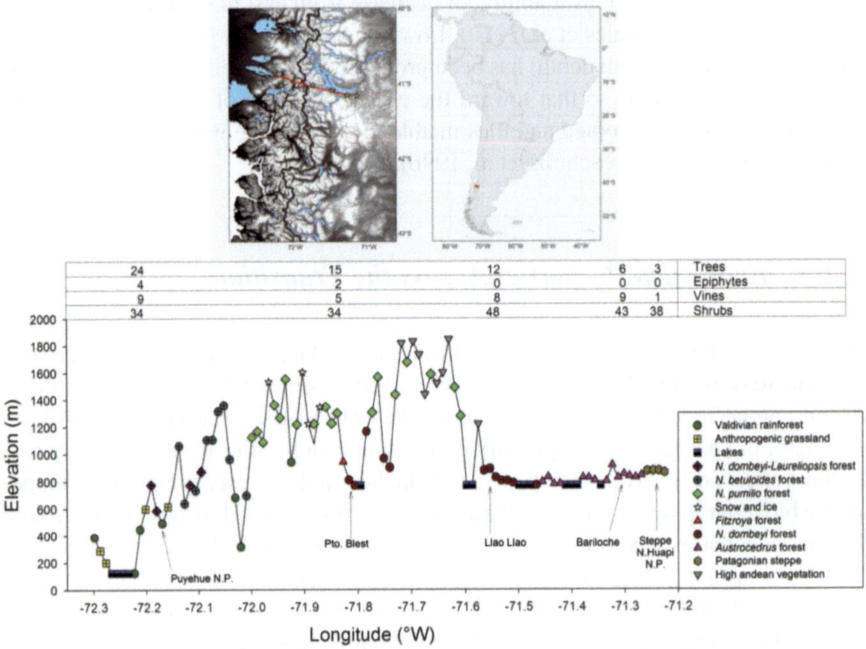

Fig. 3.1 Topographic profile across a trans-Andean gradient at c. 41°S in southern South America indicating mayor vegetation types (source: Lara et al. 2000). *Number* indicate species richness of trees, epiphytes, vines, and shrubs recorded at five localities across the transition (indicated with *stars* in the inset map): Puyehue National Park, Puerto Blest, Llao Llao Municipal Park, San Carlos de Bariloche and Eastern Nahuel Huapi National Park (after Ezcurra and Brion 2005; Aizen and Ezcurra 2008)

cold-sensitive rainforest tree components (*Aextoxicon punctatum, Saxegothea conspicua, Laureliopsis philippiana, Gevuina avellana, Podocarpus nubigena, Persea lingue*, numerous mirtaceae, etc.) typical of old-growth undisturbed forests (Veblen et al. 1996).

On the leeward slopes of the Andes in the Argentinean Lake District, a pure *Nothofagus* forest develops due to the loss of shade tolerant tree taxa at lower precipitation levels and higher elevations of the Patagonian Plateau (baseline levels of Atlantic draining lakes range from c. 500–1,000 m a.s.l.). These include evergreen *N. dombeyi*-dominated forests between lake level and 1,100 m of elevation and deciduous *Nothofagus pumilio*-dominated forests in the altitudinal range of 1,100–1,600 m. North of 40° 35′ (Sabatier et al. 2011), two deciduous trees (*Nothofagus alpina*, and *Nothofagus obliqua*) can codominate with *N. dombeyi* and more seldomly dominate the forests. In stressful sites such as frost prone/flood prone valley bottoms, xeric or disturbed slopes, the small tree/tall shrub *Nothofagus antarctica* replaces these forest types. *Nothofagus* forest may be codominated by conifers, such as the long lived *F. cupressoides* (at mesic sites > 2,500 mm year^{-1} south of c. 40° 57′S; Kitzberger et al. 2000a), *Araucaria araucana* (north of c. 40° 20′S), or

Austrocedrus chilensis. Occurrence of conifers in coexistence with *Nothofagus* in mesic sites is related to infertile conditions such as waterlogged bogs (*Fitzroya*), thin soils or rocky outcrops (*Araucaria* and *Austrocedrus*; Veblen et al. 2005). Understories are dominated by the 2–5 m tall bamboo *Chusquea culeou*.

Eastwards, with further decreases in precipitation (1,500–1,000 mm year^{-1}) *Araucaria* and *Austrocedrus*, being the relatively more drought tolerant conifers, form monospecific forests or coexist within a matrix of tall shrublands composed mainly by the bamboo (*C. culeou*), *N. antarctica* and several other small trees or tall shrubs (*Maytenus boaria*, *Lomatia hirsuta*, *Embothrium coccinueum*, *Diostea juncea*, *Schinus patagonicus*, etc.). With further decreases in precipitation (<1,000 mm year^{-1}), this matrix changes into a bunchgrass (*Stipa speciosa*, *Festuca pallescens*)/ cushion shrub (*Mulinum spinousum*)-dominated steppe where *Austrocedrus* or *Araucaria* intersperse either as isolated trees forming a woodland or small forest patches in wetter topographic positions, or rocky/stony outcrops, the former at low-mid elevations (800–1,300 m) the latter at higher elevations (1,000–1,800 m a.s.l) in association with the outskirts of high elevation *N. pumilio* forests and *N. antarctica* shrublands. Last outlier populations of *Austrocedrus* populations within the patagonian steppe are found at c. 300 mm year^{-1} (Pastorino and Gallo 2004) and of *Araucaria* at c. 450 mm year^{-1}, while easternmost angiosperm trees are small *N. pumilio* and *N. antarctica* forest clumps growing at c. 450–500 mm year^{-1} on radiation-protected south-facing slopes.

Between 40 and 42°S on western slopes of the Andes, the southern temperate rainforests host the maximum regional vascular plant species richness of southern South America with a total of 44 species of trees (Fig. 3.1; Arroyo et al. 1995). East of the continental divide Brion and Ezcurra (2005) list 15 tree species for the wet forest environment, 6 tree species for the transitional forest environment, and no trees for the steppe environment. Tree species list for specific locations along a W–E gradient (modified from Aizen and Ezcurra 2008) show a similar decline from 24 tree species at Puyehue National Park (Chile), 15 trees species in the Puerto Blest area, 12 trees in the Lla Llao Municipal Reserve area, 6 tree species in the Bariloche area, and 3 tree species in the transitional steppe in East Nahuel Huapi National Park (Fig. 3.1). Epiphytes show a similar decline with no species on the eastern slopes of the Andes, whereas climbers show a strong decline just at eastern end of the gradient. Shrub species show maximum richness in the transitional forests and woodland of the eastern slopes (Fig. 3.1). Speziale et al. (2010) show a similar patterns of plant richness variation within the eastern watershed, with a strong reduction tree species richness east of c. 71.4°W, coincident with a peak in shrub species richness (c. 71.5°W). At these intermediate portions of the transition herbs increase in richness peaking over the eastern end of the gradient (c. 71°W). Local plant species richness across the gradient is strongly controlled by precipitation and daily temperature range.

Woody species richness in particular, is strongly related to productivity. An increase of plant cover coupled with higher allocation to woody biomass, to increase the height and to have an advantage in competing for light, often occurs when water stress decreases and carbon fixation is enhanced across precipitation gradients. Tree cover (and its correlated litter cover), however has a strong negative influence on shrub and herb richness thus explaining the reduction in richness and suggesting strong competitive interactions between these life forms in the wetter portion of the transition (Speziale et al. 2010). Therefore shrub species richness is probably controlled in wetter portions of the gradient by light competition and in the drier portion of the gradient by reduced precipitation. Thus, although similar richness of shrubs is evident over the extremes of the gradient, species turnover is very high. Only five shrubs species (10–15%) are shared between the westernmost and easternmost floras in Fig. 3.1. These different shrub floras are largely different in ecological traits such as shade and drought tolerance, possibly reflecting important differences in evolutionary and biogeographic histories between forest and steppe taxa.

Based on these patterns, rather than defining a single ecotone, three sharp zones of transition can be defined across a trans-Andean gradient. First, the western transition (c. 72°W) from the tree/epiphyte species rich Valdivian rainforest to the structurally more simple species poor *Nothofagus*-dominated forests. This transition zone coincides approximately with the 600–700 m a.s.l. altitudinal limit and 3,000–2500 mm year^{-1} precipitation boundary on the western slopes of the Andes or in some low passes that cross into the eastern watershed of the Andees. A second sharp transition occurs further east (c. 71.6°W) where the continuous *Nothofagus* forest cover breaks up giving rise to first patchy but further east more extensive species rich shrublands composed of helliophylous species. This transition occurs where annual precipitation drops below c. 1,800 mm year^{-1}. Finally, a third transition takes place at c. 71.0–71.2°W where easternmost small outpost trees population (*A. chilensis and N. pumilio*) intermingle within the Patagonian steppe matrix. This transition coincides with the belt of maximum herb diversity (Speziale et al. 2010) in rainfall areas below C. 600–800 mm year^{-1}.

3.3 Disturbance Variation and Forest Dynamics Across the Transition

3.3.1 Fine-Scale Disturbances

Dominant disturbances that largely control forest dynamics vary strongly across the transition (Fig. 3.2b). Andean Valdivian rainforest is dominated by two superimposed disturbance regimes: fine-scale tree fall gap dynamics and coarse-scale disturbances of tectonic origin. Landscapes dominated by the fine-scale disturbances consisting of small openings produced by large, emergent shade intolerant trees but where shade tolerant trees have gradually colonized the understory induces an autogenic successional change towards dominance of shade tolerant trees such as *S. conspicua*, *L. philippiana*, *A. punctatum*, *P. lingue,* and several mirtaceous trees. In absence of large-scale disturbances, shade intolerant trees such as *N. dombeyi* cannot establish in dense understory populated by the shade tolerant trees and the rapidly and vigorously proliferating *Chusquea* bamboos (Veblen et al. 1996).

While gaps created by individual tree fall or the disturbance created by small groups of trees are a pervasive disturbance across the transition, the ecological role

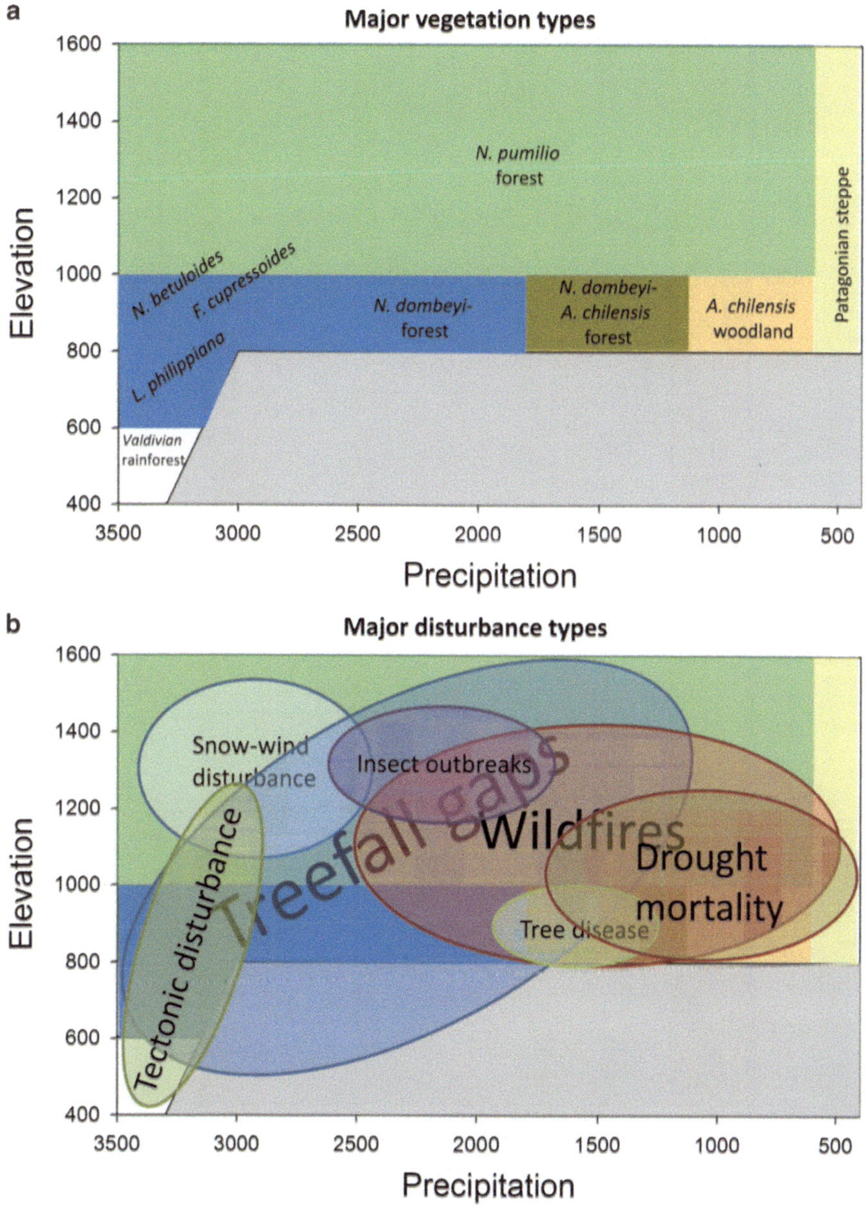

Fig. 3.2 (**a**) Potential distribution of vegetation types across a elevation (m)-precipitation (mm year^{-1}) space in at c. 41°S. Note that given the elevated plateau and base lake level on the eastern slope of the Andes, elevation <800 m a.s.l. are not available for precipitation levels <3,000 mm year^{-1}. (**b**) Distribution of dominant disturbance regimes across a elevation-precipitation space at c. 41°S

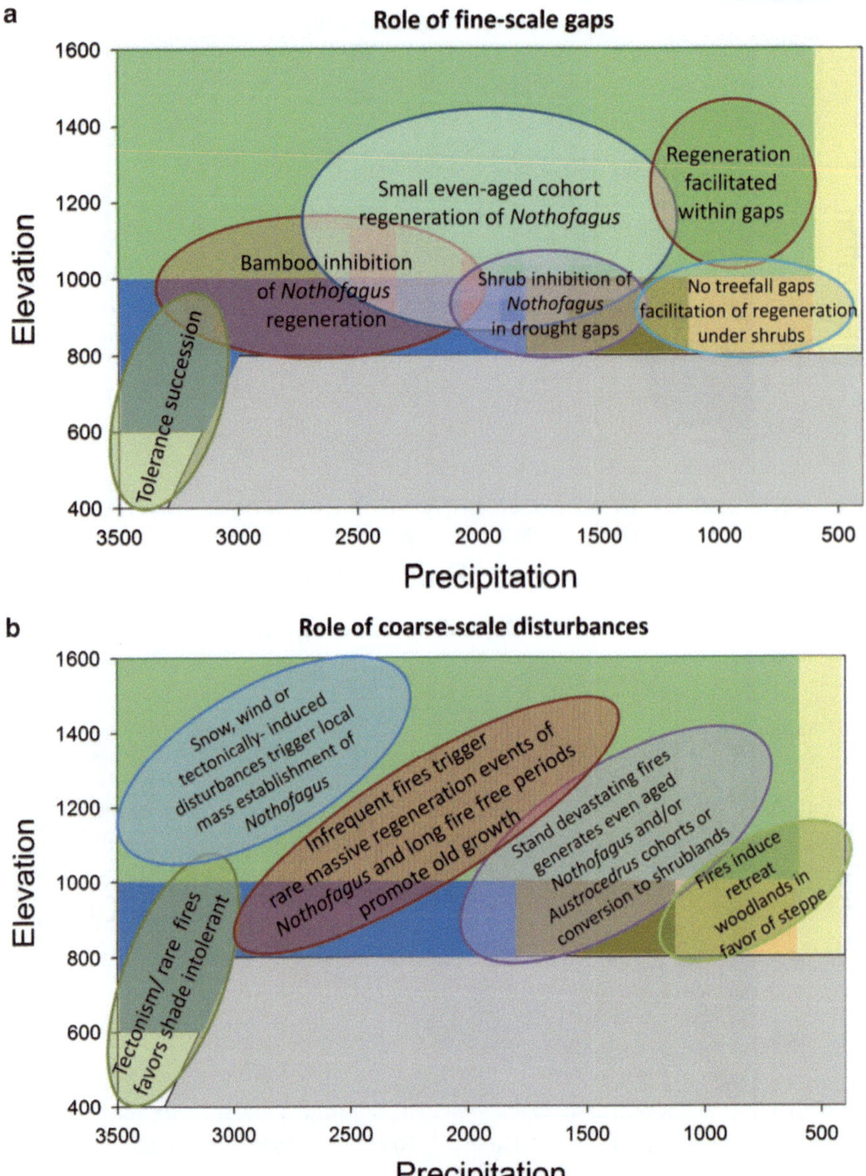

Fig. 3.3 Ecological role of fine-scale (tree-fall gap) (**a**) and coarse-scale disturbance (**b**) regimes across a elevation-precipitation space at c. 41°S

of such disturbances in the dynamics of the forest varies geographically (Fig. 3.3a; Veblen 1989). Over the wettest portion of the transition (>2,500 mm year^{-1}) and at lower altitudes (<600 m a.s.l.), tree fall gaps generate opportunities for succession towards communities dominated by shade tolerant trees. With the disappearance of shade tolerant trees on the eastern slopes of the Andes, understories at elevation <1,200 m a.s.l. are almost purely dominated by the bamboo *C. culeou*. When tree fall gaps are generated, this species aggressively responds by rapidly filling up the gaps with dense bamboo clumps that preclude *N. dombeyi* regeneration except where large nurse logs in tree fall gaps of 500–1,000 m^2 occasionally permit successful seedling establishment above much of the bamboo (Veblen 1989). At higher elevations (>1,200 m) or intermediate precipitation levels (2,000–1,600 mm year^{-1}), *C. culeou* bamboo thickets loose vigor and eventually disappear so that *N. pumilio* (Heinemann et al. 2000; Heinemann and Kitzberger 2006) and *N. dombeyi* can, respectively, regenerate as small even-aged cohorts in smaller gaps (Veblen 1989).

Importance of tree fall gaps in low elevation dry forests is lower because frequent fires have prevent the development of old-growth forests. However, given strong reduction in modern fire frequency, many tall dense *N. dombeyi* postfire cohorts are undergoing synchronous senescence opening medium-sized gaps. A similar pattern is produced in wet tall *Austrocerdus* forests that, triggered by *Pytophtora* root attack are producing massive mortality and large forest openings (LaManna et al. 2008). These larger gaps have the potential of developing even-aged cohorts; however, high cattle and exotic deer densities are effectively impeding regeneration and promoting the "emptying" of understories or the development of understory communities dominated by unpalatable/spiny shrubs (Veblen et al. 1989, 1992a; Relva et al. 2009). Another source fine-scale gaps are extreme drought events that periodically trigger *Nothofagus* canopy mortality or partial crown dieback. A modern example of this is a massive *N. dombeyi* dieback event triggered by a century record drought occurred in 1998–1999 (Suarez et al. 2004). Drought gaps created by this event are rapidly filled by preestablished shrub understories preventing the regeneration of *N. dombeyi* and favoring the relatively more shade tolerant establishment of the conifer *Austrocedrus* (Suarez and Kitzberger 2008).

In dry old-growth *N. pumilio* subalpine forests, tree fall gaps and dieback gaps regeneration dynamics is more controlled by facilitative effects than by the classic idea of competition release (Heinemann et al. 2000). Understory vegetation, mesic microclimates (e.g., shady gap boundaries, decomposed logs, etc.) are microsites where *N. pumilio* is successful within gaps. Finally at lower altitudes and lower precipitation levels in the realm of the *Austrocedrus* woodlands, the concept of gaps does not hold as these open woodlands are not limited by light but by water availability and thus, regeneration is largely controlled by rainfall variability and nurse-shrub facilitation (Villalba and Veblen 1997; Kitzberger et al. 2000b).

3.3.2 Coarse-Scale Disturbances

Being located at a boundary between two major crustal plates, the Valdivian Andean region is an area of great volcanic and tectonic activity, which results in ecological disturbances. Forest disturbances associated to tectonism such as earthquake-triggered landslides, mudflows, debris avalanches, river floods and associated volcanic lava flows, and regional volcanic sand/ash depositions promote the massive establishment of shade intolerant trees such as *N. dombeyi*, *W. trichosperma*, *E. cordifolia*, and eventually *F. cupressoides*. Composition and structure of most Andean Valdivian rainforest show the prevalence of old, large emergent shade intolerant trees, thus suggesting the importance of relatively frequent coarse-scale disturbances in promoting coexistence of shade tolerant and shade intolerant trees, and therefore, high tree species diversity (Veblen and Ashton 1978; Veblen et al. 1980, 1996).

Because of its location in the area of influence of an annually southward migrating Pacific anticyclone, the northern Patagonian transition displays relatively strong precipitation seasonality with relatively dry spring-summers and wet-cool fall-winters. Thus, fire encounters spatially variable fuel quantities and fuel desiccation conditions to become a coarse-scale disturbance with different recurrences and ecological roles across the transition (Kitzberger et al. 1997).

In the eastern steppe, fire is limited by the amount of horizontal fuel continuity. Similar to other steppes (Miller and Heyerdahl 2008), when the amount of bare ground is above a certain threshold (<c. 50%; e.g., overgrazed areas, rocky or stony outcrops), fire is not able to freely spread across the landscape. Counterbalancing this limitation, fine fuels (e.g., grasses) are desiccated and readily burnable every summer.

Conversely, in the wettest portion of the transition, large amounts of are available, yet very exceptional prolonged drought events are able to desiccate coarse fuels of rainforests and create opportunities for these communities to ignite and spread fires. Very occasionally and associated to strong droughts, and/or the massive flowering and dieback of several bamboos species, wildfires can affect hyperhumid forests. However, once spreading under these conditions, events become very large due to the high horizontal fuel continuity. Possibly the best of these two worlds for fire is found across the transition at intermediate-dry positions and lower elevations where abundant fuels are still produced in large quantities but where the water deficit is high enough (or fuels are fine enough) to generate frequent fuel desiccation opportunities for fires to spread.

Estimates of fire recurrence across the transition has been derived from annually resolved tree ring fire scar records, particularly from the conifers, the relatively fire-resistant *F. cupressoides* in the mesic forests and scar-forming *A. chilensis* at the dry end of the transition (Veblen et al. 1999). Mean composite fire intervals (i.e., the time lapses between fires dates recorded by different trees in a site) suggest that prior to the arrival of Euro-Argentineans to northern Patagonia, fires recurred at individula sites every c. 100 years in a mesic *Fitzroya-Nothofagus* forest (3,000 mm year^{-1}) and every c. 7–20 years for dry *Austrocedrus* woodlands (Veblen et al. 1992b, 1999; Kitzberger and Veblen 1997). Mean point fire intervals (i.e., the time lapses between fires dates recorded by single trees), a more conservative indication of presettlement local fire recurrence is not available for *Fitzroya-Nothofagus* rainforest but is estimated to be >3–5 centuries and was 27–42 years in the dry *Austrocedrus* woodlands (Kitzberger and Veblen 1997). Fire severity changes across the transition with high severity fires at intermediate rainfall and wet forests as well as for higher elevation forests, while dryer woodlands carry mixed severity regimes.

Temporally fire regimes have changed differently across the transition due to different histories of land occupation, cultural changes and therefore ignition frequency changes. In the dry woodlands and steppes, highest fire frequencies correspond to aboriginal times (pre-1877), a period of frequent active fire use for guanaco and rhea hunting, communication, warfare, etc. by ancient nomadic Tehuelche cultures. With military occupation of these lands and land conversion to large sheep and cattle ranches, fire frequency dropped dramatically (Veblen and Lorenz 1987; Kitzberger and Veblen 1997). In contrast, mid to wet portions of the transition (north Patagonian Lake District) was less inhabited and used less fire, thus infrequent wet forest fire during presettlement times may have primarily been naturally ignited. During the 1890s and early 1900s, colonization of forested areas was primarily driven by European settlers with a strong culture of forest opening with fire to create cattle pasture (Veblen et al. 2003). During dry enough summers, many of these forest clearing fires escaped and went out of control and thus a sharp increase in fire frequency took place during the turn of the nineteenth to twentieth centuries. By the 1940s, with the creation of the Argentinean National Park system and the cultural change towards negative connotations of wildfires, fire frequency drastically reduced up until the late 1980s after which large fires occurred related to drought and bamboo flowering/withering (Veblen et al. 2008).

Ecological consequences of fire and fire regime change across the transition (Fig. 3.3b). Similar to tectonic mass disturbances, fire in the western mesic forests induces strong colonization pulses and the development of dense even-aged cohorts of shade intolerant trees such as *N. dombeyi or N. pumilio* (Veblen et al. 1992b; Kitzberger 1994). If present, emergent thick-barked long lived conifers *F. cupressoides* or *A. araucana* (further north) may survive crown fires and endure at the site several fire cycles and generations of angiosperm (Veblen et al. unpubl.). However, given long fire free intervals mesic forest attain old-growth stages which at lower elevations on the western Andes develop into diverse Valdivian rainforest dominated by shade tolerant trees. At higher elevation or on eastern slopes of the Andes forest are typical old-growth or old even-aged pure *Nothofagus* stands with understories largely dominated by the bamboo *C. culeou*. Despite the lack of *Nothofagus* gap-phase regeneration in both Valdivian and pure wet *Nothofagus* forests, coarse-scale disturbances such as rare fires are frequent enough to ensure persistence and dominance of this shade intolerant species.

At intermediate portions of the transition and at higher elevations crown fires induce the formation of large even-aged stands of *N. dombeyi* and or *Austrocedrus* at lower altitudes (Veblen and Lorenz 1987). Given the high frequency of fires during the settlement period, this portion of the gradient is currently dominated by c. one century old even-aged stands that are undergoing cohort senescence. Contrasting with typical massive, large crowned tree falls in the old-growth forest, these forests develop tree toppling of tall self-thinned trees. In some cases when fires are too severe or too frequent, thin barked obligate-seeded *N. dombeyi* and *N. pumilio* trees fail to reestablish and the site becomes dominated by shrublands dominated by resprouting species (*N. antarctica, C. culeou, S. patagonicus, D. juncea, M. boaria, L. hirsuta*, etc.; Mermoz et al. 2005; Kitzberger et al. 2005) which in turn for being more flammable than the *Nothofagus* forest, induce alternative self-maintaining mosaic of states of shrubland and forest (Kitzberger et al. 2012).

Finally, at the eastern end of the transition, fire has an important role in regulating the relative extent of forest (mainly *Austrocedrus* in lowlands and *N. pumilio* in highlands) and steppe. During periods of higher fire frequency/severity, *Austrocedrus* retracts into suboptimal rocky/stony fire refugia where the amount of burnable biomass decreases and fires do not spread or are less lethal to adult trees (Veblen and Lorenz 1988; Kitzberger et al. 2012). Similarly, during high fire activity *N. pumilio* forests retract into moister aspects and higher elevations at the expense of steppe advances (Fig. 3.5; Gowda et al. 2012).

3.4 Direct and Disturbances-Mediated Influences of Climate Variability Across the Transition

Climatic variability can affect ecotone dynamics by directly influence on demography of dominant plant species, and by indirectly modifying the occurrence, frequency or severity of disturbances. Across the north Patagonian transition, climatic variability influences forest dynamics by affecting different key processes (Suarez and Kitzberger 2010). Near the eastern distribution limit of both *Austrocedrus* (c. 600–800 mm year^{-1}) and *N. dombeyi* (c. 1,200–1,400 mm year^{-1}), strong droughts produce massive tree mortality (Villalba and Veblen 1998; Suarez et al. 2004; Fig. 3.4a). In the conifer, episodes of massive tree mortality coincide with exceptionally dry springs and summers during the 1910s, 1942–1943, and the 1950s, whereas for the angiosperm, the exceptionally hot dry summer of 1998–1999 global change-type drought triggered the episodic death and partial crown dieback of *N. dombeyi* trees over c. 11,000 ha in Nahuel Huapi National park (Suarez et al. 2004). Interestingly, this recent drought did not affect *Austrocedrus*, and, apparently, early century droughts that affected the conifer did not induce massive mortality of *N. dombeyi* (Suarez and Kitzberger 2010). It is likely that timing and duration of drought may be having influence on different tree species across the transition (Allen et al. 2010) with multiyear droughts affecting more drought-adapted conifer and episodic seasonal droughts affecting the more moisture-demanding angiosperm.

In the mesic forests, droughts had only a minor effect on tree mortality but here, exceptionally rainy or snowy or windy winters seem to be the drivers of tree mortality by either uprooting or snow breakage events (Suarez and Kitzberger 2010). This is particularly the case of *N. dombeyi* which, due to a swallow rooting and evergreen habit, offers high aerodynamic resistance to wind and low wind firmness. Ever greenness and right branch angles in *N. dombeyi* also cause the accumulation of high snow loads which can produce high levels of branch fall and eventually entire crown breakage. Therefore, in these mesic lowland forest dynamics is largely linked to *mechanical-type disturbances* (water, wind, and snow) promoting death and regeneration, whereas dry forests dynamics tend to be controlled by *physiological-type*

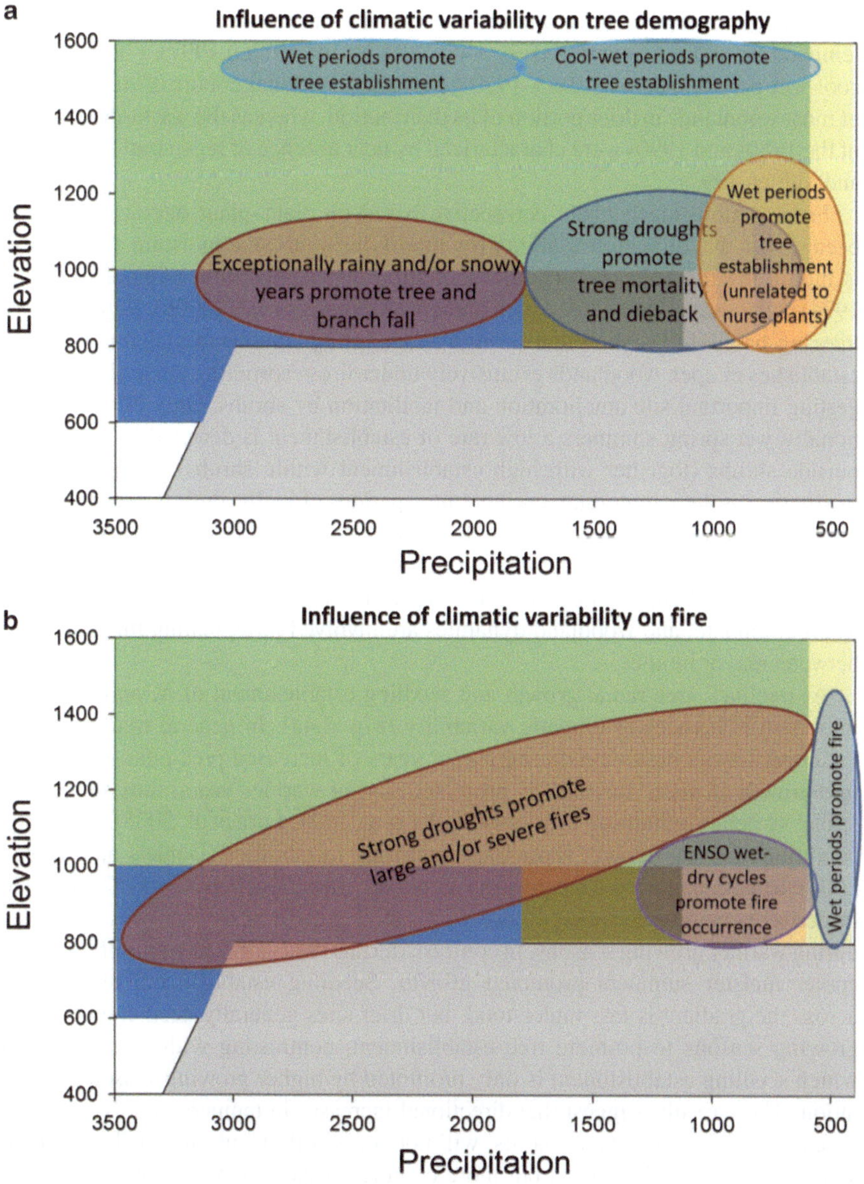

Fig. 3.4 Direct (demographic; (**a**)) and indirect (disturbance mediated; (**b**)) influence of climatic variability on forest dynamics across a elevation-precipitation space at c. 41°S

agents (e.g., soil water deficit, atmospheric demand, heat shock, etc. Suarez and Kitzberger 2010).

Climatic variability can control forest dynamics by modifying tree establishment rates. In the drier portion of the transition *Austrocedrus* seedling establishment rates are largely controlled inter annual and supra annual cycles in spring-summer temperature and precipitation (Fig. 3.4a; Villalba and Veblen 1997). For instance a cool-wet period spanning 1963–1979 generated a regional cohort of *Austrocedrus* of most woodlands in drier portion of its distribution, whereas the warm-dry decades of the 1980s and 1990s were characterized by near absence of regeneration (Villalba and Veblen 1997).

From a more mechanistic perspective based on plant–plant interactions it has been shown that climatic variability (by modifying year–year moisture availability) is modifying the balance between facilitation and competition (Aguiar and Sala 1994) and thus regulating tree establishment (Kitzberger et al. 2000b). Retrospective tree ring-based studies show that during average spring-summer rainfall *Austrocedrus* establishes in open woodlands exclusively under fire-resprouting nurse shrubs, suggesting important site amelioration and facilitation by shrubs. Only during exceptionally wet spring-summers a low rate of establishment is detected in open areas outside shrubs (together with high establishment within shrubs), suggesting that facilitation is less important during rainy periods. Finally during drought years establishment fails in all microsites (Kitzberger et al. 2000b). Because, once trees establish and overtop the shrubs, shading induces the death of heliphyllous shrubs (Veblen et al. 1992a, b), important but complex linkages between shrubland dynamics, fire, climate, and woodland dynamics are involved in regulating this boundary between mayor biomes.

At treelines, tree radial growth and seedling establishment of *N. pumilio* show important influences of climatic variability (Fig. 3.4a). In general tree growth at xeric treelines is promoted during cooler years of increased precipitation whereas tree growth at mesic treelines is promoted during warmer years, thus suggesting strong switches in limiting factors (Villalba et al. 1997; Lara et al. 2005). Similarly, conditions for tree growth show important time instabilities (Daniels and Veblen 2004). Irrespective of treeline position across the transition, between 1957 and 1976 (a period generally considered cool-wet) radial growth at treeline was promoted during warmer growing seasons. In contrast, during 1977–1996 (a warm-dry period), cooler moister summers promoted growth. Seedling establishment at treelines across the gradient is less understood, but drier sites generally need cooler-wetter growing seasons to promote tree establishment, contrasting with wetter sites at which seedling establishment is only promoted by higher growing season precipitation. These results suggest that directional increases in temperature, as predicted by current global climate scenarios, will not necessarily result in a upslope expansion of the *N. pumilio* forests but more complex responses of treelines across the transition (Daniels and Veblen 2004).

The occurrence of wildfires is largely controlled by climatic variability through its action of modifying fine fuel buildup rates and fuel desiccation (Fig. 3.4b). On the easternmost portions of the transition where steppe bunchgrasses dominate, fires are limited by fuel amounts and continuity. Because fine fuels (grasses) are highly responsive to precipitation pulses, during rainy growing seasons, ecosystems that normally do no spread efficiently due to lack of fuel loads suddenly become more prone for developing large fires (Morgan et al. 2003).

In *Austrocedrus* woodlands fine fuels coexist with coarse fuels. In these ecosystems as in other woodlands worldwide widespread fires are triggered by a sequence consisting of a pluvial year followed by a drought year. Such inter annual sequences are often tele-connected and driven by El Niño Southern Oscillation (ENSO) so that widespread fires in these regions tend to be phase-locked with ENSO (Swetnam and Betancourt 1990; Kitzberger and Veblen 1997). Thus, the strength and amplitude of the climatic oscillation is directly related to fire frequency in the northern Patagonian synchronizing power over other woodlands worldwide (Kitzberger et al. 2001).

Further west in the transition or higher at altitude, in the realm of the tall *Nothofagus* forests, fine fuels are less important and coarse fuels that require long drying periods dominate. Here fires are exclusively associated to strong droughts lasting several months, beginning during the winter, the time when soils are replenished with water. Whenever dry winter-springs associate with warm summers, wet forests ignite and spread fire without significant natural fire breaks (Mermoz et al. 2005). These strong drought events not only produce larger fires but also more severe events that create conditions that provide less regeneration opportunities to obligate seed dispersed species (such as *N. dombeyi* or *N. pumilio*; Kitzberger et al. 2005) and more opportunities for the rapid expansion of resprouting shrubland species such as *N. antarctica, C. culeou,* and other resprouting shrub species (Mermoz et al. 2005).

Climatic variability strongly controls via fire severity the landscape scale the regeneration probabilities as well as the transition probabilities from forest to shrublands (Mermoz et al. 2005; Gowda et al. 2012). Presumably many mid-slope shrublands that dominate north Patagonian forested landscapes are the result of large severe fires of the past (Veblen and Lorenz 1988; Kitzberger and Veblen 1999). An important fuel component that interacts with climatic variability in western wet and hyperhumid forests are the bamboos. Whenever drought coincides with postflowering massive die-off of bamboo culms (particularly of the dense, tall *C. culeou* or the climbing *Chusquea quila*) extremely large fires burn the wet forests. This was possibly the case of large fires reported in Puelo and Nahuel Huapi National Parks in 1944 which presumably coincided with a *C. culeou* flowering in the early 1940s in combination with a 2-year drought in 1943 and 1944 (Veblen et al. 2008). Likewise, in the mid-late 1990s large fires developed in the wet Valdivian district related to the 1993–1995 flowering of *C. quila* and associated to strong drought years (Veblen et al. 2008).

3.5 Climate, Fire, Land Use, and Long-Term Vegetation Changes Across the Transition

Vegetation changes occurred nonhomogeneously across the transition as a response to different spatio-temporal changes in human occupation, land use, and climate, so that there is a need to partition these changes by belts.

3.5.1 Xeric Steppe-Woodland Belt

High resolution pollen/charcoal from L. Mosquito a relatively dry site (mean annual precipitation 750 mm year^{-1}) shows that fire has been a prevalent disturbance that controls woodland-steppe dynamics since at least 9 ky BP (Whitlock et al. 2006). Fire episodes based on charcoal peaks suggest that the fire free interval was lower (c. 70 year) during 9–7 kyr BP, increased to 50 year during 6–2 kyr BP and decreased sharply during the last 500 years of the record to present day fire regime with fire free intervals of c. 200 years (Whitlock et al. 2006).

Although initial occupation in South America dates back c. 13 kyr BP. in the form of extinct mega faunal hunters, in northern Patagonia it was possibly not until about 8,000 years BP when indications of a new cultural assemblage utilized more specialized hunting techniques and tools (e.g., fire, boleadoras) indicative of guanaco and rhea hunting (Markgraf and Anderson 1994). Given the main habitat of guanacos hunting activities remained largely restricted the ecotonal grasslands and open woodland portion of the transition where game was more abundant and easier to hunt with substantial impacts on the landscape as recorded by early (nineteenth century) explorers and missionaries that mention large-scale fires set by Indians to hunt guanacos and rheas (Cox 1863; Musters 1871; Fonck 1896). Possibly, the extent of marginal eastern forest (*Austrocedrus*, *N. pumilio*) relative to grasslands during the last millennia may have been largely controlled by the amount of human set fires with periods of forest advance during more fire active periods (e.g., higher population density) and perhaps increased climate variability and/or an increase in convective storms (Markgraf et al. 1997; Whitlock et al. 2006).

During the early/mid-Holocene anthropological human occupation hiatus (Gil et al. 2005), despite being characterized by increased aridity (Salemme and Miotti 2008), forests ecotonal forests may have increased in extent and continuity due to reduced fire frequency. Genetic and niche modeling evidence suggests that small, isolated *Austrocedrus* forest patches in the grassland matrix are neither glacial relicts nor the results of recent migration, but rather evidence of a late Holocene more extensive forests and woodlands (Kitzberger et al. 2009).

During the mid/late Holocene as the Andean foothills were increasingly being used and populated by more organized hunter-gatherer societies, and as climate became more variable (e.g., Holocene ENSO onset), *Austrocedrus* forests may have become fragmented and confined to in rocky/stony fire refugia were fuel amount and continuity makes fire less lethal for these trees or unable to spread (Kitzberger et al. 2009). These forest retreats probably accentuated during the last stages of Aboriginal domain (seventeenth to nineteenth centuries), a period of large societal changes such as the adoption of horses and dogs in combination with fire as hunting tools, and increase in the use of fire for warfare related to strong migrations of Mapuche into the eastern slopes of the Andes forced by and earlier Spanish colonization period in south-central Chile.

With the collapse of the aboriginal domain after the Argentinean Army Desert Campaign in 1877, land was converted into large Euro-Argentinean owned sheep

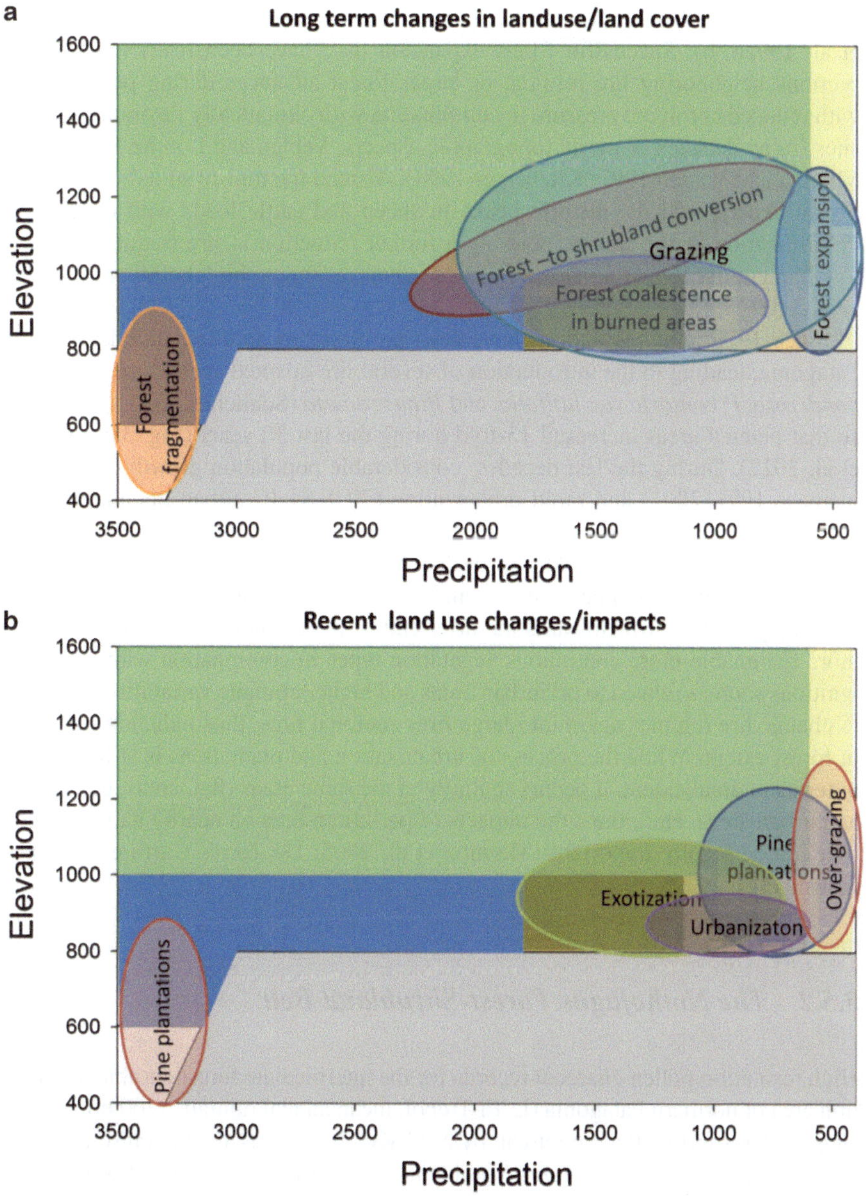

Fig. 3.5 Long-term (last centuries; (**a**)) and recent (last decades (**b**)) land cover/land use changes and related impacts across a elevation-precipitation space at c.41°S

and cattle ranches, and with this fire frequency declined abruptly and *Austrocedrus* forests began a new period of expansion (Veblen and Lorenz 1987; Kitzberger and Veblen 1997; Fig. 3.5a). This expansion however, became rapidly limited by the new set of introduced, namely herbivores sheep, cattle, exotic deer, and European hares which exert pressure on the regeneration of trees (Veblen et al. 1989, Veblen et al. 1992a, b). Successful forest expansion may have been restricted to small portions neighboring fire refugia, or larger forest advances during periods/areas with reduced herbivore pressure in combination with climatically favorable periods/sites (rainy decades or mesic topographic aspects; Veblen and Lorenz 1987, 1988; Villalba and Veblen 1997; Kitzberger 1994). Around the mid-twentieth century the region experienced the historic peaks in sheep and cattle loads with consequent desertification of the steppe, possibly a further reduction in fire frequency due to low fuel continuity but no apparent forest advances due to high herbivory pressure (Fig. 3.5a).

Since 1970, pine plantations have been promoted as an alternative land use in Patagonia, leading to the introduction of several fire-adapted species such as *Pinus ponderosa*, *P. contorta var. latifolia*, and *Pinus radiata* (Schlichter and Laclau 1998) so that planted areas increased 15-fold during the last 30 years (Fig. 3.5b; Gowda et al. 2012). During the last decades, considerable population growth (8.4–21.3% increase 1991–2001) and rapid urbanization (79.9–84.4% urban population from 1991 to 2001) has occurred in Northern Patagonia (INDEC 2001). All major cities in northern Patagonia (San Martin de los Andes, Bariloche, Esquel-Trevelin) are located in woodland steppe transition and, generally bordering westwards protected areas they tend to grow towards the drier end of the gradient. The combination of more flammable more continuous vegetation types in combination with increased ignitions sources related to periurban areas and higher climatic variability is starting to change fire regimes again into large fires ecotonal fires, thus inducing reduction in forest extent. While the process of urbanization and plantations is still incipient in terms of areal extent, it occurs spatially in the same areas (flat, areas, near roads, and urban areas, etc.), thus, the impacts of periubran fires on nearby native vegetation can be locally important (Mermoz et al. 2005; De Torres Curth et al. 2008; Gowda et al. 2012).

3.5.2 The *Nothofagus* Forest-Shrubland Belt

High resolution pollen-charcoal records for the intermediate forest-shrubland transition area of northern Patagonia (L. El Trebol, mean annual rainfall 1,700 mm year^{-1}) suggest that shortly after ice retreat, by c. 11 Kyr BP fire was already present in relation to an open *Nothofagus* forest/shrubland mosaic (Whitlock et al. 2006). During the early Holocene, fire frequency was low in wetter areas, possibly reflecting the fact that forests offered much less resources to hunter-gatherer tribes than steppe territories and were therefore much less populated and used. These intermediate *Nothofagus*-dominated forest areas had estimated fire free intervals based on charcoal peaks of c. 100–500 years (Whitlock et al. 2006) thus, old-growth forests may have been dominant features.

By 6 kyr fire interval began to shorten from c. 200 years to c. 100 years in 3.5 kyrs BP with highest fire activity during 2.5 kyr BP, possibly in response to increased inter annual climate variability has been noted in many South American records after

6,000 cal yr BP (McGlone et al. 1992) and is attributed to the onset (Markgraf and Diaz 2000) or increased importance (Rodó and Rodriguez-Arias 2004) of ENSO variability. The last two millenia reflect an increase of *Nothofagus* and reduction of *Austrocedrus* in response to wetter conditions, however fire intervals remained relatively short (c. 100 years) with high grass (or bamboo) charcoal (Whitlock et al. 2006), suggesting that forest openings were burning creating shrubland patchiness.

By the late nineteenth century European settlers arrived to this landscape and sharply increased fire frequency in the need of opening forests for cattle raising and strongly amplified vegetation shifts. During the early twentieth century fire intervals shortened approximately one order of magnitude (e.g., 1675–1920: 82 years, 1920–1950: 8 years based on fire scar data at L. Roca; Kitzberger et al. 1997; Veblen et al. 1999). Large escaped burns, well documented by early explorers and surveyors (Willis 1914; Rothkugel 1916), depending on the severity, converted large areas into shrubland dominated by resprouting shrubs and bamboo in severe fires (Mermoz et al. 2005; Gowda et al. 2012) or, when sufficient trees survived regenerated in vast even-aged stands of *N. dombeyi* in wetter areas, *N. pumilio* at higher elevations, *N. dombeyi-Austrocedrus* at intermediate rainfall, or pure *Austrocedrus* forests in drier positions (Veblen and Lorenz 1987, 1988; Fig. 3.5a). Shrublands originated from those fires became permanent features of the landscapes because they are more flammable than adjacent forests and therefore carry a self-reinforcing fire regime that maintains shrubland-forest boundaries (Kitzberger et al. 2012), because forests are unable to reestablish due to lack of near seed sources, lack of appropriate microclimatic conditions (Kitzberger et al. 2005; Gowda et al. 2012), or because dominant trees are shade intolerant and unable to establish under the shrub canopy. An exception to this is *Austrocedrus* whose establishment is facilitated by under the shrub canopy (Kitzberger et al. 2000b; Gobbi and Schlichter 1998; Letourneau et al. 2004).

Since the early twentieth century fires *Austrocedrus* forests that retreated into wet rocky fire refugia as a response to settlement period fires, is successionally replacing many shrublands at a regional scale (Gowda et al. 2012). A recent land cover comparison between historical maps and satellite imagery shows that that during the last century, northern Patagonian forests expanded as even-aged forest over almost 50% of settlement-related burned land, and on > 60% of the shrublands (Gowda et al. 2012, Fig. 3.5a). However, these trends were far from homogeneous in time and space: net forest expansion took place mainly during the wetter mid-twentieth century (Villalba and Veblen 1997), whereas since the 1970s a marginal forest retraction (and conversion to shrublands) is evident probably due to recent of recent severe fires associated to extreme droughts (Mermoz et al. 2005).

Superimposed to these fire–vegetation interactions are also important changes in mammal herbivory levels. Native herbivores in these forests are two small endangered deer species. After European colonization high herbivore loads have been introduced into the forest in the form of extensive cattle raising regimes as well as introduction of exotic game species such as red and fallow deer (*Cervus elaphus* and *Dama dama*, respectively).

These herbivores have radically changed structure, composition, and diversity of most understory plant communities and are seriously deterring or impeding gap-phase, postfire and advance regeneration of most forest types (Relva and Veblen 1998; Veblen et al. 1989, 1992a; Tercero-Bucardo et al. 2007; Relva et al. 2009; Barrios-García et al. 2012). Synergistic effects of herbivores and severe disturbances (Relva et al. 2009) are likely inducing conversion to shrublands because herbivores actively select disturbed habitat or, in the case of fire because same traits that protect plants from browsing also confer higher flammability thus making resprouting shrubs more prone than fire sensitive trees to dominate heavily browsed disturbance prone situations suggesting a self-reinforcing effect mediated by herbivores (Veblen et al. 2011; Raffaele et al. 2011).

Finally, the intermediate belt of the transition dominated of forests and shrublands is also witnessing recent, dramatic exotization of its flora (Fig. 3.5b). Shrublands at intermediate positions in the gradient are the habitats that show higher levels of exotization (c. exotics represent 19% of total cover) (Speziale and Ezcurra 2011). Important human activities in the shrublands that include plantation forestry and sheep, cattle and horse raising, high propagule pressure from roads, relatively high productivity, high radiation levels, nurse effects and empty niches all contribute to this trend (Speziale and Ezcurra 2011). Particularly worrisome is the ongoing invasion of the tree *Pseudostuga menziezii* from plantations and gardens into native forests and shrublands, a species that apparently found an empty niche related to high shade tolerance and with the potential of largely altering long-term successional trajectories of native communities, ecosystem processes, and fire regimes (Veblen et al. 2011).

3.5.3 The Wet Rainforest Belt

Despite high precipitation levels fire has been present in the rainforests since the late glacial (c. 15 kyrs BP; Moreno 2004; Whitlock et al. 2007). Until c. 11 kyrs, fire frequency was low (fire intervals based on charcoal peaks 500–1,100 years, Abarzúa and Moreno 2008). During the early Holocene (c. 11–9 kyrs BP) a relative increase in fire frequency occurred with fire intervals of c. 250 years. This shift in fire regime is coincident with a period of widespread warmth and aridity along mid-high latitudes in southern South America (Whitlock et al. 2007), increased human burning activities (Moreno et al. 2001) or past increased convective activity and fire ignition by lightning (Abarzúa and Moreno 2008) in this modern-day atmospherically stable region. Higher fire frequency periods in some areas are related to more open forest conditions and increases in shade intolerant trees such as *W. trichosperma* together light demanding understory taxa such as Poaceae (bamboos). During the mid-Holocene, fire frequency declined (500–1,000 year fire intervals) to increase again at some rainforest sites during the late Holocene (3 kyr BP) to c. 250 years intervals (Abarzúa and Moreno 2008).

Despite occurrence of local fires, tectonic and volcanic events, pollen evidence suggests that the western slopes of the Andes have supported evergreen temperate rainforests throughout the entire Holocene. Millennial-scale climate changes have however induced several events of floristic changes. Tree composition has fluctuated from north Patagonian forests dominated by cold-resistant taxa (*P. nubigena*, *F. cupressoides*, *Pilgerodendron nubigena*, *S. conspicua*) during cool late glacial conditions to Valdivian forests with thermophilous, summer-drought resistant species (*E. cordifolia*, *Caldcluvia paniculata*) during warming trend from the early to mid-Holocene. Coincident with ENSO onset and the strengthening of the southern westerlies, several cooling trends during the mid-Holocene led to several reexpansion events of cold-resistant rainforest trees at the expense on thermophyllous taxa. Since c. 3 kyrs warming has led to the decline of cold tolerant trees and the establishment of modern vegetation (Moreno 2004).

Humans inhabited this region concurrently with the onset of the postglacial climate warming and the period of maximum fire activity in the past 15,000 years. Starting in the mid-Holocene, a shift occurs from nomadic to social more organized sedentary communities that trigger an agricultural expansion in the form of slush-and-burn agriculture (Dillehay 1997). Local openings created by these activities may have generated extensive clearings and agricultural fields especially along the main river basins which later with the demise of aboriginal during Colonial time may have presumably been reinvaded by *Nothofagus* (Armesto et al. 2010; Fig. 3.5a).

During the Colonial time 1500s–1800s large-scale deforestation was associated with the expansion of wheat farming and cattle introduction (Armesto et al. 2010). This trend intensified during the late 1800s and early 1900s with extensive burning practices by settlers as ownership of the land was gained by clearing forests (Rozzi et al. 2000). Enormous extension of forest that had remained as such since their arrival after the last glacial period were extirpated within a few years during the late 1800s in the Chilean Lake District, one of the most rapid deforestation events to be recorded in Latin America (Veblen 1983). These areas remained deforested due to limited conditions for tree reestablishment.

Finally, starting in the mid-twentieth century industrial forestry developed based on the extensive replacement of native forests, secondary forests, and agricultural land with fast-growing exotic tree plantations (Fig. 3.5b). Fueled by government subsidies and market economy the area planted with *P. radiata* and *Eucalyptus* tripled between the years 1980 and 2000 (Armesto et al. 2010). Replacement to plantation of exotic species is currently the greatest threats of native temperate rainforest (Neira et al. 2002) without an adequate reserve network (Wilson et al. 2005).

3.6 Conclusions

The north Patagonian forest-steppe transition has usually related to a classic rain shadow effect. It is however increasingly clear that vegetation is responding not only to a simple gradient of water but also to whole complex multivariate set of correlated physical conditions such as cloudiness, frost frequency, thermal regime, soil physical, and chemical fertility among other factors. Adding further complexity, the physical

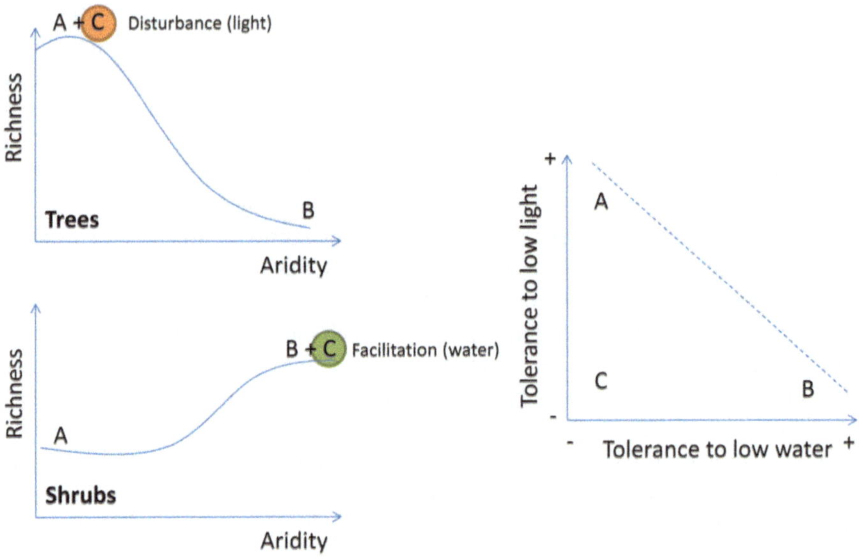

Fig. 3.6 Proposed hypothetical explanation of tree and shrub richness patterns across the trans-Andean north Patagonian transition from Valdivian rainforest to Patagonian steppe based on species trade-offs between shade tolerance and drought tolerance (modified from Smith and Huston 1989)

microscale conditions experienced by plant across the transition are themselves modified by other plants so that neighboring plants can exert competitive or facilitative effects based on resources, physical, soil amelioration, etc. (Holmgren 2000).

Plant physiological and demographic responses to these sets of factor are also complex and in some case counterintuitive. Invariance in rooting depth along the gradient is one example. Contrasting establishment vs. growth requirement may also create counterintuitive patterns such as the existence of large portions of the transition were certain species (or functional types) can grow but do not exist due to limitation during critical establishment phases (Schulze et al. 1996).

Community responses are even more complex. Richness patterns along the transition are largely dependent on the functional/structural group. Species or functional trait sorting along the transition has been shown to generally respond to changes in major resources such as light availability and water availability and the fact species display life history trade-offs so that single species cannot be at the same time tolerant to two limiting resources (Smith and Huston 1989). In this transition, tree species richness peak at high precipitation (Figs. 3.1 and 3.6) can be partially explained by the coexistence of shade tolerant/drought intolerant species (functional group A, Fig. 3.6) with shade intolerant/drought intolerant species (functional group C) that has opportunities of establishment mediated by frequent during coarse-scale disturbances that open up the closed forest canopy (Veblen et al. 1980). Species richness declines towards the dry end because only shade intolerant/drought tolerant (group B) can establish under these aridity/high light conditions. Shrubs instead peak in richness at intermediate to dry conditions along the transition. At the wet extreme closed shady conditions or rapid canopy closure after disturbance does not allow the existence of shade intolerant/drought intolerant species (functional group C) and only shade tolerant/drought intolerant shrubs are able to survive under the canopy (functional group A). High radiation conditions at the arid end allow the existence of shade intolerant/drought tolerant shrubs (group B) which may have an ameliorating facilitative microclimatic effect (Holmgren 2000; Raffaele and Veblen 1998) thus promoting the coexistence with shade intolerant/drought intolerant shrubs (functional group C, Fig. 3.6).

Ecotones can be defined at multiple hierarchical spatial scales from local patch boundary scales driven by dispersal, demography etc., meso-scale transitions defined with species turnover and beta diversity, to the classic biome boundary conception. Northern Patagonian ecotone forest-steppe patch boundary dynamics occurs where forest islands contact the grass/shrubland matrix or where grass/shrub patches perforate the forest matrix and an intermediate state where neither forest patches nor forest holes are discernible (Gruyère structure, sensu, Rapoport 1982).

Small-scale ecotone dynamics in northern Patagonia is driven by the balance between forest-promoting and grassland/shrubland-promoting factors. Factors that promote forests expansion (both by tree island expansion or forest gap filling) are increases in spring-summer precipitation and reductions in summer temperature over multiannual periods, fire frequency reductions, heterogeneity in the form of rock/stony fire refugia, herbivore pressure reductions and facilitate effects of shrubs on trees. Factors that favor shrub expansion (either by forest island reduction or forest gap expansion) are increased frequency of droughts and warm spells promoting forest die-off and severe large wildfires, increased ignition and fire frequency reducing forest resilience to subsequent fires or promoting self-reinforcing shrubland/grassland states, increased herbivory reducing tree establishment (directly or by affecting facilitation) and promoting more flammable vegetation. Interactive effects of climate fire and herbivory can further promote shrublands/grassland expansion in detriment of forest.

At the meso-scale the northern Patagonian ecotone can be defined from W–E by three main transitions, (1) a wet transition with loss of structural complexity and tree richness that defines the limit of Valdivian and pure *Nothofagus* forest, (2) an intermediate transition with loss of forest continuity at expense of extensive shrublands, and (3) a dry transition where tree and elements gradually fade into the steppe matrix. Each of these transitions is likely to be controlled by different sets of factors. The loss of shade tolerant elements is probably regulated by temperature and coarse disturbance frequency, the second transition a function of fire severity which in turn relates to drought frequency, and the third transition is governed by climate, herbivory, and long-term fire regimes.

Finally the macro scale defines the classic concept of boundaries between major biomes in this case temperate forests and grasslands. At this scale the boundary is defined by some average lines that divides functional type dominance in this case tree vs. grass dominance can be broadly predicted from temperature and precipitation and correlate well with water balance (Woodward 1987). However, the role of fire in globally preventing ecosystems from achieving the potential height, biomass, and the dominant functional types is increasingly being acknowledged (Bond et al. 2005). For this ecotone, "fire off" Dynamic Global Vegetation model simulations show that the forest biome boundary shifts eastwards, thus suggesting a strong fire control on the current northern Patagonian forest-steppe boundary (Bond et al. 2005). Even at these coarse scales it is evident that other factors besides average climate (e.g., disturbance, climate variability, anthropogenic impacts) control ecotones dynamics.

Clearly the nature regime and consequences of disturbances vary markedly across the northern Patagonian ecotone. Tectonism and tree fall gaps has a decreasing W–E influence while fire has the reverse pattern. Consequences of tree falls on forest dynamics are radically different along the transition. On the most productive species rich end fine-scale gaps release suppressed trees from light competition mobilizing succession towards shade tolerant species or resource preempting understory species (e.g., bamboos). On the arid end, fine-scale gaps, when present, do not trigger immediate responses but slow reestablishment in biotic (tree shade, nurse shrub protection, decomposing coarse woody debris) or abiotic (south slopes) "safe" microsites (Heinemann et al. 2000; Heinemann and Kitzberger 2006).

The role of coarse-scale disturbances is also variable across the transition. In the core Valdivian rainforest, fires or tectonic disturbances favors the opportunities of less competitive shade intolerant trees promoting higher tree diversity. At mid-sections of the transitions, fires generate landscape heterogeneity in terms of producing a mosaic of even-aged stands or bi-phase mosaics of forests and shrublands. Finally, at the xeric end, frquent fires have a homogenizing role in promoting the retraction of forests and advance of grasslands.

Climatic variability has also a variable role across the transition both directly and mediated by fire. Direct effects of climatic fluctuations have clearly more marked effects at the drier end of the transition. Forest die-off related to inter annual or multiannual droughts tends to occur at eastern limits of the dominant trees species (*Austrocedrus*: Villalba and Veblen 1998, *N. dombeyi*: Suarez et al. 2004) with important consequences on plant community composition (Suarez and Kitzberger 2008). Rainfall variability is also largely controlling tree establishment rates of dominant tree species across local forest-steppe and forest-shrubland boundaries (Villalba and Veblen 1997; Heinemann 2007) as well as in postfire situations (Tercero-Bucardo 2009).

Direct effects of climatic extremes in more productive portions of the transition are related mechanical disturbance (extreme snowfall, wind, or rainfall events) driving mortality and subsequent growth and establishment (Suarez and Kitzberger 2010).

Interestingly, in opposition to direct effects rainfall variability, indirect effects of droughts mediated by fire are more important in wet than in dry extremes of transition. Much deeper/longer droughts are necessary to generate fires in the coarse fuel dominated hyperhumid western forests or high altitude subalpine, while average summers generate conditions for burning in fine fuel—dominated shrublands, grasslands, or dry woodlands (Kitzberger et al. 1997). Therefore, increases in rainfall variability may have a dual role across transitions, generating *more demographic dynamism* at the xeric end (establishment pulses during extremely wet periods, and mortality events during deep or persistent droughts). In the wetter portion of the transition climatic variability increases *physical disturbance dynamisms* more fine-scale tree falls during extremely windy, snowy, rainy periods, and more coarse-scale disturbances in the form of large fires during deep seasonal droughts.

Long-term records across the transition suggest that both climate and fire has fluctuated markedly throughout the Holocene. In addition, patterns of pre-Hispanic human occupation have been widely different between the westerns slope of the Andes with relatively more inhabited eastern slopes and widely used woodland areas. Patterns of settlement are also largely different with a long settlement history on the Chilean slopes, little impact on the eastern Andes and a relatively young settlement history of the eastern slopes. Modern patterns of land use vary wildly across the transition. Highest degrees of human intervention and impacts are being witnessed at both the wet and the dry extremes of the transition but based on radically different mechanisms. Western rainforest have been converted to agriculture and more recently to plantations based on short period of brute-force events fragmentation by burning, agriculture, cattle raising, and industrial tree planting. In contrast, the dry extreme of the transition is witnessing a high levels of human intervention based on a more diffuse pattern of land degradation by fire, exotic herbivory, exotization, and local plantations and urbanization. Possibly produced by the inaccessibility of the Andes and the creation of an important National Park system, mid-sections of the gradient have witnessed the least amount of anthropogenic changes and impacts.

The northern Patagonian transition between one of the most structurally and biological diverse temperate forest and treeless Patagonian steppe develops over only 70 km. Despite being this one of the steepest vegetation transitions on earth I have attempted to show providing the model of the northern Patagonian forest-steppe transition that (1) no single environmental factor explains biological variation, (2) ecotones are not boundary lines but a complex entity of hierarchically structured spatial patterns, (3) ecotones have a complex set of natural disturbance regimes with spatially varying roles, (4) different portions of ecotones are affected by climatic variability in rather different and finally (5) ecotones are areas differentially used by humans and thus differently impacted by different land use. All these characteristics identified call for a more integrated understanding of the complex internal dynamics to gain better understanding of worldwide consequences of anthropogenic global changes on these important biological spatial transitions.

References

Abarzúa AM, Moreno PI (2008) Changing fire regimes in the temperate rainforest region of southern Chile over the last 16,000 yr. Quat Res 69:62–71

Aguiar MR, Sala OE (1994) Competition, facilitation, seed distribution and the origin of patches in a Patagonian steppe. Oikos 70:26–34

Aizen MA, Ezcurra C (2008) Do leaf margins of the temperate forest flora of southern South America reflect a warmer past? Glob Ecol Biogeogr 17:164–174

Allen CD, Macalady AK, Chenchouni H, Bachelet D, McDowell N, Vennetier M, Kitzberger T, Rigling A, Breshears DD, Hogg EH, Gonzalez P, Fensham R, Zhang Z, Castro J, Demidova N, Lim J, Allard G. Running SW, Semerci A & N Cobb (2010) A global overview of drought and heat-induced tree mortality reveals emerging climate change risks for forests. Forest Ecology and Management 259:660–684

Armesto JJ, Manuschevich D, Mora A, Smith-Ramirez C, Rozzi R, Abarzúa AM, Marquet PA (2010) From the Holocene to the Anthropocene: a historical framework for land cover change in southwestern South America in the past 15,000 years. Land Use Policy 27:148–160

Arroyo MK, Cavieres L, Peãloza A, Riveros M, Faggi AM (1995) Relaciones fitogeorgáficas y patrones regionales de riqueza de especies en la flora del bosque lluvioso templado de Sudamérica. In: Armesto JJ, Villagrán C, Arroyo MK (eds) Ecología de los bosques nativos de Chile. Editorial Universitaria, Santiago, Chile, pp 71–92

Barrios-García MN, Relva MA, Kitzberger T (2012) Patterns of use and damage by exotic deer on native plant communities in northwestern Patagonia. Eur J Wildl Res 58:137–146

Besoaín E (1985) Los suelos. In: Toso J (ed) Suelos volcánicos de Chile. Instituto de Investigaciones Agropecuaria (INIA), Ministerio de Agricultura, Santiago, Chile, pp 25–95

Bond WJ, Woodward FI, Midgley GF (2005) The global distribution of ecosystems in a world without fire. New Phytol 165:525–538

Burgos JJ (1985) Clima del extremo sur de Sudamerica. In: Boelcke O, Moore DM, Roig FA (eds) Transecta Botánica de la Patagonia Austral, CONICET. Argentina, Inst. Patagonia Chile & Royal Society Great Britain, Buenos Aires, pp 10–40

Cox G (1863) Viajes a las regiones septentrionales de Patagonia 1862–1863. An Univ Chile 23(3–239):437–509

Curth De Torres, Mónica I, Ghermandi L, Pfister G (2008) Los incendios en el noroeste de la Patagonia: su relación con las condiciones meteorológicas y la presión antrópica a lo largo de 20 años. Ecol Austral 18:153–167

Daniels LD, Veblen TT (2004) Spatiotemporal influences of climate on altitudinal treeline in northern Patagonia. Ecology 85:1284–1296

Di Castri F, Hansen AJ, Holland MM (1988) A new look at ecotones. Emerging international projects on landscape boundaries. Special Issue 17- Biology International. IUBS, p 161

Dillehay T (ed) (1997) The archaeological context, vol. II of Monte Verde: a late Pleistocene settlement in Chile. Smithsonian, Washington, DC

Ezcurra C, Brion CAM (2005) Plantas del Nahuel Huapi. Catálogo de la Flora Vascular del Parque Nacional Nahuel Huapi, Argentina. In: Bavaria (ed) San Carlos de Bariloche, pp 1–70

Fonck F (1896) Libro de los diarios de Fray Francisco Menéndez. In: Niemeyer CF Valparaiso, Chile

Gaitán JJ, López CR (2007) Análisis del gradiente edáfico en la región andinopatagónica. Cienc Suelo 25:53–63

Gil A, Zárate M, Neme G (2005) Mid-Holocene paleoenvironmentsand the archeological record of southern Mendoza, Argentina. Quat Int 132:81–94

Gobbi M, Schlichter T (1998) Survival of *Austrocedrus chilensis* seedlings in relation to microsite conditions and forest thinning. For Ecol Manage 111:137–146

Gowda JH, Kitzberger T, Premoli AC (2012) Landscape responses to a century of land use along the northern Patagonian forest-steppe transition. Plant Ecol 213:259–272

Heinemann K (2007) Aspectos espaciales y temporales del establecimiento de Nothofagus pumilio en claros de bosques maduros en los extremos de un gradiente ambiental en el Noroeste de la Patagonia. Doctoral Thesis Biology. Universidad Nacional del Comauhe

Heinemann K, Kitzberger T (2006) Effects of position, understorey vegetation and coarse woody debris on tree regeneration in two environmentally contrasting forests of north-western Patagonia: a manipulative approach. J Biogeogr 33:1357–1367

Heinemann K, Kitzberger T, Veblen TT (2000) Influences of gap microheterogeneity on the regeneration of *Nothofagus pumilio* in a xeric old-growth forest of northwestern Patagonia, Argentina. Can J For Res 30:25–31

Holmgren M (2000) Combined effects of shade and drought on tulip poplar seedlings: trade-off in tolerance or facilitation? Oikos 90:67–78
Hueck K (1978) Los bosques de Sudamérica. GTZ. Hoehl Druck, Bad Hersfeld, Alemania, p 476
INDEC (2001) Censo Nacional de Poblacion, Hogares y Viviendas 2001. http://www.indec.gov.ar/webcenso/publicaciones/publicaciones.asp
Kitzberger T (1994) Fire regime variation along a northern Patagonian forest-steppe gradient: stand and landscape response. Ph.D. Thesis, University of Colorado, Boulder, CO, USA
Kitzberger T, Veblen TT (1997) Influences of humans and ENSO on fire history of *Austrocedrus chilensis* woodlands in northern Patagonia, Argentina. Ecoscience 4:508–520
Kitzberger T, Veblen TT (1999) Fire-induced changes in northern Patagonian landscapes. Landsc Ecol 14:1–15
Kitzberger T, Veblen TT, Villalba R (1997) Climatic influences on fire regimes along a rainforest-to-xeric woodland gradient in northern Patagonia, Argentina. J Biogeogr 23:35–47
Kitzberger T, Pérez A, Iglesias G, Premoli AC, Veblen TT (2000a) Distribución y estado de conservación del alerce (*Fitzroya cupressoides* (Mol.) Johnst.) en Argentina. Bosque 21:79–89
Kitzberger T, Steinaker DF, Veblen TT (2000b) Effects of climatic variability on facilitation of tree establishment in northern Patagonia. Ecology 81:1914–1924
Kitzberger T, Swetnam TW, Veblen TT (2001) Inter-hemispheric synchrony of forest fires and the El Niño-Southern Oscillation. Glob Ecol Biogeogr 10:315–326
Kitzberger T, Raffaele E, Heinemann K, Mazzarino MJ (2005) Effects of fire severity in a north Patagonian subalpine forest. J Veg Sci 16:5–12
Kitzberger T, Souto C, Premoli A (2009) The ghost of fragmentation past: habitat suitability models and genetic markers help unveil dry forest non-equilibrium landscapes in northern Patagonia. 2009 Latin American IALE Conference, Campos de Jordao, Brazil
Kitzberger T, Aráoz E, Gowda JH, Mermoz M, Morales JM (2012) Decreases in fire spread probability with forest age promotes alternative community states, reduced resilience to climate variability and large fire regime shifts. Ecosystems 15:97–112
Kleinebecker T, Holzel N, Vogel A (2007) Gradients of continentality and moisture in South Patagonian ombrotrophic peatland vegetation. Folia Geobot 42:363–382
Kupfer JA, Cairns DM (1996) The suitability of montane ecotones as indicators of global climatic change. Prog Phys Geogr 20:253–272
LaManna L, Mateucci SD, Kitzberger T (2008) Abiotic factors related to the incidence of the *Austrocedrus chilensis* disease syndrome at a landscape scale. Forest Ecol Manage 256:1087–1095
Lara A, Rutherford P, Montory C, Bran D, Pérez A, Clayton S, Ayesa J, Barrios D, Gross M, Iglesias G (2000) Vegetación de la Eco-región de los Bosques Valdivianos Escala 1:500.000. Informe Final Proyecto Binacional Chile–Argentina UACh—INTAAPN–FVSA: Boletín Técnico Fundación Vida Silvestre Argentina, N°51
Lara A, Villalba R, Wolodarsky-Franke A, Aravena JC, Luckman BH, Cuq E (2005) Spatial and temporal variation in *Nothofagus pumilio* growth at tree line along its latitudinal range (35°40′–55°S) in the Chilean Andes. J Biogeogr 32:879–893
Letourneau FJ, Andenmatten E, Schlichter T (2004) Effect of climatic conditions and tree size on *Austrocedrus chilensis*-shrub interactions in northern Patagonia. Forest Ecol Manage 191:29–38
Markgraf V, Anderson L (1994) Fire history of Patagonia: climate versus human cause. Rev Inst Geogr S Paulo 15:33–47
Markgraf V, Diaz HF (2000) The past ENSO record: a synthesis. In: Diaz HF, Markgraf V (eds) El Nino and the Southern Oscillation; multiscale variability and global and regional impacts. Cambridge University Press, Cambridge, pp 465–488
Markgraf V, Betancourt J, Rylander KA (1997) Late-Holocene rodent middens from Rio Limay, Neuquen Province, Argentina. Holocene 7:325–329
McGlone M, Kershaw PA, Markgraf V (1992) El Nino/Southern Oscillation climatic variability in Australasian and South American paleoenvironmental records. In: Diaz HF, Markgraf V (eds)

El Nino, Historical and Paleoclimatic aspects of the Southern Oscillation. Cambridge University Press, Cambridge, UK, pp 435–462
Mermoz M, Kitzberger T, Veblen TT (2005) Landscape influences on occurrence and spread of wildfires in Patagonian forests and shrublands. Ecology 86:2705–2715
Miller RF, Heyerdahl EK (2008) Fine-scale variation of historical fire regimes in sagebrush-steppe and juniper woodland: an example from California, USA. Int J Wildland Fire 17:245–254
Morales D, LaManna L, Buduba C (2010) Propiedades químicas de suelos desarrollados sobre distintos materiales originales bajo bosques de *Austrocedrus chilensis*. Rev FCA UN Cuyo 42:185–200
Moreno PI (2004) Millennial-scale climate variability in northwest Patagonia over the last 15 000 yr. J Quat Sci 19:35–47
Moreno PI, Jacobson GL, Lowell TV, Denton GH (2001) Interhemispheric climate links revealed by a late-glacial cooling episode in southern Chile. Nature 409:804–808
Morgan P, Defoseé GE, Rodriguez NF (2003) Management implications of fire and climate in Western Americas. In: Veblen TT, Baker W, Montenegro G, Swetnam TW (eds) Fire and climatic changes in temperate ecosystems of the Western Americas. Springer, New York, pp 413–440
Musters GC (1871) At home with the Patagonians: a year's wanderings over untrodden ground from the straits of Magellan to the Rio Negro. John Murray, London
Neira E, Verscheure H, Revenga C (2002) Chile's frontier forests: conserving a global treasure. World Resources Institute, Comité Nacional Pro Defensa de la Fauna y Flora, Universidad Austral de Chile
Nielson RP (1991) Climatic constraints and issues of scale controlling regional biomes pp. 31–51. In: Holl MM, Risser PG, Naiman RJ (eds) Ecotones. The role of landscape boundaries in the management and restoration of changing environments. Chapman and Hall, New York, p 142
Paruelo JM, Aguiar MR, Golluscio RA, Leon RJC, Pujol G (1993) Environmental controls of the NDVI dynamics in Patagonia based on NOAA-AVHRR satellite data. J Veg Sci 4:425–428
Paruelo JM, Beltrán A, Jobbágy E, Sala OE, Golluscio RA (1998) The climate of Patagonia: general patterns and controls on biotic processes. Ecol Austral 8:85–101
Pastorino MJ, Gallo L (2004) Los cipreses de Pilcaniyeu. El extremo más árido de la distribución natural del ciprés de la cordillera. Presencia 16:17–19
Raffaele E, Veblen TT (1998) Facilitation by nurse shrubs of resprouting behavior in a post-fire shrubland in northern Patagonia, Argentina. J Veg Sci 9:693–698
Raffaele E, Veblen TT, Blackhall M, Tercero-Bucardo N (2011) Synergistic influences of introduced herbivores and fire on vegetation change in northern Patagonia, Argentina. J Veg Sci 22:59–71
Rapoport EH (1982) Areography: geographical strategies of species. Pergamon Press, Oxford
Relva MA, Veblen TT (1998) Impacts of introduced large herbivores on *Austrocedrus chilensis* forests in northern Patagonia, Argentina. Forest Ecol Manage 108:27–40
Relva MA, Westerholm CL, Kitzberger T (2009) Effects of introduced ungulates on forest understory communities in northern Patagonia are modified by timing and severity of stand mortality. Plant Ecol 201:11–22
Rodó X, Rodriguez-Arias MA (2004) El Nino-Southern oscillation: absent in the early Holocene? J Clim 17:423–426
Rothkugel M (1916) Los bosques patagónicos. Ministerio de Agricultura, Buenos Aires
Rozzi J, Silander JJ, Armesto PF, Massardo F (2000) Three levels of integrating ecology with the conservation of South American temperate forests: the initiative of the Institute of Ecological Research Chiloé. Chile Biodiv Conserv 9:1199–1217
Sabatier Y, Azpilicueta MAM, Marchelli P, González-Peñalba M, Lozano L, García L, Martinez A, Gallo LA, Umaña F, Bran D, Pastorino MJ (2011) Distribución natural de *Nothofagus alpina* y *Nothofagus obliqua* (nothofagaceae) en Argentina, dos especies de primera importancia forestal de los bosques templados norpatagónicos. Bol Soc Argent Bot 46:131–138
Salemme MC, Miotti L (2008) Archeological hunter-gatherer landscapes since the latest Pleistocene in Fuego-Patagonia. In: Rabassa J (ed) The Late Cenozoic of Patagonia and Tierra del Fuego.

Elsevier B.V. ISBN: 9780444529541, pp 437–483 [Developments in Quaternary Sciences 11:1–513]
Schlichter TM, Laclau P (1998) Ecotono estepa-bosque y plantaciones forestales en la Patagonia norte. Ecología Austral, 8(2), 285–296
Schulze ED, Mooney HA, Sala OE, Jobbagy E, Buchmann N, Bauer G, Canadell J, Jackson RB, Loreti J, Oesterheld M, Ehleringer JR (1996) Rooting depth, water availability, and vegetation cover along an aridity gradient in Patagonia. Oecologia 108:503–511
Smith T, Huston M (1989) A theory of the spatial and temporal dynamics of plant communities. Plant Ecol 83:49–69
Speziale KL, Ezcurra C (2011) Patterns of alien plant invasions in northwestern Patagonia, Argentina. J Arid Environ 75:890–897
Speziale KL, Ruggiero A, Ezcurra C (2010) Plant species richness-environment relationships across the Subantarctic-Patagonian transition zone. J Biogeogr 37:449–464
Suarez ML, Kitzberger T (2008) Recruitment patterns following a severe drought: long-term compositional shifts in Patagonian forests. Can J For Res 38:3002–3010
Suarez ML, Kitzberger T (2010) Differential effects of climate variability on forest dynamics along a precipitation gradient in northern Patagonia. J Ecol 98:1023–1034
Suarez ML, Ghermandi L, Kitzberger T (2004) Factors predisposing episodic drought-induced tree mortality in *Nothofagus*: site, climatic sensitivity and growth trends. J Ecol 92:954–966
Swetnam TW, Betancourt JL (1990) Fire–southern oscillation relations in the southwestern United States. Science 249:1017–1020
Tercero-Bucardo N (2009) Trayectoria sucesional posfuego con clima alterado e historia de regeneración en bosques templados de Nothofagus de Argentina. Doctoral Thesis Biology. Universidad Nacional del Comahue
Tercero-Bucardo N, Kitzberger T, Veblen TT, Raffaele E (2007) A field experiment on climatic and herbivore impacts on post-fire tree regeneration in north-western Patagonia. J Ecol 95:771–779
Veblen TT (1983) Degradation of native forest resources in southern Chile. In: Steen HK (ed) History of sustained-yield forests: a symposium. Forest History Society, Durham, NC
Veblen TT (1989) Tree regeneration responses to gaps along a transandean gradient. Ecology 70:541–543
Veblen TT (2007) Temperate Forests of the Southern Andean Region Pp. 217–231. In: Veblen TT, Young KR, Orme AR (eds) The Physical Geography of South America. Oxford University Press, Oxford, p 361
Veblen TT, Ashton D (1978) Catastrophic influences on the vegetation of the Valdivian Andes, Chile. Plant Ecol 36:149–167
Veblen TT, Lorenz DC (1987) Post-fire stand development of *Austrocedrus–Nothofagus* forests in Patagonia. Vegetatio 73:113–126
Veblen TT, Lorenz DC (1988) Recent vegetation changes along the forest/steppe ecotone in northern Patagonia. Ann Assoc Am Geogr 78:93–111
Veblen TT, Schlegel FM, Escobar B (1980) Structure and dynamics of old-growth *Nothofagus* forests in the Valdivian Andes, Chile. J Ecol 68:1–31
Veblen TT, Mermoz M, Martin C, Ramilo E (1989) Effects of exotic deer on forest regeneration and composition in northern Patagonia. J Appl Ecol 26:711–724
Veblen TT, Mermoz M, Martin C, Kitzberger T (1992a) Ecological impacts of introduced animals in Nahuel Huapi National Park, Argentina. Conserv Biol 6:71–83
Veblen TT, Kitzberger T, Lara A (1992b) Disturbance and vegetation dynamics along a transect from rain forest to Patagonian shrublands. J Veg Sci 3:507–520
Veblen TT, Donoso C, Kitzberger T, Rebertus AJ (1996) Ecology of southern Chilean and southern Argentinean *Nothofagus* forests. In: Veblen TT, Hill RS, Read J (eds) Ecology and biogeography of *Nothofagus* forests. Yale University Press, New Haven
Veblen TT, Kitzberger T, Villalba R, Donnegan J (1999) Fire history in northern Patagonia: the roles of humans and climatic variation. Ecol Monogr 69:47–67

Veblen TT, Kitzberger T, Raffaele E, Lorenz DC (2003) Fire history and vegetation changes in northern Patagonia, Argentina. In: Veblen TT, Baker W, Montenegro G, Swetnam TW (eds) Fire and climatic changes in temperate ecosystems of the Western Americas. Springer, New York, pp 265–295

Veblen TT, Armesto JJ, Burns BR, Kitzberger T, Lara A, León B, Young KR (2005) The coniferous forests of South America Pp. 701–725. In: Andersson FA (ed) Coniferous Forests. Ecosystems of the World. Elsevier, Brussels, p 600

Veblen TT, Kitzberger T, Raffaele E, Mermoz M, González ME, Sibold JS, Holz A (2008) The historical range of variability of fires in the Andean-Patagonian *Nothofagus* forest region. Int J Wildland Fire 17:724–741

Veblen T, Holz A, Paritsis J, Raffaele E, Kitzberger T, Blackhall M (2011) Adapting to global environmental change in Patagonia: what role for disturbance ecology? Austral Ecol 36:891–903

Villalba R, Veblen TT (1997) Regional patterns of tree population age structures in northern Patagonia: climatic and disturbance influences. J Ecol 85:113–124

Villalba R, Veblen TT (1998) Influences of large-scale climatic variability on episodic tree mortality in northern Patagonia. Ecology 79:2624–2640

Villalba R, Boninsegna JA, Veblen TT, Schmelter A, Rubulis S (1997) Recent trends in tree-ring records from high elevation sites in the Andes of northern Patagonia. Clim Chang 36:425–454

Whitlock C, Bianchi MM, Bartlein PJ, Markgraf V, Marlon J, Walsh M, McCoy N (2006) Postglacial vegetation, climate, and fire history along the east side of the Andes (lat 41–42.5°S), Argentina. Quat Res 66:187–201

Whitlock C, Moreno PI, Bartlein P (2007) Climatic controls of Holocene fire patterns in southern South America. Quat Res 68:28–36

Willis B (1914) El norte de la Patagonia. Dirección de Parques Nacionales, Buenos Aires

Wilson K, Newton A, Echeverría C, Weston C, Burgman M (2005) A vulnerability analysis of the temperate forests of south central Chile. Biol Conserv 122:9–21

Woodward FI (1987) Climate and plant distribution. Cambridge University Press, Cambridge, UK

Chapter 4
Woody–Herbaceous–Livestock Species Interaction

Temuulen Tsagaan Sankey

4.1 Introduction

Woody and herbaceous species interaction in disturbed and natural environments has attracted a great deal of research attention due to its implications for land cover change, land surface–atmosphere interaction, global carbon budget (House et al. 2003), biodiversity, primary and secondary productivity, and the associated land use management (Archer 1994). Ecosystems of mixed woody and herbaceous plants comprise 15–35% of the terrestrial surface area and are distributed from hot tropical to cold temperate climates across varying topography and soils (House et al. 2003). Mixed woody–herbaceous ecosystems are often heavily impacted by natural and anthropogenic factors such as fire and grazing (House et al. 2003).

When woody–herbaceous species balance is disturbed, one of the two life-forms is likely to dominate the other and a shift occurs in the density of woody and herbaceous plants and the location of woody–herbaceous boundaries, known as a kind of ecotone. Proximate causes of shifts in woody–herbaceous ecotones have been studied in many different parts of the world to understand the dynamics and balance between woody and herbaceous species. A few of these causes have been widely agreed upon to be the main driving factors. Most conceptual models of woody–herbaceous balance shifts acknowledge the interactive effects of multiple factors rather than a single driving force (Daly et al. 2000; House et al. 2003; Kupfer and Miller 2005). Among them are climate change, increased CO_2, nitrogen pollution, drought, fire suppression, and grazing (Dando and Hansen 1990; Archer 1994; Bachelet et al. 2000; Bartolome et al. 2000; Asner et al. 2004). Topographic slope and aspect, snow accumulation, and soil texture and depth further influence changes

T.T. Sankey (✉)
Idaho State University, 322 E. Front Street, Boise, ID 83712, USA
e-mail: sankteki@isu.edu

R.W. Myster (ed.), *Ecotones Between Forest and Grassland*,
DOI 10.1007/978-1-4614-3797-0_4, © Springer Science+Business Media New York 2012

in the balance and determine spatial patterns of the balance shift (Brown 1994; Walsh and Butler 1994; Kupfer and Cairns 1996).

Livestock grazing is known as one of the major factors that influence woody and herbaceous species interaction (Werner 1990a, b, c, d, e) throughout the world, as livestock grazing occupies 25% of the global land surface (Asner et al. 2004). Over the last 300 years, livestock grazing systems have increased 600% in extent and are projected to continue to increase with growing global human population and the associated increase in demand for meat and dairy products (Asner et al. 2004). A recent review of livestock grazing effects and ecosystem responses by Asner et al. (2004) identifies three major responses of ecosystems to livestock grazing observed at regional scales: desertification, woody species encroachment, and deforestation. In Asner et al.'s definition, desertification refers to grassland and steppe conversion to desert shrubland in arid regions of the world, while woody encroachment refers to grassland conversion to savanna and woodland in semiarid regions. In this chapter, I discuss and combine both ecosystem responses into a single-term woody encroachment.

Livestock grazing interacts with multiple other physical and biological factors at various spatial and temporal scales while influencing woody–herbaceous species balance. The complex interaction of livestock grazing with other factors such as climate, topography, fire, and soils has often made it difficult to quantitatively assess livestock grazing effects on woody–herbaceous species interaction at decadal and centennial scales (Archer 1994). To fully understand livestock grazing effects, all variables need to be controlled simultaneously in various environments and at a range of spatial and temporal scales. Due to the logistics involved in such a study and the lack of quantitative historical data, most ecological research is not able to do this. The current knowledge of woody–herbaceous –livestock species interaction is largely based on short-term studies and ecological investigations of only one or two variables with limited control on other potential factors.

4.2 Woody–Herbaceous Species Interactions and Associated Models

Mixed woody–herbaceous communities are diverse in composition, structure, functional forms, and spatial patterns due to their wide-spread distribution across the world (House et al. 2003). The interaction of woody and herbaceous species and the key driving factors that facilitate the interaction have been well studied (Fig. 4.1), although different conceptual models emphasize different driving factors (Belsky 1990). Woody and herbaceous species can influence each other in many different ways and the effects can be expressed in various forms.

The effects of woody plants on herbaceous species can be positive, neutral, or negative depending upon the characteristics of the woody and herbaceous growth-forms, ecophysiological features, photosynthetic pathway (C_3 vs. C_4) and habit (deciduous vs. evergreen), and water and nutrient requirements (Scholes and Archer

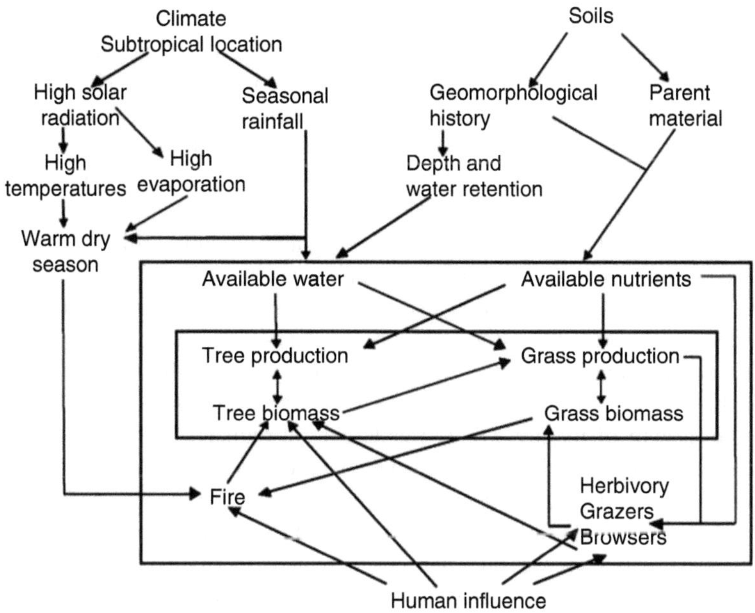

Fig. 4.1 Key driving factors for mixed woody-herbaceous systems (from House et al. 2003). Water, nutrients, fire, and herbivory are defined as the determinants of structure and function in woody-herbaceous systems and they collectively affect the innermost level, the balance between tree and grass

1997 and references therein). The effects of woody plants can also be expressed in varying forms. Firstly, woody plants can affect herbaceous species composition (Burrows et al. 1990). In mixed woody–herbaceous communities, herbaceous species composition under a tree canopy might be very different compared to that in the inter-tree space. C_3 grasses might be found mostly under the tree canopy, while C_4 grasses might dominate in-between trees in subtropical and temperate regions. Furthermore, herbaceous species composition can vary under the canopy from the tree trunk to the edge of the canopy (Scholes and Archer 1997). Secondly, woody species can influence herbaceous species production, biomass allocation, and phenology. Trees can often reduce herbaceous species biomass production (Burrows et al. 1990). However, herbaceous biomass production under tree canopies can also increase (Burrows et al. 1990) due to improved nutrient supply, reduced evapotranspiration (Reid and Ellis 1995), and increased water availability (Walker et al. 1981). Alteration of the geologic parent material and soil characteristics and the improvement of harsh environmental conditions are considered other facilitation effects of woody species for herbaceous plants. These facilitation effects might not be observed for many years after tree establishment, because the effects are dependent on tree size, age, and density and are not obvious until woody species reach a critical size and age (Scholes and Archer 1997). In other cases, trees/shrubs can have facilitation effects for herbaceous plants only when they are young. As the trees/shrubs grow bigger, the facilitation effects might be outweighed by their competition effects on herbaceous plants. This competition often results in a strong, negative correlation between tree density or cover and grass cover or biomass (Stuart-Hill and Tainton 1989). Herbaceous production and diversity, therefore, might be low at high tree/shrub density (Burrows et al. 1990). The negative correlation might be due to the tree/shrub litter accumulation (which might increase soil acidity), canopy shading, reduced rainfall under the canopy, and root competition.

The effects of herbaceous species on woody plants are most critical during woody seedling establishment stage, although the effect can be variable. Firstly, herbaceous species can impact woody seedling establishment and recruitment directly by effectively competing for light, water, and nutrients (Knoop and Walker 1985). The competition can prevent woody seedling emergence, increase the mortality of newly established woody seedlings, and reduce woody seedling growth and recruitment. Even the growth of mature woody plants can be reduced by herbaceous species competition for water in wetter years, when herbaceous biomass is high (Knoop and Walker 1985). Secondly, herbaceous species can influence woody seedling recruitment indirectly (Scholes and Archer 1997 and references therein). Herbaceous species biomass can increase fine fuel loads, which increases fire frequency and intensity, leading to increased mortality of small woody seedlings that are especially vulnerable to fire (Dando and Hansen 1990; Archer 1994). However, the direct and indirect influences of herbaceous species on woody plants are often not enough to completely exclude woody plants and to prevent woody encroachment. Woody plants still might be able to expand into adjacent grassland with a wide range of herbaceous species composition and production (Scholes and Archer 1997). Woody plants can establish during wet periods, when competition from herbaceous species are limited. Once woody seedlings establish and grow beyond the height of the herbaceous layer, they can establish vertical dominance and herbaceous species might have little or no influence on them. Scholes and Archer (1997) summarize that experimental studies in savanna environments largely found no significant effects on woody species, when herbaceous plants were cut and cleared. Only on fine textured soils with greater clay content herbaceous species appeared to limit water recharge from rainfall deeper in the soil profile where tree roots uptake water.

In mixed woody–herbaceous communities, the interaction between woody plants themselves has been considered important. Tree–tree interaction or shrub–shrub interaction can lead to competition for belowground resources such as water and nutrients as well as competition for light. This intraspecific competition is often assumed to lead to self-thinning and ultimately a regular spatial pattern of woody plants. Clumped and random spatial patterns are also possible in savanna tree distribution due to fire effects, topography, soils, and resource patchiness (Scholes and Archer 1997 and references therein). Clumped and random spatial patterns can be associated with some level of facilitation effects such as increased seed dispersal and improved environmental conditions under canopies and nearby existing trees/shrubs.

Three different types of models in the literature describe woody-herbaceous species interaction and coexistence, particularly in savanna ecosystems: niche separation models, balanced competition models, and disequilibrium models (Scholes and

Archer 1997 and references therein). Niche separation models are based on the assumption that a variable, such as water, is a limiting factor and woody and herbaceous species, therefore, have to use resources at different times or places (House et al. 2003). For example, grasses and shrubs can have different root systems at different depths in the soil profile so that they can use water at different soil depths to coexist (Walker et al. 1981; Knoop and Walker 1985).

Balanced competition models are based on the concept of intraspecific competition, which is assumed to be stronger than interspecific competition (House et al. 2003). In other words, competition between herbaceous species is assumed to be stronger than competition between woody and herbaceous species. Likewise, competition between woody species is assumed to be stronger than competition between woody and herbaceous species. The result would be woody species that outcompete herbaceous species and establish dominance or herbaceous species that outcompete newly establishing woody seedlings and prevent woody establishment and encroachment. Balanced competition models, therefore, predict two stable states: woodland and grassland. Similar to the balanced competition models, Walker and Noy-Meir's 1982 model predicts the two stable states after adding grazing as a factor to a niche separation model centered on soil water, which was initially proposed by Walter in 1971 (Jeltsch et al. 2000).

Following the simpler models that predict stable equilibrium of woody and herbaceous vegetation, newer concepts emerged modeling and predicting nonequilibrium dynamics in the woody-herbaceous interactions (Jeltsch et al. 2000). Equilibrium models might explain the coexistence of woody and herbaceous species at smaller scales, whereas disequilibrium models are more appropriate for describing landscape- and decadal-scale dynamics (Sharp and Whittaker 2003). Disequilibrium models predict cycles and oscillations in the relative abundance of woody and herbaceous species at larger scales (Sharp and Whittaker 2003). They suggest that mixtures of woody and herbaceous species only exist due to disturbances such as fire and grazing and, therefore, represent a transitional state between the possible stable states (Jeltsch et al. 2000). Further development of the disequilibrium models predicts multiple stable states with varying tree-grass ratio (House et al. 2003). Disequilibrium models have also been extended to include a spatial aspect and a concept of patches of disequilibrium which result from stochastic processes such as gap dynamics (Jeltsch et al. 2000).

4.3 Woodland and Grassland Stable States and Conceptual Models

Ecosystems respond in different ways when external conditions change over time. Some ecosystems might respond gradually in a smooth, continuous manner, while others respond abruptly, especially after a certain threshold is passed in external conditions (Fig. 4.2). Many different ecosystem studies in the literature have demonstrated the existence of alternative stable states and multiple stable states in different

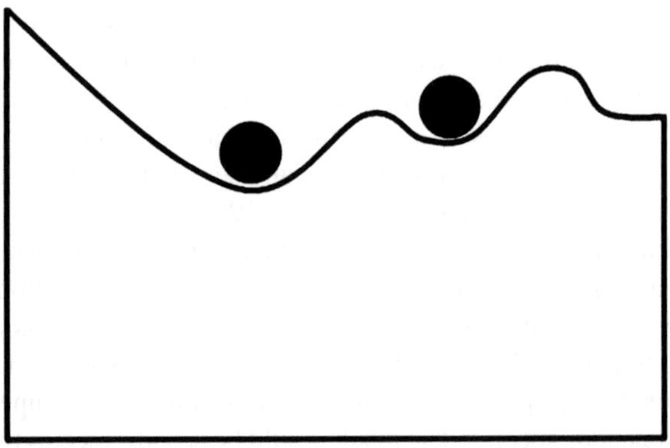

Fig. 4.2 Alternative stable states and their basins of attraction (modified from Scheffer et al. 2001). Stable equilibria correspond to the valleys or attraction basins, while unstable transitional periods correspond to the hill between the valleys. If the size of the attraction basin is small, ecosystem resilience is small and even small changes in the external conditions might move the system into an alternative stable state

environments (Werner 1990a, b, c, d, e; Scheffer et al. 2001 and references therein). Studies of woody–herbaceous interactions have been common among such research demonstrating two possible alternative states: woodland and grassland. Scheffer et al. (2001) term the changes between the two alternative stable states "catastrophic shift," because shifts occur very rapidly and there is often no "early warning signals." Moreover, it is extremely difficult to recover an ecosystem after such shifts and many ecosystems remain in the new alternative state (Walker et al. 1981; Sharp and Whittaker 2003), even if previous environmental conditions are restored.

Numerous studies documented a shift from a grassland stable state to a woodland stable state (Fig. 4.3) in Africa, southwestern USA, drier parts of India, and in Australia (Walker et al. 1981), while shifts from a woodland state to a grassland state have also been observed (Burrows et al. 1990). African grasslands, for example, were kept open by herbivory and fire until herbivore numbers drastically declined allowing successful establishment of woody plants. Once successfully established and recruited, the woody species were no longer kept in balance by herbivores and their canopy shade reduced herbaceous biomass accumulation, which then reduced fire frequency. Reduced fire frequency further increased successful woody establishment and encroachment (Scheffer et al. 2001). Returning this ecosystem to the grassland stable state would require drastic measures taken at substantial spatial and temporal scales. In contrast, conditions in dry environments can enable a shift from a woodland stable state to a grassland stable state (Scheffer et al. 2001 and references therein). In dry environments, if well-established tree populations are heavily disturbed and killed due to fire and other factors, conditions might be too harsh to allow woody seedlings to establish in the absence of nurse trees and herbaceous species might dominate. Restoring the woodland stable state might require a rare combination of adequate precipitation and reduced grazing effects.

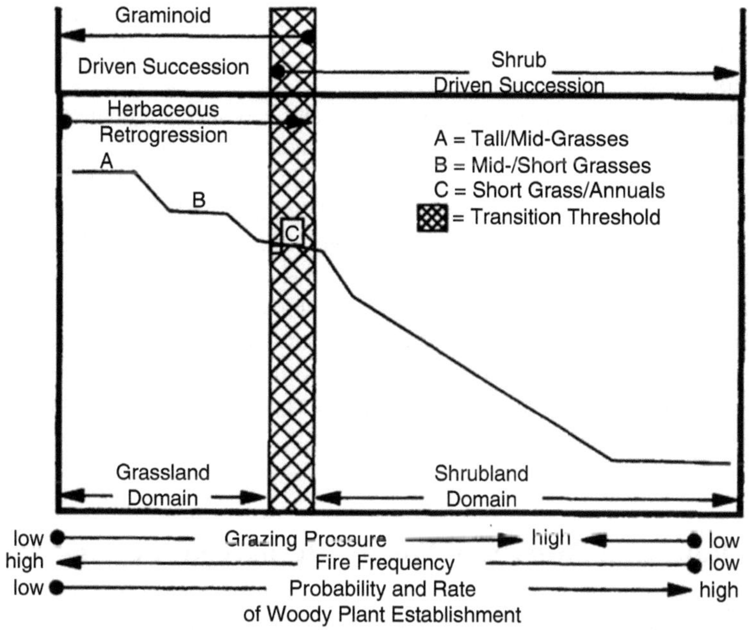

Fig. 4.3 A conceptual model of shrubland and grassland stable states and a transition between the two states (from Archer 1994). This model demonstrates the conversion from a grassland stable state to a shrubland stable state and the existence of a threshold in livestock grazing pressure

The dynamics and shifts between woodland and grassland are dependent upon processes and mechanisms that influence the resistance, resilience, and persistence of the associated woodland and grassland ecosystems. Both ecosystems create positive feedbacks to persist. Such positive feedback mechanisms in woodlands include tree suppression of grass through shading (Menaut et al. 1990), increased seed input within areas around tree patches (Archer 1990), and reduced fire frequency (Archer 1990; Menaut et al. 1990). Woody–herbaceous ecotones exist as a result of a balance in such feedback mechanisms. Equally strong persistence and feedback mechanisms of the two ecosystems create a stable ecotone that does not change rapidly in space over time. In contrast, imbalance in the feedback systems might result in unstable conditions over time (Werner 1990a, b, c, d, e) and constant fluctuations in ecotones.

4.4 Woody–Herbaceous Ecotones

Ecotones play a vital role in understanding the interaction between woody and herbaceous species. The ecological importance of ecotones and their roles in understanding global environmental changes have long been recognized (Holland and Risser 1991). Ecotone characteristics, including their location, size, shape, and composition are more sensitive to global environmental changes than those of homogenous landscape units (Turner et al. 1991). Ecotones thus provide good early indicators of such changes (Gosz 1991).

Biophysical characteristics of forest-grassland ecotones are defined by a complex interaction of biotic and abiotic factors, including plant interactions, disturbance regime, physiography, topography, geologic parent materials, soil properties, and climate variables (Alverson et al. 1988; Tilghman 1989; Smit and Olff 1998; Carmel and Kadmon 1999; Mast and Veblen 1999; Zald 2002). Changes in these factors can have a substantial impact on woody–herbaceous ecotones and cause a shift in their location (Camarero et al. 2000; Taylor 1995). Shifts in forest-grassland ecotones impact carbon sequestration and land surface–atmosphere interactions and have important implications for biodiversity, primary and secondary productivity, soil development, and populations and carrying capacity of both domestic and wild animals (Archer 1994).

4.5 Rates and Patterns of Woody–Herbaceous Ecotone Shift

Varying rates and patterns of forest encroachment into the adjacent grassland can be observed. I documented encroachment by a dominant tree species along northern Mongolian forest-grassland ecotones, Siberian larch (*Larix sibirica*) (Sankey et al. 2006a, b). My results indicated that the ecotones shifted into the adjacent grassland in different patterns compared to aspen (*Populus tremuloides*) and Douglas-fir (*Pseudotsuga menziesii*) observed in my other study in southwestern Montana of western USA that used dendrochronological data with individual tree maps along a lower ecotone (Sankey et al. 2006a, b). I also documented western juniper encroachment into sagebrush steppe vegetation in southwestern Idaho, USA using satellite imagery and airborne lidar data along with dendrochronological data (Sankey et al. 2010). In this latter study, I documented encroachment patterns observed in both of the previously mentioned studies thus demonstrating that different patterns of encroachment can occur simultaneously. The patterns observed in my studies as well as other woody encroachment studies can be divided into three general patterns of tree encroachment into the adjacent grassland (Fig. 4.4).

Type I change is a shift in the forest-grassland boundary location into the adjacent grassland. This type of ecotone change results from a mechanism where new trees establish in the adjacent grassland advancing the ecotone location towards the grassland (Sankey et al. 2006a, b). Type I ecotone change might mostly occur in systems where the dominant tree species regenerates through seed dispersal, although it can be observed in systems with vegetatively-reproducing species. For example, Douglas-fir is a seed-dispersed species. Its seeds are dispersed through wind, animals, and birds (Hermann and Lavender 1965). Seeds usually fall within 100 m from a seed tree or a stand edge, but they can fall 1–2 km away from the seed sources (Hermann and Lavender 1965).

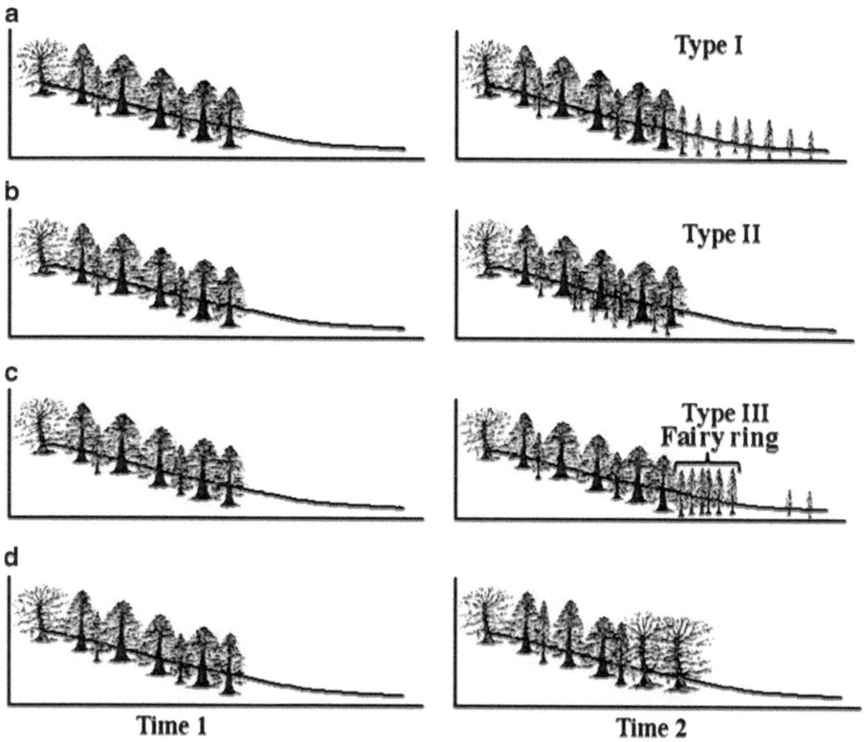

Fig. 4.4 Patterns of tree encroachment into the adjacent grassland observed in grazed areas. Trees might encroach into the adjacent grassland in three different patterns. These patterns are not mutually exclusive and can occur simultaneously. Type I pattern is a shift in the forest-grassland ecotone location over time into the adjacent grassland (**a**). Type II pattern is tree density increase within the same forest-grassland boundary location (**b**). Type III pattern is fairy ring establishment that advances the forest-grassland boundary into the adjacent grassland (**c**). Forest-grassland ecotones might also retreat and herbaceous species might expand into the adjacent forest over time (**d**)

Type II change is an increase in tree density at the forest-grassland boundary (Arno and Gruell 1986). During this change, new trees establish within the same boundary location and do not advance the boundary towards the adjacent grassland. Type II ecotone change occurs in systems where tree establishment sites are available under the forest canopy, but no establishment occurs outside of the forest boundary due to unfavorable site conditions and disturbance. Type II ecotone change can occur with Type I ecotone change at the same time, if conditions outside of the forest boundary change allowing Type I ecotone change. Type II change can be observed in systems with both tree species that regenerate vegetatively and through seeds. For example, Douglas-fir, aspen, western juniper, and Siberian larch all can result in Type II ecotone change.

Type III change is an establishment of a fairy ring in the grassland along the edge of the forest (Sankey 2007). This change occurs when new trees establish as a fringe in the grassland, adjacent to the forest boundary. Fairy rings consist of new stands of densely distributed new stems of similar age. Fairy ring establishment is often associated with wave regeneration mechanisms. During this mechanism, new regenerations occur as pulses advancing the forest boundary into the adjacent grassland. Both seed-dispersed and vegetatively-reproducing species can regenerate in pulses. For example, aspen regenerates in pulses after a fire disturbance event (DeByle and Winokur 1985), whereas Siberian larch can regenerate in pulses following reductions in grazing disturbance forming a fringe (Didier 2001).

4.6 Woody–Herbaceous–Livestock Species Dynamics

Most ecological processes in woody–herbaceous species dynamics can be impacted by herbivory, an important local control over vegetation. Vertebrate and invertebrate herbivores can regulate plant cover types, their composition, structure, and productivity (Alverson et al. 1988; Tilghman 1989; Mast et al. 1997; Carmel and Kadmon 1999; Mast and Veblen 1999; Bachelet et al. 2000; Bartolome et al. 2000; Scheffer et al. 2001; Wahungu et al. 2002). Herbaceous and woody plant species react to herbivory differently due to their differences in tolerance to grazing and palatability (Archer 1994). This makes their interaction in grazed environments more complex than in undisturbed environments. This complexity has generated abundant interest in the interaction between herbaceous and woody plant species and, in particular, the changes from herbaceous vegetation cover to woody species cover due to grazing. Some studies show that the processes of woody species seedling emergence, growth, and survival are facilitated by grazing (Walker et al. 1981; Dando and Hansen 1990; Reid and Ellis 1995; Archer 1994; Sharp and Whittaker 2003), while others suggest a conflicting result that these processes are inhibited by grazing (Reid and Ellis 1995; Carmel and Kadmon 1999; Bartolome et al. 2000). In most cases, the assumption is that coexisting herbaceous species and woody species within an ecosystem show the opposite trends under livestock grazing effects. If woody species increase with increasing grazing effects, herbaceous species are assumed to decrease and vice versa.

Woody species encroachment due to grazing has been demonstrated throughout the world, including southern Asia, Australia, Africa, South America, and North America (Walker et al. 1981; Archer 1989). Ecological processes described in studies that proposed woody species encroachment due to grazing are (1) Grazing decreases seed production, seedling establishment, biomass, and basal area of palatable herbaceous species and increases their mortality; (2) Reduced herbaceous species ground cover increases sunlight levels on the ground, which increases seed germination and early establishment of woody species seedlings; (3) Reduced herbaceous species biomass decreases fine fuel accumulation and reduces fire frequency, which increases woody species invasion; (4) Invading woody species are less palatable than herbaceous species and are not browsed enough to be eliminated; (5) Grazing makes herbaceous species less able to compete for resources and unable to limit woody species growth and their seedling establishment; and (6) Livestock

Table 4.1 Long-term averages of grazing pressure estimated in AUM ha^{-1} in the Sankey et al. (2006a, b) study in Montana, USA

Grazing levels	AUM ha^{-1}
Grazing level 1	0.00
Grazing level 2	0.11
Grazing level 3	0.12
Grazing level 4	0.17
Grazing level 5	0.28
Grazing level 6	0.41
Grazing level 7	0.50
Grazing level 8	0.79
Grazing level 9	1.00
Grazing level 10	2.00

disperse woody species seeds across the landscape, which facilitates woody species expansion (Archer 1994).

The opposite effects of grazing on woody species establishment have also been demonstrated (Carmel and Kadmon 1999; Bartolome et al. 2000). Studies of these effects suggest that grazing can inhibit tree seedling establishment, survival, and growth. Grazing, therefore, might be expected to control woody species encroachment into grasslands. Ecological processes described in studies of negative effects of grazing on woody species include (1) Slow growth rate of most woody species allows repeated grazing in their seedling stage when they are most vulnerable to grazing (Alverson et al. 1988; Tilghman 1989); (2) Intense grazing causes shoot loss, tissue damage, and biomass loss for woody species (Hjalten et al. 1993), which decreases their seedling growth (Alverson et al. 1988; Tilghman 1989) and increases seedling mortality (Hjalten et al. 1993); (3) Increased seedling mortality reduces recruitment into the tree population (McInnes et al. 1992; Rooney et al. 2002); and (4) Trampling and rubbing against the bark by grazing animals damage woody species and their seedlings (Kay and Bartos 2000).

The current literature indicates two seemingly conflicting linear relationships between grazing effects and woody species (i.e., increasing woody species with increasing grazing effects and thus decreasing herbaceous species or vice versa). The vast majority of grazing impact studies in the current literature, however, has compared only two or three levels of grazing intensity. Such comparisons of limited number of grazing levels might provide only a simple linear relationship between woody species establishment and grazing effects. The potential variability in woody establishment due to varying levels of grazing intensity needs to be investigated across a wider gradient of multiple levels of grazing intensity.

I examined aspen and Douglas-fir tree encroachment into the adjacent grassland in Montana, USA under ten different levels of livestock grazing intensity using nonexperimental observational data of 60 years (Table 4.1) (Sankey et al. 2006a, b). The objective of my study was to determine if tree establishment-livestock grazing relationship always had a simple linear trend as suggested by previous studies or if

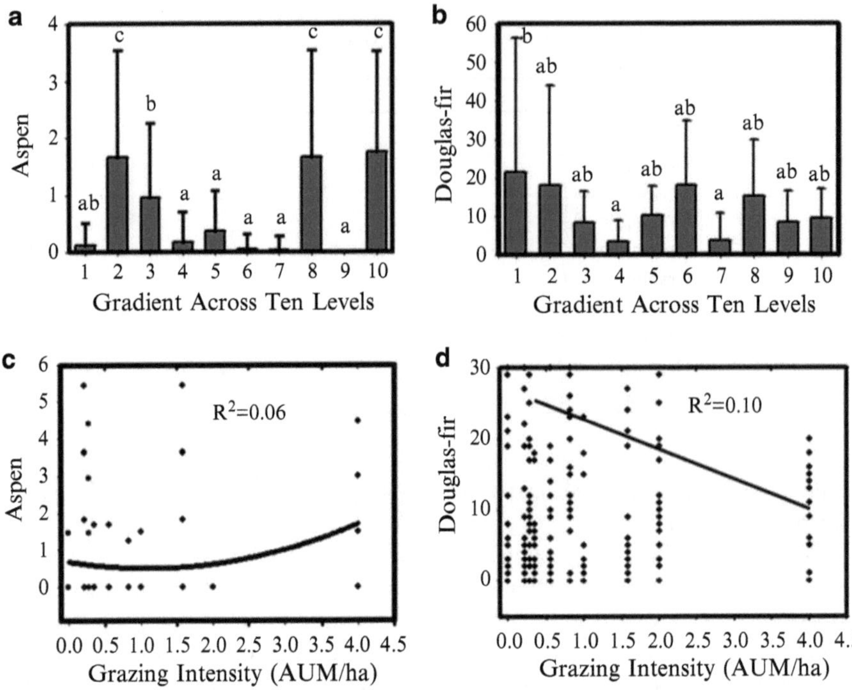

Fig. 4.5 Aspen (*Populus tremuloides*) and Douglas-fir (*Pseudotsuga menziesii*) tree establishment with varying grazing pressure in Sankey et al. (2006a, b) study in Montana, USA. Number of new tree stems established (expressed in percent) under each grazing level is on the *Y*-axis. The ten grazing levels on the *X*-axis in (**a, b**) correspond with categorical classes of grazing pressure, while the grazing levels in (**c, d**) are livestock grazing pressure estimated in animal unit months per hectare (AUM ha^{-1}). Grazing pressure was significantly ($\alpha = 0.10$) correlated with aspen and Douglas-fir tree establishment ($p = 0.01$ and $p = 0.07$, respectively)

patterns were different at decadal time scales. When aspen tree establishment was analyzed using an analysis of variance (ANOVA) model with all pair-wise comparisons with a gradient of all ten grazing levels, increasing in intensity from 1 to 10 (Table 4.1), grazing levels 2, 8, and 10 had significantly greater aspen establishment than all other grazing levels (*p*-value <0.001) (Fig. 4.5a). Grazing level 3 had significantly greater aspen establishment than grazing levels 4, 5–7, and 9 (*p*-value <0.001). There were no other significant differences in aspen establishment along this gradient. A similar test indicated that grazing level 1 had significantly greater Douglas-fir establishment compared to grazing levels 4 and 7 (*p*-value of 0.002) (Fig. 4.5b). There were no other significant differences in Douglas-fir establishment. There was no apparent trend of linear increase or decrease in aspen and Douglas-fir establishment with increasing grazing intensity along this gradient. In addition to the ANOVA model, a regression model of all 10 grazing levels (AUM ha^{-1}) (Table 4.1) and aspen establishment was built with a significant squared term (Fig. 4.5c). The statistically significant squared term might suggest a possibility of a curvilinear relationship between aspen establishment and grazing levels. The regression model of all ten grazing levels (Table 4.1) and Douglas-fir establishment did not have a statistically significant squared term, but indicated a linear relationship between Douglas-fir and grazing levels with a trend of decrease (Fig. 4.5d). However, the regression model produced a low correlation coefficient and did not suggest a strong relationship.

Collectively, my results indicated varying relationships between tree establishment and livestock grazing intensity. The relationships were not always simple linear increase or decrease in tree establishment with increasing livestock grazing intensity, although linear relationships were observed in some cases. At decadal time scales, simple linear trends of inhibition and facilitation effects as suggested by previous studies did not appear to hold across varying gradients of grazing levels and two different tree species. Indeed, complex curvilinear trends might be possible at decadal time scales across wider gradients of grazing intensity (Sankey et al. 2006a, b). This is consistent with other studies that suggest that mixed woody–herbaceous systems can have nonlinear trends in woody plant abundance and rates of change in tree abundance (McPherson 1992; Archer et al. 1988; Miller and Wigand 1994).

Taken together, previous studies indicate that multiple trajectories might be observed in livestock grazing effects on woody species. Most of these trajectories might be difficult to discern in long-term studies with limited control on grazing treatments and other potential factors that influence woody-herbaceous species interaction. Better-controlled experimental designs and process-based studies might allow conclusive tests of the possible trajectories that have been hypothesized in the current literature. Currently proposed, but not exclusively tested, trajectories in the relationship between woody establishment and livestock grazing effects include four possible trends: (1) facilitation effect or a simple linear increase in woody species with increasing grazing intensity (Fig. 4.6a, Archer 1994), (2) inhibition effect or a simple linear decrease in woody species with increasing grazing intensity (Fig. 4.6b) (Carmel and Kadmon 1999; Bartolome et al. 2000), (3) a curvilinear relationship in which inhibition effects dominate at low and high grazing intensities, but facilitation effects dominate at medium grazing intensities (Fig. 4.6c) (Cairns and Moen 2004), and (4) a curvilinear relationship in which facilitation effects dominate at low and high grazing intensities, but inhibition effects dominate at medium grazing intensities (Fig. 4.6d) (Sankey et al. 2006a, b).

The first two trajectories can be explained by the ecological processes described in the positive and negative effects of livestock grazing discussed earlier in this section. The first trajectory, facilitation effects, predicts decreasing herbaceous species and increasing woody species with increasing grazing intensity. This trend might be observed in systems where herbaceous species are palatable and are preferred by the livestock species over the woody species. The second trajectory, inhibition effects, predicts increasing herbaceous species and decreasing woody species with increasing grazing intensity. This trend might be observed in systems where woody species are palatable and are commonly grazed or browsed by livestock. In this case, woody establishment can be greatest under low grazing intensity. In the third trajectory, woody and herbaceous species are balanced at medium grazing intensity, but woody

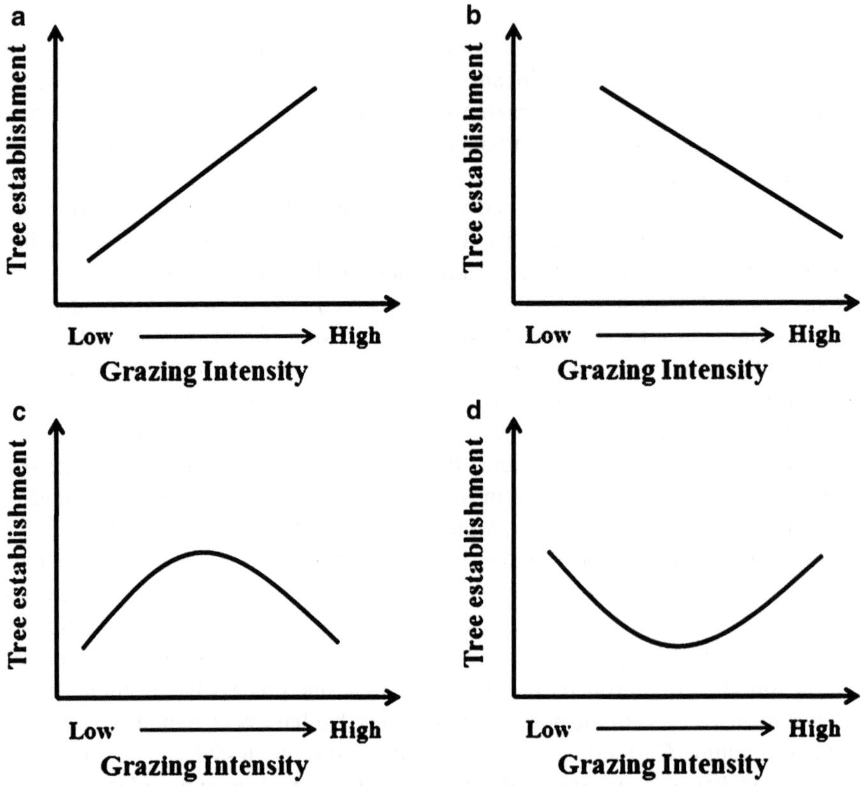

Fig. 4.6 Currently proposed trajectories in the relationship between tree establishment and livestock grazing effects. Herbaceous species compete for similar resources and are assumed to show the opposite trend compared to trees. Livestock grazing can have facilitation (**a**) or inhibition (**b**) effects on tree establishment resulting in simple linear trends. Facilitation and inhibition effects can also dominate at varying levels of grazing pressure resulting in complex curvilinear trends (**c** and **d**)

species are expected to decrease at low and high grazing intensities. This trend is expected to occur where grazing effects largely include woody species trampling, seed dispersal, and seed predation, but woody species foliage consumption is relatively low (Cairns and Moen 2004). It is also expected to occur in systems where multiple grazing animal species are present (Cairns and Moen 2004).

In the fourth trajectory, woody and herbaceous species are hypothesized to be balanced at medium grazing intensity, but woody species are expected to increase at low and high grazing intensities. This trajectory might be explained by the grazing optimization hypothesis. The grazing optimization hypothesis, developed for herbaceous species, states that herbaceous species productivity increases with increasing grazing intensity at low grazing levels due to overcompensation (McNaughton 1979). This trend continues up to a point called "the level of optimal grazing" and then declines with increasing grazing intensity at high grazing levels. Since coexisting herbaceous and woody species compete for largely similar resources, woody species might be expected to show the opposite trend. Therefore, at lower levels of grazing intensity, we might expect tree establishment to be relatively high. Even if woody species are grazed/browsed at this intensity, they can show increasing competitiveness due to stimulatory effects of grazing (Stuart-Hill and Tainton 1989). Tree establishment might then decrease with increasing grazing intensity as herbaceous species productivity increases due to overcompensation. At medium levels of grazing intensity, tree establishment might reach its minimum because herbaceous species productivity is highest at these levels and, thus, tree competition with herbaceous species is greatest. At higher levels of grazing intensity, however, we expect tree establishment to increase because herbaceous species at these levels show a trend of decrease.

Walker et al. (1981) proposed conceptual models of woody-herbaceous–livestock interaction that the four trajectories in Fig. 4.6 don't fully describe. Walker et al.'s conceptual models describe the stability of woody-herbaceous species balance under livestock grazing effects in semiarid savanna ecosystems. In their definition, semiarid savanna includes regions in which scattered to numerous trees/shrubs are distributed across continuous grass cover. They first describe the effects of livestock grazing on grass recruitment curve (Fig. 4.7a) based on McNaughton's (1979) grazing optimization hypothesis and suggest that grazing has the greatest stimulating effect at intermediate values of grass. They also describe the effects of woody vegetation on grass recruitment curve (Fig. 4.7b), because woody and herbaceous species compete for the same water resource in the top-soil in semiarid regions (Walker et al. 1981). Their conceptual model suggests that small and medium amounts of woody species (W) have greater effects on high amounts of grass (G) than small amounts of grass. Walker et al. (1981) then describe the effects of grass on woody vegetation (Fig. 4.7c) and suggest that high values of grass (G) do not have strong effects on the woody vegetation recruitment curve, because woody species have exclusive access to subsoil water. Lastly, Walker et al. (1981) describe the zero-isocline or zero recruitment of grass as related to grass, woody species, and the effects of grazing (II) (Fig. 4.7d). The zero grass recruitment curve indicates the equilibrium between woody plants and grass under grazing effects. Similar to some of the previously discussed trajectories, this model suggests a curvilinear relationship in the woody-herbaceous–livestock species interaction. However, this model suggests varying curves with increasing grazing pressure. The equilibrium always indicates a similarly shaped curve at varying grazing pressure, but the curve falls at varying amounts of total vegetation and varying woody:grass ratios. Unlike the other models, which largely assume unvarying total vegetation amount, this model assumes decreasing amounts of total vegetation with increasing grazing pressure. The assumption might depend on the relative palatability and tolerance of the plant species involved to grazing as well as the forage preferences of the grazing animal species. Total vegetation amount might decrease with increasing grazing pressure in some systems, while in other systems only the palatable species might decrease and unpalatable species might remain constant or increase with increasing grazing pressure.

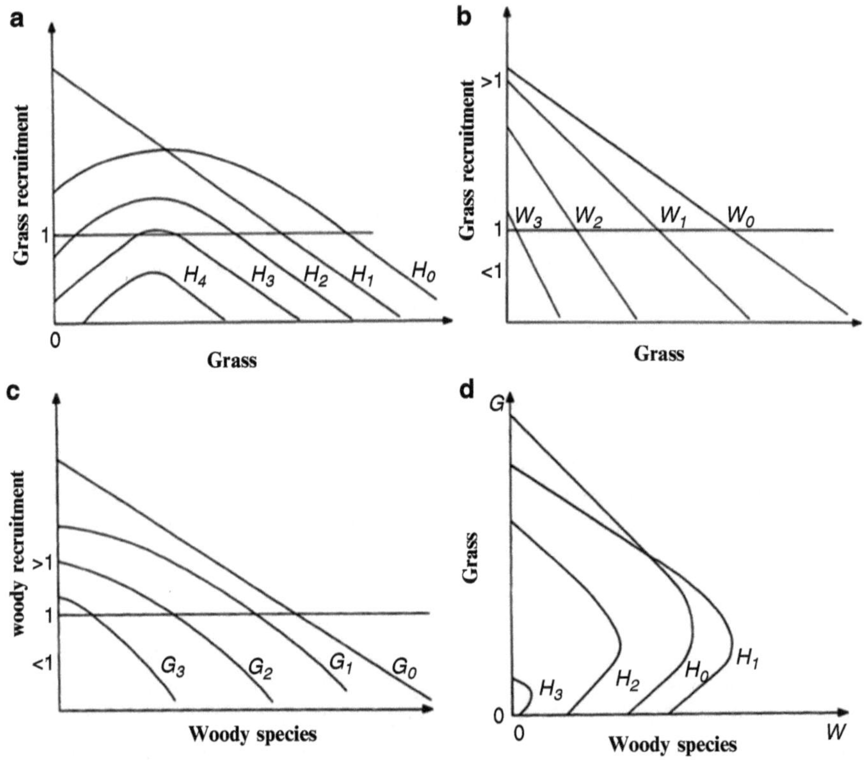

Fig. 4.7 Livestock grazing effects, grass and woody species recruitment, and their interactions (from Walker et al. 1981). (**a**) Shows the hypothesized effects of increasing grazing pressure on grass recruitment, where grazing pressure ranges from none (H_0) to heavy (H_4). (**b**) Shows the hypothesized effects of woody species on grass recruitment, where woody plant density ranges from W_0 or no woody plants to high density or W_3. (**c**) Demonstrates the potential effects of grass on woody species recruitment, grass density ranges from G_0 or no grass to G_4 or high density of grass. (**d**) Shows the zero grass recruitment in relation to increasing grazing pressure, grass, and woody species simultaneously

I also examined the relationship between Siberian larch forest-grassland ecotone shift and a gradient of five different livestock grazing regimes dominated by different livestock species (Table 4.2) in northern Mongolia (Sankey et al. 2006a, b). The five grazing regimes varied in overall grazing intensity from 3.3 to 5.9 AUM ha^{-1}. They varied in species composition such that each site represented either sheep-dominance, sheep-goat mix, sheep-goat-cattle mix, or cattle-dominance (Table 4.2). Forage preferences between these animal species are known to be substantially different (Vallentine 2001). Cattle are grazers and consume mostly graminoids. Sheep are intermediate feeders and consume grasses, forbs, and woody species. Goats are browsers and prefer leaves and tender twigs of new growth on trees and shrubs. My results indicated that Siberian larch forest-grassland ecotone response to grazing varied among different grazing regimes and tree establishment varied statistically significantly. The number of new trees established varied significantly between Site

Table 4.2 Long-term averages of grazing intensity observed in the Sankey et al. (2006a, b) study in northern Mongolia

Sites	Grazing level	AUM ha^{-1}	Dominant livestock species	Number of households at the site
Site 1	Very low	3.3	Sheep	20
Site 2	Low	4.2	Goat-sheep	32
Site 3	Medium	4.8	Cattle-sheep-goat	56
Site 4	Medium	4.9	Cattle	36
Site 5	High	5.9	Sheep	43

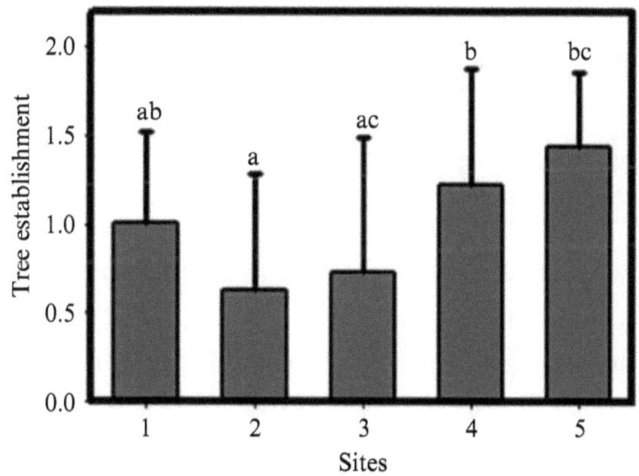

Fig. 4.8 Mean (SE) Siberian larch (*Larix sibirica*) tree establishment with different grazing regimes in Sankey et al. (2006a, b) study in northern Mongolia. The significant differences are indicated by different letters in the figure. Site 2 (goat-sheep-dominated and low overall grazing intensity) had significantly lower tree establishment than Site 5 (sheep-dominated and high overall grazing intensity) (Table 4.2). Sites 2 and 3 (cattle-sheep-goat mix and medium overall grazing intensity) also had significantly lower tree establishment than Site 4 (cattle-dominated and medium overall grazing intensity) (Table 4.2)

2 (goat-sheep-dominated and low overall grazing intensity) and Site 4 (cattle-dominated and medium overall grazing intensity) and between Site 2 (goat-sheep-dominated and low overall grazing intensity) and Site 5 (sheep-dominated and high overall grazing intensity) (Fig. 4.8). The number of new stems established also varied between Site 3 (cattle-sheep-goat mix and medium overall grazing intensity) and Site 4 (cattle-dominated and medium overall grazing intensity). There was no statistical difference between sheep-dominance at low and high overall grazing intensities. This might indicate that sheep, in general, do not have substantial negative effects on tree establishment regardless of sheep grazing intensity, which might be explained by their lower consumption of woody species compared to herbaceous species. The sites with high numbers of goats had lower tree establishment than all other sites with lower numbers of goats, regardless of overall grazing intensity. The implications of this study are important for future studies of woody-herbaceous–livestock species interaction and future land resource management. It is not only the

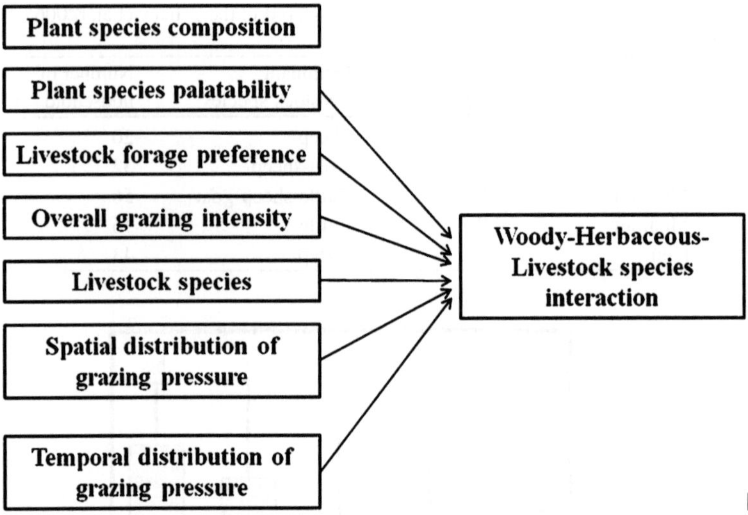

Fig. 4.9 Potential factors influencing grazing effects on woody-herbaceous dynamics

overall grazing intensity that researchers and land managers should be concerned with, but also the types of grazing animal species and the different combinations of varying grazing intensities and livestock species (Sankey et al. 2006a, b).

I also found that many new trees established during the decades when livestock were distributed in numerous small herds in northern Mongolia (Sankey et al. 2006a, b). Prior to these decades (1930–1950), livestock were distributed in a few large herds owned by a few religious leaders. A major political regime shift occurred in Mongolia causing drastic socio-economic changes. Due to the resulting land management changes, the large herds were redistributed into small herds. A pulse of tree regeneration appears to have established following the herd redistribution (Sankey et al. 2006a, b). This indicates that herd distribution at the landscape scale and the driving socio-economic changes and policy changes are important variables to consider when studying woody-herbaceous–livestock interaction. Human legacies can have lasting effects on this interaction at varying spatial and temporal scales. Several variables appear important to consider in livestock grazing effects on woody-herbaceous dynamics. They include overall grazing intensity, grazing animal species and their forage preferences (herbaceous vs. woody species), spatial distribution of grazing pressure (few large herds vs. many small herds), and temporal distribution of grazing pressure (winter grazing vs. summer grazing), tree and herbaceous species composition, their palatability, and tolerance to grazing (Fig. 4.9). Accurate understanding of these variables has important implications for modeling and managing woody-herbaceous–livestock species interaction. Different combinations of varying grazing intensities and livestock species composition, for example, can be used to either facilitate or inhibit directional changes in the woody-herbaceous species balance. Forest-grassland ecotone shift can be influenced by both overall grazing intensity and different grazing animals, if a shift is occurring. Different gradients of overall grazing intensity and grazing animals might correspond with different trajectories of change in woody-herbaceous balance. Grazers, for example, might facilitate tree encroachment, while browsers might inhibit tree encroachment and facilitate increased herbaceous species distribution. Furthermore, different levels of grazing effects might be observed in different tree species due to their differences in palatability and their response to grazing. Douglas-fir, for example, is unpalatable to most livestock species, while aspen is highly palatable to livestock species. Siberian larch can also be highly palatable to livestock species. Changes in woody-herbaceous species balance, therefore, might be facilitated or inhibited to different levels depending upon the dominant tree species.

4.7 Other Potential Factors Influencing Woody-Herbaceous Species Dynamics

In addition to livestock grazing, several other variables have been proposed as potential factors in woody-herbaceous dynamics that cause a shift in the ecotone (Fig. 4.10). In savanna environments, four factors have been acknowledged as the main determinants that create and maintain the coexistence. These four determinants are water, nutrients, herbivory, and fire (Werner 1990a, b, c, d, e). In other mixed woody-herbaceous systems, atmospheric CO_2 increase has been suggested as one of the potential factors that can cause woody-herbaceous shift. Carbon dioxide affects plant photosynthetic rates, stomatal conductance, water-use efficiency, resource allocation, growth, and architecture (Bazzaz 1996). An increase in CO_2 from 270 to 370 ppm in the last 200 years has been proposed as a possible cause of woody species expansion (Archer 1994 and references therein). Increased CO_2 is expected to change competitive abilities of different plant species, altering the interaction among species and, consequently, species composition in the community (Bazzaz 1996). Archer (1994) summarizes that increased atmospheric CO_2 is hypothesized to favor woody plants over herbaceous species due to the following specific reasons: (1) woody plants have C_3 photosynthetic pathway, while many grass species have C_4 photosynthetic pathway, (2) C_3 species have greater advantage for growth and competition with increased CO_2, (3) C_4 grasses evolved at lower CO_2 concentrations (~200 ppm) and woody encroachment into C_4 grasslands occurred during a 30% increase in atmospheric CO_2 over the last 200 years. However, Archer (1994) argues that C_3 grasses would also be expected to invade C_4 grasses, if atmospheric CO_2 increase was a probable cause of C_3 woody species encroachment. C_3 grasses, however, have not invaded C_4 grasses. Furthermore, Archer (1994) suggests that C_3 cold desert and temperate grasses have also been invaded by woody species.

Changes in climate variables such as mean annual temperature, rainfall, and evapotranspiration are expected to influence the balance between herbaceous and woody species, because the distribution of many grasslands and savannas throughout

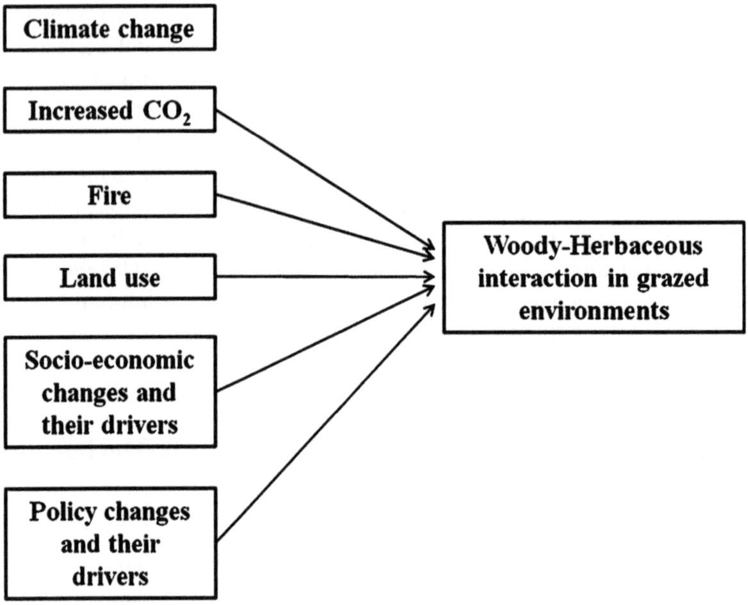

Fig. 4.10 Multiple interacting factors that are commonly proposed as causes of shifts in the woody-herbaceous species balance. In grazed woody-herbaceous systems, the effects of these factors might reduce or increase the effects of livestock grazing

the world are closely related to these variables (Archer 1994 and references therein). Potential changes that might lead to woody encroachment at the expense of herbaceous species include increased or decreased rainfall, shifts in the seasonality of rainfall, shifts in the distribution of precipitation events, increased temperature, and drought (Archer 1994 and references therein). Increased rainfall might facilitate increased woody species establishment in mixed herbaceous-woody communities. Under normal precipitation conditions, herbaceous and woody species would utilize soil moisture at different depths and herbaceous species would not competitively eliminate woody species. However, when precipitation increases, woody species have the advantage to utilize deeper and more abundant soil moisture. This might increase successful establishment of more woody plants, after which woody and herbaceous species would continue to exploit soil moisture from different depths and herbaceous species would not eliminate newly established woody species. When precipitation decreases, herbaceous species mortality might increase leading to gaps that can be colonized by woody species that are more drought tolerant (Archer 1994).

Archer (1994) summarizes that shifts in seasonality of rainfall in the last century might have contributed to shrubland expansion in southwestern USA and future shifts from summer to winter precipitation associated with increased atmospheric CO_2 might make the current grasslands vulnerable to woody encroachment. In arid and semiarid environments, cool season moisture favors woody plants, while warm season precipitation favors grasses. When precipitation falls during the cool season, soil moisture percolates down the soil horizons and accumulates at deeper depths. Woody species with deeper root systems are able to utilize such soil moisture. Grasses, however, are unable to exploit such soil moisture and can be especially vulnerable during summer drought. Small frequent precipitation events would favor herbaceous species with shallow root systems, while larger precipitation events would benefit woody species with deeper roots. Therefore, a shift in precipitation distribution would have important implications for herbaceous-woody species balance. In arid environments, changes in extreme climatic events might have more profound effects on vegetation than the gradual change in mean conditions.

Increased atmospheric CO_2-increased temperature models predict that woodland distribution would increase in extent in tropical, subtropical, and cool temperate regions of the world under increasing temperature conditions, potentially because woody species are more stress-tolerant than herbaceous species (Archer 1994 and references therein). Similarly, during drought periods woody species are better able to persist, while herbaceous species decline. The resulting gaps can be occupied by woody species. Periodic droughts, therefore, might be associated with episodic woody plant establishment.

Fire suppression is another explanation that has been proposed for increased woody species distribution. Fire has been shown to be a primary factor that creates and maintains grasslands (DeByle 1981; Arno and Gruell 1986; Dando and Hansen 1990; Covington and Moore 1994; Mast et al. 1997). When fire is suppressed, woody species encroach into grasslands (DeByle 1981; Arno and Gruell 1986; Dando and Hansen 1990; Covington and Moore 1994) through increased woody seedling establishment (Mast et al. 1997; Archer 1994) and survival (Dando and Hansen 1990; Archer 1994). Once seedlings reach a sufficient size and age, they are able to tolerate fires and dominate grasslands (Archer 1994). When fire was suppressed, many different grasslands of varying composition were encroached by woody species of *Juniperus*, *Artemisia*, and *Prosopis* in western and southwestern USA (Burkhardt and Tisdale 1976; Blackburn and Tueller 1970; Young and Evans 1981; Johnson 1987; Brown and Archer 1989; Miller and Wigand 1994; Miller and Rose 1995, 1999; Baker and Shinneman 2004), and by *Pseudotsuga* and *Pinus* in other parts of the USA (Arno and Gruell 1986; Dando and Hansen 1990; Mast et al. 1997; Mast and Veblen 1999).

Mixed woody-herbaceous ecosystems are sensitive to land use changes (Werner 1990a, b, c, d, e). However, land use in many savanna environments are intensifying around the world (Werner 1990) and the extent of area modified by human use is continually increasing (House et al. 2003). The above described factors, especially changing climate and increasing carbon dioxide accelerate the effects of land use and changes in land use on woody-herbaceous–livestock species interaction (House et al. 2003). Land use policies and socio-economic interests of the pastoral livestock industry, the most common land use in mixed tree-herbaceous ecosystems, are increasingly focused on improved pasture production through additions of fertilizers and supplements, improved breeds and types of livestock, and manipulations of the plant species composition through introducing new species (e.g., legumes) and removing trees (Werner 1990a, b, c, d, e). Lastly, many different local factors have been proposed, in addition to the proximate factors discussed above, as important variables influencing the woody-herbaceous interaction and leading to a transition into grassland or a transition to woodland. Jeltsch et al. (2000) reviewed these local factors from studies around the world to propose ecological buffering mechanisms as their unifying theory that explains long-term tree–grass interaction.

4.8 Current and Future Research on Woody-Herbaceous–Livestock Species Interaction

Current studies provide detailed, field-based observations of woody-herbaceous–livestock species interactions (Sankey et al. 2006a, b). Tree/shrub age and distribution are often characterized with livestock grazing information and the data are used to make inferences regarding woody-herbaceous–livestock species interaction and to build empirical models (Burrows et al. 1990). However, models of interactions have not been explicitly tested outside of the regions and sites for which they were developed (House et al. 2003). Further studies need to use such data in quantitative models and simulation-based approaches (McKeon et al. 1990) and test the validity of empirical models that are based on site-specific data and relationships. Future research can also include process-based studies with carefully designed experiments. Such studies, although they are likely to be short-term, would provide important details on ecological processes involved in the woody-herbaceous–livestock interaction. Using the detailed understanding of the processes involved, accurate empirical relationships and simulation models could be built to observe potential patterns and changes at longer time scales (Daly et al. 2000). This would further enhance our understanding of woody-herbaceous–livestock species dynamics and their changes at different spatial and temporal scales. Process-based studies should also have more controlled experiments, where effects of different grazing intensities and the effects of varying grazing animals can be statistically separated. This can further improve our understanding of the effects of overall grazing intensities and different grazing animals and allow an understanding of the importance of overall grazing intensity vs. grazing animal species.

Another trend in current studies of woody-herbaceous species dynamics with or without livestock grazing effects is the use of digitally available data such as satellite imagery and digital aerial photography to map and monitor changes in woody-herbaceous plants. In northwestern USA, for example, several studies have used digital imagery to detect the commonly field-observed expansion of juniper and pinyon-juniper woodlands into adjacent shrublands and grasslands (Strand et al. 2006; Weisberg et al. 2007; Sankey et al. 2010). Field-based approaches for detecting woody cover increase provide highly accurate and valuable results, but they can be labor-intensive, time-consuming, and limited in the spatial extent they can cover. In comparison, the application of remote sensing methods can be more cost-effective and timely due to the large areal extent they cover. Digital satellite imagery also provides opportunities for more robust and comprehensive analysis of change, as the imagery can be easily integrated with other sources of digital data, such as digital maps of grazing lands and topography. Moreover, data from satellite platforms, such as Landsat, can be acquired in retrospect to examine past changes or past vegetation distribution and to compare with current distribution in order to quantify the extent and rates of change. Such analysis of remote sensing data along with detailed field data could provide information on indicators of global environmental changes and enhance our understanding of processes, signals, extent, and rates of woody-herbaceous vegetation changes under herbivory effects. The information would also be useful in grazing management and land use decision making regarding desired vegetation patterns across the landscape.

Quantitative models woody-herbaceous–livestock species interaction could also include data of other important factors that contribute to changes in woody-herbaceous species such as climate change, CO_2 increase, and fire suppression. This would enhance our ability to quantitatively describe the combined effects of multiple interactive factors on ecotone shift. This would also improve our predictive ability and forecasting skills regarding when, where, and under what conditions changes occur leading to a regional environmental change. Finally, currently proposed hypotheses and empirical models should be quantitatively tested in varying regions with different plant and livestock species.

References

Alverson WS, Waller DW, Solheim SL (1988) Forests too deer: edge effects in Northern Wisconsin. Conserv Biol 2:348–358

Archer S (1989) Have Southern Texas savannas been converted to woodlands in recent history? Am Nat 134:545–561

Archer S (1990) Development and stability of grass/woody mosaics in a subtropical savanna parkland, Texas, USA. J Biogeogr 17:453–462

Archer S (1994) Woody plant encroachment into southwestern grasslands and savannas: rates, patterns, and proximate causes. In: Vavra M, Laycock WA, Pieper RD (eds) Ecological implications of livestock herbivory in the west. Society for Range Management, Denver

Archer S, Scifres CJ, Bassham CR, Maggio R (1988) Autogenic succession in a subtropical savanna: conversion of grassland to thorn woodland. Ecol Monogr 58:111–127

Arno SF, Gruell GE (1986) Douglas-fir encroachment into mountain grasslands in southwestern Montana. J Range Manage 39:272–275

Asner GP, Elmore AJ, Olander LP, Martin RE, Harris T (2004) Grazing systems, ecosystem responses, and global change. In Annual Review of Environment and Resources (Eds Matson PA, Gadgil A, Kammne DM) 29:261–299. Stanford, CA, USA; Annual Reviews, Palo Alto, CA, USA

Bachelet D, Lenihan JM, Daly C, Neilson RP (2000) Interactions between fire, grazing and climate change at Wind Cave National Park, SD. Ecol Model 134:229–244

Baker WL, Shinneman DJ (2004) Fire and restoration of pinon-juniper woodlands in the western United States: a review. For Ecol Manage 189:1–21

Bartolome J, Franch J, Plaixats J, Seligman NG (2000) Grazing alone is not enough to maintain landscape diversity in the Montseny Biosphere Reserve. Agric Ecosyst Environ 77:267–273

Bazzaz FA (1996) Plants in changing environments: linking physiological, population, and community ecology. Cambridge University Press, Cambridge
Belsky AJ (1990) Tree/grass ratios in east African savannas: a comparison of existing models. J Biogeogr 17:483–489
Blackburn WH, Tueller PT (1970) Pinyon and juniper invasion in black sagebrush communities in east-central Nevada. Ecology 51:841–848
Brown DG (1994) Comparison of vegetation-topography relationships at the alpine treeline ecotone. Phys Geogr 15:125–145
Brown JR, Archer S (1989) Woody plant invasion of grasslands: establishment of honey mesquite (*Prosopis glandulosa var. glandulosa*) on sites differing in herbaceous biomass and grazing history. Oecologia 80:19–26
Burkhardt JW, Tisdale EW (1976) Causes of juniper invasion in southwestern Idaho. Ecology 57:472–484
Burrows WH, Carter JO, Scanlan JC, Anderson ER (1990) Management of savannas for livestock production in north-east Australia: contrasts across the tree-grass continuum. J Biogeogr 17:503–512
Cairns DM, Moen J (2004) Herbivory influence tree lines. J Ecol 92:1019–1024
Camarero JJ, Gutierrez E, Fortin M (2000) Spatial pattern of subalpine forest-alpine grassland ecotones in the Spanish Central Pyrenees. For Ecol Manage 134:1–16
Carmel Y, Kadmon R (1999) Effects of grazing and topography on long-term vegetation changes in a Mediterranean ecosystem in Israel. Plant Ecol 145:243–254
Covington WW, Moore MM (1994) Southwestern ponderosa forest structure: changes since Euro-American settlement. J For 40:39–47
Daly C, Bachelet D, Lenihan JM, Neilson RP, Parton W, Ojima D (2000) Dynamic simulation of tree-grass interactions for global change studies. Ecol Appl 10:449–469
Dando LM, Hansen KJ (1990) Tree invasion into a range environment near Butte, Montana. Great Plains Rocky Mount Geogr J 18(1):65–76
DeByle NV (1981) Clearcutting and fire in the larch/douglas fir forests of western Montana-A multifaceted research summary. General Technical Report INT-99. U.S. Forest Service. Fort Collins, Colorado
DeByle NV, Winokur RP (1985) Aspen: ecology and management in the Western United States. General Technical Report RM-119. Rocky Mountain Forest and Research Station. Fort Collins, Colorado
Didier L (2001) Invasion patterns of European larch and Swiss stone pine in subalpine pastures in the French Alps. For Ecol Manage 145:67–77
Gosz JR (1991) Fundamental ecological characteristics of landscape boundaries. In: Holland M, Risser PG, Naiman RJ (eds) Ecotones, the role of landscape boundaries in the management and restoration of changing environments. Chapman & Hall, New York
Hermann RK, Lavender DP (1965) Douglas-Fir. Silvics Manual, vol 1. United States Department of Agriculture. Forest Service. Agriculture Handbook 654
Hjalten J, Danell K, Ericson L (1993) Effects of simulated herbivory and intra-specific competition on the compensatory ability of birches. Ecology 74:1136–1142
Holland M, Risser PG (1991) The role of landscape boundaries in the management and restoration of changing environments: introduction. In: Holland M, Risser PG, Naiman RJ (eds) Ecotones, the role of landscape boundaries in the management and restoration of changing environments. Chapman & Hall, New York
House JI, Archer S, Breshears DD, Scholes RJ, NCEAS Tree-Grass Interactions Participants (2003) Conundrums in mixed woody-herbaceous plant systems. J Biogeogr 30:1763–1777
Jeltsch F, Weber GE, Grimm V (2000) Ecological buffering mechanisms in savannas: a unifying theory of long-term tree-grass coexistence. Plant Ecol 161:161–171
Johnson KL (1987) Sagebrush over time: a photographic study of rangeland change. In: McArthur ED, Welch BL (eds) Biology of Artemisia and Chrysothamnus. USDA Forest Service General Technical Service Report INT-200. Ogden, Utah

Kay EC, Bartos DL (2000) Ungulate herbivory on Utah aspen: assessment of long-term exclosures. J Range Manage 53:145–153
Knoop WT, Walker BH (1985) Interactions of woody and herbaceous vegetation in a southern African savanna. J Ecol 73:235–253
Kupfer JA, Cairns DM (1996) The suitability of montane ecotones as indicators of global climatic change. Prog Phys Geogr 20:253–272
Kupfer JA, Miller JD (2005) Wildfire effects and post-fire responses of an invasive mesquite population: the interactive importance of grazing and non-native herbaceous species invasion. J Biogeogr 32:453–466
Mast JN, Veblen TT (1999) Tree spatial patterns and stand development along the pine-grassland ecotone in the Colorado Front Range. Can J For Resour 29:575–584
Mast JN, Veblen TT, Hodgson ME (1997) Tree invasion within a pine-grassland ecotone: an approach with historic aerial photography and GIS modeling. For Ecol Manage 93:181–194
McInnes PF, Naiman RJ, Pastor J, Cohen Y (1992) Effects of moose browsing on vegetation and litter of the boreal forest, Isle Royle, Michigan, USA. Ecology 73:2059–2075
McKeon GM, Day KA, Howden SM, Mott JJ, Orr DM, Scattini WJ, Weston EJ (1990) Northern Australian savannas: management for pastoral production. J Biogeogr 17:355–372
McNaughton SJ (1979) Grazing as an optimization process: grass-ungulate relationships in the Serengeti. Am Nat 113:691–703
McPherson GR (1992) Comparison of linear and non-linear overstory-understory models for ponderosa pine: a conceptual framework. For Ecol Manage 55:31–34
Menaut JC, Gignoux J, Prado C, Clobert J (1990) Tree community dynamics in a humid savanna of the Cote-d'Ivoire: modelling the effects of fire and competition with grass and neighbors. J Biogeogr 17:471–481
Miller RF, Rose JA (1995) Historic expansion of *Juniperus occidentalis* (western juniper) in southeastern Oregon. Great Basin Nat 55:37–45
Miller RF, Rose JA (1999) Fire history and western juniper encroachment in sagebrush steppe. J Range Manage 52:550–559
Miller RF, Wigand PE (1994) Holocene changes in semiarid pinyon-juniper woodlands. Bioscience 44:465–474
Reid RS, Ellis JS (1995) Impacts of pastoralists on woodlands in South Turkana, Kenya: livestock-mediated tree recruitment. Ecol Appl 5:978–992
Rooney TP, Solheim SL, Waller DM (2002) Factors affecting the regeneration of northern white cedar in lowland forests of the Upper Great Lakes Region, USA. For Ecol Manage 163: 119–130
Sankey TT (2007) Spatial patterns of Douglas-fir and aspen forest expansion. New For 35:1–12
Sankey TT, Montagne C, Graumlich L, Lawrence R, Nielsen J (2006a) Twentieth century forest-grassland ecotone shift in Montana under differing livestock grazing pressure. For Ecol Manage 234:282–292
Sankey TT, Montagne C, Graumlich L, Lawrence R, Nielsen J (2006b) Lower forest-grassland ecotones and 20th century livestock herbivory effects in northern Mongolia. For Ecol Manage 233:36–44
Sankey TT, Glenn N, Ehinger S, Boehm A, Hardegree S (2010) Characterizing western juniper expansion via a fusion of Landsat 5 Thematic Mapper and lidar data. Rangel Ecol Manage 63:514–523
Scheffer M, Carpenter S, Foley JA, Folke C, Walker B (2001) Catastrophic shifts in ecosystems. Nature 413:591–596
Scholes RJ, Archer SR (1997) Tree-grass interaction in savannas. Annu Rev Ecol Syst 28:517–544
Sharp BR, Whittaker RJ (2003) The irreversible cattle-driven transformation of a seasonally flooded Australian savanna. J Biogeogr 30:783–802
Smit R, Olff H (1998) Woody species colonization in relation to habitat productivity. Plant Ecol 139:203–209

Strand EK, Smith AMS, Bunting SC, Vierling LA, Hann DB, Gessler PE (2006) Wavelet estimation of plant spatial patterns in multitemporal aerial photography. Int J Remote Sens 27:2049–2054

Stuart-Hill GC, Tainton NM (1989) The competitive interaction between *Acacia karroo* and the herbaceous layer and how this is influenced by defoliation. J Appl Ecol 26:285–298

Taylor AH (1995) Forest expansion and climate change in the mountain hemlock zone, Lassen Volcanic National Park, California, USA. Arctic Alpine Res 27(3):207–216

Tilghman NG (1989) Impacts of white-tailed deer on forest regeneration in Northwestern Pennsylvania. J Wildl Manage 53:524–531

Turner MG, Gardner RH, O'Neill RV (1991) Potential responses of landscape boundaries to global environmental change. In: Holland M, Risser PG, Naiman RJ (eds) Ecotones, the role of landscape boundaries in the management and restoration of changing environments. Chapman & Hall, New York

Vallentine JF (2001) Grazing management. Academic, San Diego

Wahungu GM, Catterall CP, Olsen MF (2002) Seedling predation and growth at a rainforest-pasture ecotone, and the value of shoots as seedling analogues. For Ecol Manage 162:251–260

Walker BH, Ludwig D, Holling CS, Peterman RM (1981) Stability of semi-arid savanna grazing systems. J Ecol 69:473–498

Walsh SJ, Butler DR (1994) Influence of snow patterns and snow avalanche on the alpine treeline ecotone. Phys Geogr 15:181–199

Weisberg PJ, Lingua E, Pillai RB (2007) Spatial patterns of pinyon-juniper woodland expansion in central Nevada. J Range Manage 60:115–124

Werner PA (1990) Biological mosaics and tree/grass ratios (introduction). J. Biog 17:451

Werner PA (1990a) Biological mosaics and tree/grass ratios (introduction). J Biogeogr 17:451

Werner PA (1990b) Savanna management for pastoral industries. J Biogeogr 17:501–502

Werner PA (1990c) Introduction. J Biogeogr 17:343–344

Werner PA (1990d) Conclusions. J Biogeogr 17:553–557

Werner PA (1990e) Ecological determinants of savannas: abiotic and biotic (introduction). J Biogeogr 17:401–402

Young JA, Evans RA (1981) Demography and fire history of a western juniper stand. J Range Manage 34:501–505

Zald HSJ (2002) Physiographic and reproductive components of treeline response to climate variation in the Alaska Range. Glob Glimpses Center Glob Change Arctic Syst Res 10(1):7–9

Chapter 5
Woody Plant Invasions in Pampa Grasslands: A Biogeographical and Community Assembly Perspective

Enrique J. Chaneton, Noemí Mazía, William B. Batista, Andrés G. Rolhauser, and Claudio M. Ghersa

5.1 Introduction

Grasslands provide a number of ecosystem services that have been essential to human well-being since pre-historical times (Gibson 2009). Yet they also represent the most endangered terrestrial biome due to conversion into agricultural systems (Sala 2001). The grassland biome covers 15 million km^2 (11 %) of the Earth's surface with non-woody vegetation, excluding deserts and savannas (Sala 2001). Natural grasslands encompass vast regions including the North American Great Plains, the Eurasian steppes of Russia, China and Mongolia, and the South American Pampas (Gibson 2009).

Temperate grasslands occur in areas with *c.* 250–1,200 mm of annual precipitation, being replaced by deserts and forests at lower and higher precipitation, respectively. With mean annual temperatures above *c.* 16 °C and 500–1,200 mm/year, grasslands give way to savannas and open woodlands (Whittaker 1975). Nevertheless, grasslands often occupy areas that are wet enough to support forests (Bond 2008). The existence of grassy—treeless—vegetation where climate would permit forests

E.J. Chaneton (✉) • C.M. Ghersa
IFEVA–CONICET, Universidad de Buenos Aires, Avenida San Martín 4453,
1417 Buenos Aires, Argentina
e-mail: chaneton@ifeva.edu.ar

N. Mazía
Cátedra de Dasonomía, Universidad de Buenos Aires, Avenida San Martín 4453,
1417 Buenos Aires, Argentina

W.B. Batista
IFEVA–CONICET and Departamento de Métodos Cuantitativos,
Universidad de Buenos Aires, Avenida San Martín 4453, 1417 Buenos Aires, Argentina

A.G. Rolhauser
Departamento de Biología, Facultad de Ciencias Exactas, Físicas y Naturales, Universidad Nacional de San Juan, Av. Ignacio de la Roza (Oeste) 599, Rivadavia, San Juan, Argentina

R.W. Myster (ed.), *Ecotones Between Forest and Grassland*,
DOI 10.1007/978-1-4614-3797-0_5,

has long intrigued researchers (Christy 1892; Parodi 1942; Ellenberg 1962; Walter 1967). While climatic grassland/forest boundaries have been maintained for millennia, modern grasslands appear to be changing as a result of woody invasions (Bond 2008; Simberloff et al. 2010; Lamarque et al. 2011).

Woody plant encroachment reflects the synergistic influence of various drivers of global environmental change (Hobbs 2000). Land-use changes associated with agriculture and domestic grazing have caused the widespread loss or fragmentation of large areas of temperate grasslands (Sala 2001; Gibson 2009). These landscape-level modifications have often been followed by the introduction of exotic tree species, many of which became invasive of natural or seminatural grasslands (Richardson and Rejmánek 2011). In addition, anthropogenic changes in disturbance regimes created novel opportunities for woody species establishment within herbaceous vegetation (Hobbs and Huenneke 1992; Bond 2008). These changes may interact with altered precipitation patterns, affecting expansion rates of native or exotic tree species (Dukes and Mooney 1999; Sankaran et al. 2004). Successful invasions by exotic trees draw attention towards biogeographic and phylogenetic constraints that might have limited the presence of trees in the original grasslands (Ricklefs and Schluter 1993; Mack 2003). Hence, questions regarding the historical existence of treeless grasslands as well as present shifts in woody cover may be integrated within a common framework, by emphasising the role of multiple processes that control vegetation structure at different spatio-temporal scales.

In this chapter, we examine the ongoing process of tree invasion in South American Pampa grasslands. In doing so, we consider two broad phenomena, namely, the dynamics of native woody species across grassland/forest boundaries and the spread of exotic trees over heterogeneous habitat mosaics. We argue that the latter occurs primarily through "diffuse" ecotones created by anthropogenic landscape fragmentation. Overall, grassland invasion by either native or exotic tree species represents the assembly into a local community of a novel life form that managed to pass various ecological "filters" (Funk et al. 2008; Fig. 5.1). Our central thesis is that human activity facilitates woody species establishment in grasslands by altering assembly filters for either exotic or native tree taxa. We shall use the term "invasion" to denote the successful addition of a new species to a recipient system, regardless of whether it was introduced from other continents or expanded its range from an adjacent biogeographic zone (see Vermeij 2005).

We focus on the humid Pampas of Argentina, the largest portion of the Río de la Plata grasslands, which also comprise the Campos of Uruguay and southern Brazil (Soriano 1992). Woody invasions in the Campos have recently been assessed by others (De Patta Pillar et al. 2009). We begin by discussing grassland invasion by woody plants in the light of a hierarchical model of community assembly. Second, we introduce the main biogeographical features of the study region, before revisiting hypotheses for the absence of forests in the Pampas. We close the section with a consideration of multi-causal constraints to the presence of trees in pristine grassland. Third, we summarise empirical evidence from different scenarios of tree invasion in Pampa grasslands. We conclude by suggesting how facilitative interactions between exotic and native woody taxa could play a key role in accelerating a state shift towards a novel, woody-dominated ecosystem.

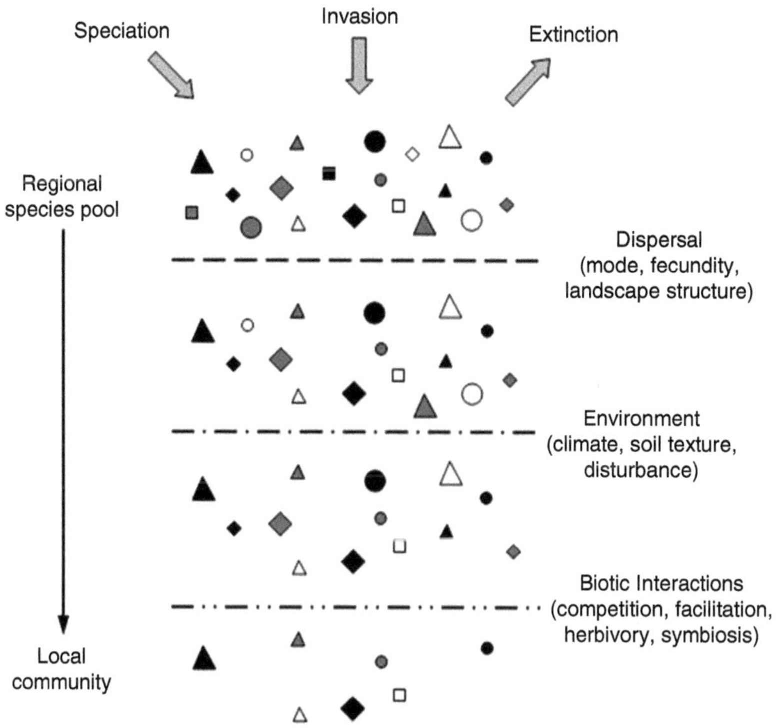

Fig. 5.1 Hierarchical model of community assembly. *Horizontal lines* represent ecological filters for species addition to a local grassland community. *Small symbols* denote herbaceous species; *large symbols* denote tree species, including native species of wet subtropical origin (*white*), native species from dry savannas (*grey*), and introduced exotic species (*black*)

5.2 Woody Invasions as Hierarchical Assembly Processes

The assembly of new species into a local community is hierarchically constrained by dispersal from the regional species pool and niche limitations created by habitat conditions and biotic interactions (Ricklefs and Schluter 1993; Funk et al. 2008; Myers and Harms 2009). Factors affecting the chances of successful species establishment can be seen as a series of ecological "filters" (Fig. 5.1). Only species in the regional pool having certain adaptive traits will be able to pass these filters (Keddy 1992; Funk et al. 2008; see also Theoharides and Dukes 2007). Assembly filters roughly correspond to processes operating at different spatio-temporal scales.

Species may be added to a regional pool as a result of speciation processes or by immigration from another region (Ricklefs and Schluter 1993). Biotic exchanges at large biogeographical scales are aided by human transport (Williamson 1996; Theoharides and Dukes 2007), which may act to enrich the phylogenetic background of a regional flora (Mack 2003). Exotic invasions are a manifestation of the re-shuffling of regional species pools caused by human activities on a global scale. In this context, the absence of woody life forms in a grassy ecosystem may just reflect the evolutionary history of the flora, including major events deep in the regional past (Ricklefs and Schluter 1993; Mack 2003). It also suggests that native trees growing in nearby forest habitats may not have the "right" functional traits to establish successfully in a grassland environment (Bredenkamp et al. 2002; see Fig. 5.1).

For a species to occur at a given locality, it must first arrive via dispersal from available propagule sources (Fig. 5.1). The chance of seed arrival is a function of landscape configuration, including site isolation, as well as species fecundity and dispersal mode (anemochory, zoochory, etc.) of the focal species. The existence of appropriate vectors (e.g. frugivorous birds) can be critical for animal-dispersed species. Seed dispersers interact strongly with landscape structure in determining the spatial pattern of dispersal, both within and across biomes (Nores et al. 2005). Propagule "pressure" (i.e. the quantity and frequency of propagule arrival) has been indicated as a primary determinant of exotic invasions that is highly influenced by human activity (Williamson 1996). Low propagule pressure may increase the chances that an invading species will be eliminated by stochastic events (Simberloff 2009). Exotic tree species that were planted at several points within a region, produce large seed crops with a short age at maturity, and are readily dispersed by wind or animals (e.g. cattle, birds) may exhibit high invasion potential (Rejmánek and Richardson 1996; Lamarque et al. 2011). Conversely, native woody species found in grassland/forest ecotones might have yet to disperse in sufficient numbers to become established in grassland habitats.

Species lacking the functional traits to cope with local habitat conditions will fail to become established or to persist in the long term (Fig. 5.1). Therefore, successful invaders should be pre-adapted to abiotic conditions in the recipient system (Williamson 1996; Mack 2003). This involves the ability to tolerate various environmental stresses (e.g. frost, drought, waterlogging), which can be recurrent features of certain localities. For instance, soil aridity or low temperatures may limit the presence of trees in natural grasslands (Bredenkamp et al. 2002). Climate fluctuations may create resource opportunities for episodic seedling recruitment as observed for resident savanna trees (Sankaran et al. 2004). Natural disturbances may act to facilitate or prevent exotic species establishment (D'Antonio et al. 1999; Chaneton et al. 2002). For instance, effects of burning and grazing on woody encroachment are mostly idiosyncratic and context-dependent (Hobbs 2000; Bond 2008), although some invasive taxa like *Pinus* spp. appear to thrive in frequently disturbed habitats (Simberloff et al. 2010).

Only a small fraction of the species arriving at a site do establish persistent populations without being excluded by interactions with other species (Fig. 5.1). Biotic limitations to establishment can reflect either negative impacts from competitors and consumers (e.g. seed predators) or the absence of specific mutualists (Crawley 1987; Levine et al. 2004). Elton (1958) regarded the network of resident species interactions as the prime barrier against invaders (cf. Levine et al. 2004). Tree seedling growth in grasslands can be strongly limited by competition from herbaceous plants (Davis et al. 1998). Selective removal of the grass cover may release resources and increase woody recruitment (Sankaran et al. 2004). Exotic trees with high growth rates at the seedling stage appear to be better equipped for invading undisturbed grasslands (Rejmánek and Richardson 1996; Lamarque et al. 2011). Lastly, positive interactions (facilitation) between woody invaders, or between tree seedlings and resident plants, may be important as a means of ameliorating the effects of physical or biotic stress during tree establishment (e.g. Tecco et al. 2006; Smit et al. 2006).

It follows that grassland invasions by woody species (native or exotic) should be seen as multi-causal processes that integrate the effects of several putative factors acting at various scales. The existence of treeless grasslands such as the humid Pampas could be traced back to the influence of historical processes that shaped the composition of the regional pool as well as to the roles of propagule pressure, environmental stress, disturbance, and biotic interactions (Fig. 5.1). Tree invasion of primary grasslands begs the question of which filters were modified and what were the intervening causes.

5.3 The Humid Pampas: A Phytogeographic Overview

The temperate humid Pampas extend over *c.* 400,000 km^2 between 31 and 38 °S in east-central Argentina (Fig. 5.2). The Pampean phytogeographic province is surrounded by the Espinal province, a belt-shaped system of open woodlands and savannas (Cabrera 1958, 1976; Cabrera and Willink 1980). Native forests also occur along riparian corridors of the Uruguay and Paraná rivers, which extend from the subtropical Paranaense province into the northern part of the Pampas (Nores et al. 2005). Precipitation decreases from the north-east to the south-west of the region (1,000–500 mm/year). Rainfall occurs throughout the year, with mild summer water deficits in the east, and more severe, winter dry periods in the west. Mean annual temperature is 18 °C, although frosts are common in winter (Soriano 1992). The landscape developed on Andean loess deposits and alternates vast flat areas with gently rolling hills. Rocky outcrops form the Tandilia and Ventania ranges (1,500 m a.s.l.) in the southern Pampas (Fig. 5.2). The dominant soils in the wetter, eastern areas are Mollisols with a heavy argillic B horizon, while towards the west, soils are mostly Entisols with a coarse, sandy texture (Soriano 1992).

The origin of the Pampas flora reflects the influence of the Neotropical and Antarctic realms (Cabrera and Willink 1980; Ribichich 2002). The primary vegetation corresponds to mesophytic grasslands with a mix of C_3 and C_4 perennial grasses (Burkart 1975; Soriano 1992). Common genera include *Stipa, Poa, Piptochaetium, Melica, Briza, Bothriochloa, Aristida, Panicum,* and *Paspalum.* Exotic herbaceous forbs are common throughout the region (Ghersa and León 1999; Perelman et al. 2001; Burkart et al. 2011). The Pampas comprise five phytogeographic districts (Fig. 5.2), differing in the degree of landscape transformation by humans (Baldi et al. 2006):

Mesopotamic pampa—This is most humid and warmer section of the Pampas. The vegetation structure and composition have strong affinities with the Espinal and Chaco provinces (Cabrera 1958, 1976; Soriano 1992). This district is surrounded by

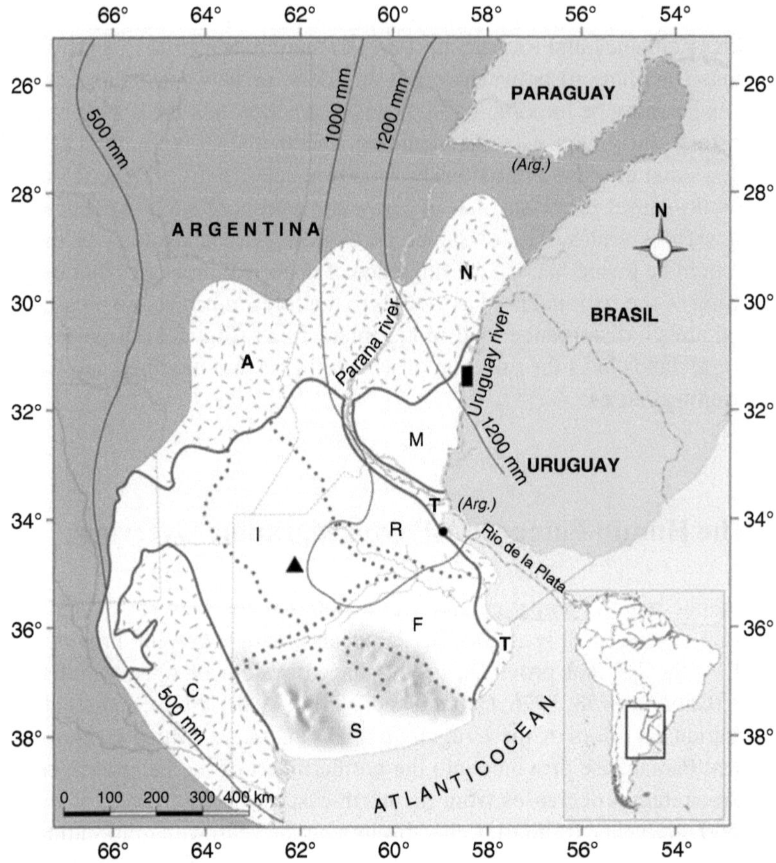

Fig. 5.2 Limits of the Pampa grasslands and Espinal woodlands in east-central Argentina. Phytogeographic districts (*dotted lines*) of the Pampa province (*thick line*) are denoted by different letters: *M* Mesopotamic pampa; *R* Rolling pampa; *F* Flooding pampa; *I* Inland pampa; *S* Southern pampa. Districts of the Espinal province (*stippled area*) are denoted by *bold-type letters*: *Ñ* Ñandubay; *A* Algarrobo; *C* Caldén; *T* Tala. *Lightly shaded areas* within S show the Ventania and Tandilia mountain ranges. *Symbols* indicate the location of long-term research sites: El Palmar National park (*rectangle*) and Estancia San Claudio (*triangle*)

different types of forest formations, including subtropical gallery forests along the Uruguay and Parana riverbanks. Grasslands are dominated by C_4 tussock grasses and several native shrubs. Palm trees (*Butia yatay*) are scattered across large grassland areas, creating a conspicuous savanna-like physiognomy. Wildfires likely were a recurrent disturbance agent in the past, but nowadays have been largely suppressed.

Rolling pampa—This section of the Pampas has been the most deeply transformed by rowcrop agriculture, cattle husbandry, and human settlement since the Europeans arrived in the mid-1500s. Darwin (1893) described the grasslands as being heavily invaded by exotic thistles. The potential vegetation is a "flechillar" with a species-rich mixture of C_3 and C_4 tussock grasses (Cabrera 1976; Soriano 1992). Yet seminatural grasslands occupy only a small fraction of the area, and no relicts of native grassland exist today (Parodi 1947; Burkart et al. 2011).

Inland pampa—The eastern side of the district is a mosaic of croplands and rangelands. Native grasslands are confined to relictual fragments extending along abandoned railway tracks and secondary roadsides (Soriano 1992). In contrast, the western side comprises extensive natural grasslands, which are mainly used for cattle grazing. Prescribed burning is used as a management tool in the drier western areas. The dominant community is a steppe-like "flechillar" with tall-tussock grasses forming dense, species-poor, homogeneous stands (Burkart et al. 2011).

Flooding pampa—The area comprises extremely flat, flood-prone landscapes devoted to domestic grazing on natural grasslands and sown pastures (Soriano 1992). Crops have been sown only in the most elevated areas. Grazed grasslands are dominated by short tussock and prostrate grasses and harbour high numbers of low-growing exotic forbs. The C_3/C_4 native grass ratio increases southward (Perelman et al. 2001). Periodic floods and soil salinity limit plant species richness in lowlands areas.

Southern pampa—This is the cooler and drier section of the Pampas. The landscape mosaic is dominated by winter crops, pastures, and natural grassland remnants. In unploughed, well-managed grazed sites the grassland is made up of a species-rich ensemble of C_3 grasses and forbs, thriving within a matrix of tall bunchgrasses (Soriano 1992). Montane grasslands occurring on rocky outcrops and elevated hills (>500 m) contain high species richness and many endemisms.

All five pampean districts are in contact with the Espinal province (Fig. 5.2). The Espinal comprises xerophytic forests, parklands, and savannas (Cabrera 1976). The climate is warm and humid in the north, and temperate semiarid in the west and south sections. Annual precipitation varies from 1,200 mm in the north to less than 500 mm in the south-west. Mean temperatures range between 20 and 14 °C. The tree overstorey has varying densities, being dominated by *Prosopis* and *Acacia* species (Fabaceae, Mimosoideae). Tree species richness declines southwards. Woody species have a neotropical lineage, including *Celtis*, *Schinus*, *Jodina*, and *Geoffroea*. Much of the original vegetation has been modified by logging and livestock management, and presumably by fire suppression as well. The Espinal can be divided into four main districts (Cabrera 1976).

Ñandubay district—The northern section of the Espinal (Fig. 5.2) comprising low forests (<10 m height) dominated by *Prosopis algarrobillo* (ñandubay) and *Prosopis nigra* (algarrobo negro). Other common woody species include *Acacia caven* (espinillo), *Geoffroea decorticans* (chañar), *Aspidosperma quebracho-blanco*, *Celtis tala* (tala), *Schinus molle* (molle), *Jodina rhombifolia* (sombra de toro), and *Scutia buxifolia* (coronillo). There is a well-developed shrub understorey. The herbaceous layer comprises a diverse community of C_4 grasses, forbs, and bromeliads (see Cabrera 1976).

Algarrobo district—The western section of the Espinal corresponds to a once extensive xerophytic woodland (see Fig. 5.2), which has been largely lost to logging and agriculture. There remain a few relicts of native forest in which the dominant trees are *Populus alba* (algarrobo blanco), *P. nigra*, and several other species shared with the Ñandubay district, except *P. algarrobillo* (Cabrera 1976).

Caldén district—The southern section intergrades with the western Inland pampa along a wide ecotone (Fig. 5.2). The dominant species is *Prosopis caldenia* (caldén), which forms open woodlands and savannas with a sparse shrub layer and 35–50 % grass cover. This area has been severely degraded by cattle grazing and wood extraction, and therefore the actual limits with the Pampas are a source of controversy (cf. Cabrera 1976; León and Anderson 1983). Common subdominant woody species include *Prosopis flexuosa* (algarrobo), *P. nigra*, *G. decorticans*, *J. rhombifolia*, and *Schinus* spp. (incienso). The grass layer (flechillar) comprises several bunchgrasses including *Elyonurus muticus*, *Piptochaetium stipoides*, *Poa ligularis*, *Poa lanuginosa*, and various species of *Setaria* and *Stipa* (Cabrera 1976).

Tala district—It extends southward as a narrow coastal forest along the lower section of the Paraná river and the coast of Buenos Aires on the Atlantic (Parodi 1940; Cabrera 1958, 1976; Fig. 5.2). The so-called "talares" are xerophytic hardwood forests growing on well-drained coastal dunes or calcareous soils formed on ancient marine (shell) deposits. Forest patches intermingle with lowland grasslands (Soriano 1992). The community is dominated by a deciduous thorny tree, *C. tala* (Ulmaceae), and contains a mixture of woody natives such as *S. buxifolia*, *A. caven*, *Stellaria longifolia*, *J. rhombifolia*, *P. alba*, *Phytolacca dioica* (ombú), and *Sambucus australis* (sauco). The original forest has been deeply fragmented by logging and few patches remain undisturbed (Cabrera 1976). In some areas, the forest was cleared during colonial times (Garavaglia 1999a). Forest remnants are being colonised by nearly 20 exotic tree species, of which the evergreen *Ligustrum lucidum* is the worst invader (Cagnoni et al. 1996; Ribichich and Protomastro 1998).

5.4 Why Were There No Trees in the Pristine Pampas?

The existence of treeless natural grasslands in virtually all continents has attracted much controversy (Bredenkamp et al. 2002; Bond 2008). Broadly speaking, explanations for the lack of trees in grassy steppes and prairies can be divided into "bottom–up" and "top–down" arguments. The former emphasise the influence of climatic conditions and abiotic stress on tree growth potential, whereas the latter focus on the role of woody biomass consumption by fire and herbivores (Bond 2008). On the one hand, the global distribution of forests would be limited by macroclimatic aridity and soil texture (Walter 1967; Whittaker 1975). On the other hand, there is little doubt that human-induced fire suppression and reduction of native herbivore populations provide opportunities for woody encroachment on grassy vegetation (Sankaran et al. 2004; Bond 2008; Bond et al. 2005). We notice, however, that both these perspectives implicitly assume that ill-adapted tree genotypes would be available in the regional pool, and hence focus on environmental and biotic filters to species assembly (see Fig. 5.1). We shall come back to this point later in this section.

Vegetation scientists have often regarded the Pampa grasslands as an anomaly, since other regions with similar temperate humid climates typically support deciduous or mixed forests. Were the Pampas a climatically determined (primary) grassland or an anthropogenic one? (Schmieder 1927; Parodi 1942; Ellenberg 1962; Walter 1967). It has also been questioned why is that forests do not occur naturally in the Pampas when many introduced tree species grow spontaneously in old fields, along road verges and fencelines, and within forest plantations (Parodi 1942; Ellenberg 1962; Ghersa et al. 2002). In the following, we briefly revisit traditional ideas for the paucity of trees in the pristine Pampas. We stress that hypotheses thus far have largely been articulated around single causes. We then integrate the various putative causes for the existence of treeless Pampa grasslands within a multi-causal, hierarchical framework.

5.4.1 Single-Cause Explanations for Treeless Grasslands

Perhaps the most frequently cited cause for the paucity of trees in temperate grassy systems is the occurrence of seasonal drought (Bredenkamp et al. 2002). For instance, the absence of trees in Eurasian primary grasslands and North American prairies has been associated with an aridity gradient. Where conditions become too dry, forests and woodlands give way to grassland. It is then common to find trees where the local moisture regime is more favourable for tree growth, like in river valleys or flood plains immersed within steppe or prairie zones (Bredenkamp et al. 2002). This is indeed a common feature of the Rolling pampa landscape, where woody species typical of the Espinal occur along tributaries of the Paraná river (Walter 1967; Ghersa and León 1999).

A related hypothesis states that the root systems of resident grasses exert strong competition for water, limiting the establishment of tree seedlings (Parodi 1942). There is much experimental evidence that grass competition can significantly reduce seedling or sapling growth under water-limiting conditions (Davis et al. 1998; Sankaran et al. 2004; Riginos 2009). This suggests that the regeneration niche of the trees would overlap with that of resident grasses (Facelli and León 1986; Bond 2008). Taken together, a negative water balance during part of the year and the higher competitive ability of grasses would determine the existence of natural "climatogenic" grasslands, and the Pampas would be no exception (Walter 1967). Forestry practices provide stress-free microsites for seedlings, which explains why many planted tree species grow well in the humid Pampas.

Some researchers have pointed out the role of soil texture in tipping the competitive balance for water between grasses and trees (Walter 1967). Heavy, clayed soils may favour grass dominance by limiting tap-root penetration, whereas trees would become dominant in lighter, sandy soils. In the Pampas, the ecotones between grassland and woodland (Espinal) roughly correspond with transitions from heavy loessic soils to coarser-textured soils (Cabrera 1976; Yazawa 1989). The combination of fine-textured soils and summer drought would make it difficult for tree seedlings to establish in bare-soil spaces among grass tussocks (Parodi 1942). This seemingly strong effect of soil texture on life-form balance may account for the sharp boundaries between xerophytic forests (talares) and grasslands on the north-eastern side of the Rolling and Flooding pampas (Ribichich and Protomastro 1998; see Fig. 5.2).

Temperature has been proposed to control the presence of trees in some regions. In particular, frost periods may limit seedling recruitment for subtropical woody taxa. For instance, according to Bredenkamp et al. (2002), it is not aridity that excludes trees from Southern African grasslands. The cooler climates prevailing in some temperate grasslands, especially during the winter, would exclude subtropical trees. Many cold- and frost-tolerant tree species from the northern hemisphere and Australia do grow well when planted in these grasslands (e.g. *Eucalyptus camaldulensis, Platanus acerifolia, P. alba*). Some of these species even become invasive (e.g. some pines and acacias). However, native trees do not survive when planted, as they are killed by winter frosts. This suggest that while aridity excludes cold-resistant trees from temperate steppes, low temperatures (and frost) may exclude subtropical trees (Bredenkamp et al. 2002). Such a causal scenario might explain why, when occurring at higher latitudes, trees originating in the subtropical Paranaense province remain confined to mild riparian habitats, without invading adjacent Pampa grasslands (Cabrera 1958; Nores et al. 2005).

Fire has been invoked as a major determinant for the absence of woody plants in grassy biomes worldwide (Bond et al. 2005). In particular, frequent burning used by early human societies for land clearing, hunting, and agriculture has been implicated in the origin of anthropogenic (secondary) grasslands in western Europe (see Bredenkamp et al. 2002). Some authors have suggested that fires set by native American Indians were the cause for the existence of treeless North American prairies (Christy 1892). This hypothesis was extended to South American grasslands, on the basis that burning was found to be a common practice among native Indians, at least, during the early period of European settlement in the Pampas (Schmieder 1927; Ellenberg 1962; Henning 1988). In this light, the Pampa grasslands would be anthropogenic, rather than a natural vegetation type in equilibrium with climate and soil conditions. This notion has been rejected for reasons that will soon become clear (Parodi 1942; Walter 1967; Soriano 1992).

The other top–down force with the potential to control vegetation structure is herbivory, especially by large-bodied grazers and browsers (Johnson 2009). Temperate South American grasslands developed without supporting high densities of large vertebrate herbivores since the late Pleistocene extinctions (Burkart 1975; Barnosky and Lindsey 2010). However, a rich fauna of megaherbivores roamed the region for much of the Tertiary (Webb 1978). Did plant consumers play a role in precluding the expansion of trees onto grassy plains in the Pampas? May be, although megaherbivores coexisted with open woodlands and gallery forests in areas like the Great Chaco and Patagonia. Medial Miocene to Pliocene fossil faunas became dominated by pampas and steppe adapted forms, while gallery forests likely persisted along stream borders (Webb 1978). Since the megafauna extinctions of southern South America coincided with the drying of the climate and occurred after the arrival of early humans (some 13,000 year BP; Barnosky and Lindsey 2010), and even so the region remained covered by grassland up to the present (Tecchi 1983; Prieto 1996), there is no compelling evidence that extinct herbivores were a primary driver of the prehistoric paucity of trees in the Pampas (cf. Johnson 2009).

A final hypothesis for the absence of forests emphasises the geomorphological youth and changing climatic history of the Pampas (Parodi 1942; Iriondo 1999; Iriarte 2006). There is strong fossil evidence from phytoliths (Tecchi 1983), pollen records (Prieto 1996), and faunal assemblages (Tonni et al. 1999) that grassy vegetation occupied the Pampas from the early Miocene and through the Quaternary. The late Pleistocene (*c.* 14,810–10,000 ^{14}C year BP) of the Pampas was defined by drier and cooler conditions (Iriondo 1999), as shown by the presence of C_3-dominated psamophytic steppes (Iriarte 2006). These conditions prevailed until the onset of a humid period in the early Holocene (*c.* 10,000–6,600 ^{14}C year BP), which saw the emergence of extensive wetlands and the replacement of C_3 pooid by C_4 panicoid grasses (Iriarte 2006). Dry steppes were displaced by humid grasslands (Prieto 1996, 2000). During the mid-Holocene, around 6,600 ^{14}C year BP, began a new period of dryness, which led to the expansion of halophytic vegetation in flat lowlands. After *c.* 4,000 ^{14}C year BP, a decline in halophytes signals the onset of more humid and stable climatic conditions, characteristic of the late Holocene to the present (Prieto 1996; Tonni et al. 1999). Overall, evidence suggests that the Pampas are a relict of drier past climates. Recurrent drying–wetting cycles might have also acted to constrain the presence of tall woody plants.

5.4.2 A Multi-causal Perspective: Putting History and Ecology Back Together

Hypotheses for the paucity of trees in the humid Pampas have referred to either proximate causes (e.g. tree seedlings cannot successfully compete with grasses) or distant causes (e.g. the Pampas are a relict of a drier past climate). Yet single-cause explanations may fail to account for the absence of tree life forms in grassland (Bond 2008), as for biotic invasions in general (Theoharides and Dukes 2007). Moreover, univariate models of vegetation structuring are inconsistent with modern ecological theory, which embraces the role of long-term evolutionary and historical processes as well as local niche relations in shaping present-day community patterns (Ricklefs and Schluter 1993; Valiente-Banuet et al. 2006; Fig. 5.1). We stress that the former absence of trees in the Pampas likely resulted from multiple interacting factors (i.e. filters) operating at disparate scales. Limitations to tree occurrence were probably fourfold.

Species pool—It appears that the regional species pool for the Pampas did not contain autochthonous trees with evolved adaptations to the local environment. Yet the region was surrounded by forests of subtropical affinity since pre-historic times (Webb 1978; Cabrera and Willink 1980). It is unknown why the dry woodland and wet forest lineages did not give rise to a "Pampas tree genotype". We may think of two reasons. First, there might have been strong phylogenetic constraints to the evolution from northern South American ancestors of temperate tree genotypes adapted to cool, subhumid climates (see Mack 2003). Southern lineages from the Antarctic domain were confined to the Andes after the last glacial period and the emergence of vast semiarid steppes in Patagonia (Markgraf et al. 1995). Second, the unstable climatic and geological conditions that characterised the Pampas from the late Tertiary through the Quaternary (Webb 1978; Prieto 1996, 2000; Iriarte 2006) could make it difficult for any marginal or founder long-lived tree population to develop the "right" suite of adaptive traits. Moreover, prior to the late Pleistocene mass extinctions, resident megaherbivores also probably helped to maintain a treeless grassy biome (Webb 1978; Johnson 2009).

Seed dispersal—Even if pre-adapted tree genotypes were available in past species pools along the margins of the Pampas, successful invasion by trees could have been precluded by low dispersal rates and hence low propagule pressure. Dispersal limitations to native trees could reflect lack of seed sources, as determined by the northward retreat of tropical forests during the late Tertiary–early Quaternary period of increased aridity in southern South America (Markgraf et al. 1995). Further, low seed arrival may result from reduced activity of dispersal agents. Birds may act as long-distance dispersers, but may fail to transport tree seed into grassland areas with few perch structures (Milton et al. 2007).

Abiotic stress—Although several tree species were actually present in the broader regional pool of the Pampas (encompassing the ecotones with Espinal woodlands and Paranaense riparian forests), most seemed to lack the functional traits to establish and reproduce in a temperate subhumid environment (Fig. 5.1). Provided enough seed of either xeromorphic or subtropical tree genotypes was dispersed into adjacent grasslands, then colonisation would have been impeded by heavy soils, drought, or frosts, especially during periods when climate was much drier than at present (Prieto 1996, 2000). Such niche limitations should create demographic bottlenecks during early life stages, thus reducing the chances of invasion (Bond 2008). Indeed, native trees from savanna and gallery forests generally do not survive when planted, as they are often killed by winter frosts.

Biotic interactions—The potential for woody species from adjacent communities to establish in the Pampas was also probably limited by competition from resident grasses (Parodi 1942; Walter 1967). Tussock grasses strongly suppress tree seedling growth and survival in humid grasslands (Mazía et al. 2001; Chaneton et al. 2004), and may indirectly prevent seedling emergence by providing cover to granivorous rodents (Mazía et al. 2010). These factors alone only rarely repel invasion, unless immigrating species are propagule-limited (Levine et al. 2004; Lockwood et al. 2005), which is the case at increasing distance from seed sources. In addition, root symbionts such as mycorrhizal fungi must be available for successful recruitment of many trees. Dominant tree taxa in the Espinal belong to the Fabaceae and require certain *Rhizobium* strains for nitrogen fixation. Thus, absence of specific soil microbial partners can add another, often underestimated barrier to tree invasion (see Simberloff et al. 2010).

Overall, we expect that low propagule pressure, abiotic stresses, biotic resistance, and a paucity of specific symbionts might have exerted a synergistic influence in slowing tree invasion rates (Crawley 1987; Lockwood et al. 2005). These factors operated under a higher-level constraint, that is, the palaeoclimatic history of the region and its long-term influence on species pool evolution.

5.5 Contemporary Patterns and Mechanisms of Tree Invasion

The original treeless physiognomy of the Pampas began to change dramatically with the arrival of European settlers to the Río de la Plata in the mid-1500s. Human activity was instrumental in shaping all four stages of the process of invasion by tree species alien to South America, from transport through colonisation and establishment, to landscape spread (Theoharides and Dukes 2007). Species pool limitations were overridden by the introduction of temperate tree genotypes pre-adapted to the Pampas environment. Since the seventeenth century trees and other woody species became conspicuous features of "estancias" in the northern (Rolling) pampas. Frequency of planted trees in urban and countryside areas increased between the eighteenth and nineteenth centuries (Garavaglia 1999b), with both native and exotic species being introduced for living fences, shade, wind barriers, and as ornamental plants (Ghersa and León 1998, 1999). Native woody species included *Acacia bonariensis* (añapindá), *Parkingsonia aculeata* (cina-cina), *P. dioica* (ombú), and *C. tala* (tala), while early exotics were *L. lucidum*, *Gleditsia triacanthos* (acacia negra), *Melia azedarach* (paraíso), *Morus alba* (mora blanca), and *Phoenix canariensis* (palmera), and later *Eucalyptus* spp. Pollen records for *G. triacanthos* in the eastern Rolling pampa date back to the late 1700s (Prieto et al. 2004). This process reshaped the regional pool by adding species with novel functional traits (Fig. 5.1).

Widespread plantation of woodlots and tree lines increased the number of seed sources and hence propagule pressure. Main roads served as corridors for the early dissemination of introduced tree species. Planted trees created new habitat for wildlife, including native birds and mammals, which likely contributed to accelerate the spread of woody plants (Ghersa and León 1998). By the early 1800s, native trees of the Espinal like *G. decorticans*, *J. rhombifolia*, and *Prosopis* spp. could be found in small forest patches in northern and eastern sites of the Pampas (Garavaglia 1999a). On the northwestern Pampas ecotone, Espinal woody species were dispersed into dry grasslands by cattle herds driven by indigenous people (León and Anderson 1983). Rapid tree encroachment was facilitated by the land-use changes that dominated the Pampas in the late 1800s after the Indian wars (Tapson 1962). Before that, burning of tall tussock grasslands by the aborigines (Darwin 1893) possibly slowed the spread of invasive woody plants.

During the 1900s, the profound anthropogenic transformation of the original landscape involved the division of the land by linear elements such as roads, fence-lines, channels, and railways. This created a network of seminatural vegetation corridors that now harbour many exotic tree species (Ghersa et al. 2002). Bird species (e.g. *Turdus rufiventris*, zorzal) find perching sites along these corridors and in domestic woodlots, thus transporting the seeds of introduced woody plants across the landscape (Ghersa and León 1998). Riparian zones have also been invaded by frost-tolerant exotic trees such as *G. triacanthos*, *L. lucidum*, and *M. alba* (Ghersa et al. 2002), probably induced by directed, bird-mediated dispersal (Nores et al. 2005). Lastly, the modern Pampas landscape contains many human-disturbed areas of noncultivated or waste land, which are invaded by exotic trees like *G. triacanthos* and *Ulmus pumila* (olmo) forming dense forest patches (Facelli and León 1986; Mazía et al. 2001). Only the vast grazed plains of the Flooding pampa remained relatively free of woody invasions (Fig. 5.3).

Overall, anthropogenic landscape fragmentation generated a diffuse, reticulated "ecotone" between modified habitat patches and remnant grasslands. Whereas historical (biogeographical) boundaries with adjacent forested regions remained mostly stable, the ecotones created by human activity in the Pampas interior promoted encroachment of introduced woody species onto once treeless landscapes. In the remaining of this section, we review evidence for contemporary patterns of woody plant invasion in the Pampas. We highlight two separate phenomena, namely, woody encroachment across geographic ecotones and spread of exotic trees onto human-disturbed grasslands.

5.5.1 Dynamics Across Natural Grassland/Forest Ecotones

5.5.1.1 Humid Palm Savannas

Savannas dominated by the palm tree *B. yatay* occur on sandy soils in the eastern Chaco and Espinal provinces (Martinez-Crovetto and Piccinini 1950; Carnevali 1994). The largest area of *B. yatay* palm savannas extends over the watersheds of several streams that flow eastward into the Uruguay River, on the northern limit of the Pampas (Fig. 5.2). Stands with widely varying densities of *B. yatay* occur on the highlands and river terraces, while alluvial plains support tall grasslands and riparian subtropical forests (Cabrera 1976). Since the creation of El Palmar National Park in 1970, a large track of palm savannas has been protected under a policy of cattle exclusion and fire control. The area had previously been used for cattle ranching, and the savannas were structured by a tall palm tree layer and a short species-rich herbaceous layer (Martinez-Crovetto and Piccinini 1950). Scarcity of juvenile palms suggested that cattle were preventing palm recruitment (Dimitri and Rial 1955; Cabrera 1976). Hence, cattle exclusion should allow palm recruitment,

Fig. 5.3 Views of the Pampas. (**a**) Treeless landscape in the central Flooding pampa, (**b**) trees invading a roadside corridor in the Rolling pampa, (**c**) invasion of an old-field grassland by *Ulmus pumila* in the Inland pampa, (**d**) a thicket formed by *Gleditsia triacanthos* and *Ligustrum lucidum* (*dark tree*) in an old pasture field, Inland pampa (photos by E. Chaneton (**a**, **c**, **d**) and F. Weyland (**b**))

while maintaining the open savanna/grassland physiognomy. Yet palm savannas were rapidly encroached by native bushy species, whereas denser savannas have been invaded by both native and exotic tree species (Ciccero and Balabusic 1994; Goveto 2006; Fig. 5.4).

Tree encroachment on *B. yatay* savannas has been examined at the landscape and local (stand) scales. In 2000, a total of 26 tree species were recorded in the understorey at El Palmar (Table 5.1). Most invading trees are native to riparian forests and are distributed over the Paranaense or Espinal provinces (Cabrera 1976). In addition, palm savannas were invaded by tree species exotic to South America, especially *M. azedarach*. Among the encroaching tree species, the majority have their seeds dispersed by birds (Table 5.1). High-palm density savannas on lower terraces have been encroached by many tree species from nearby riparian forests, while sparser savannas on uplands have fewer trees and include hardwood species from the *Ñandubay* forests (Table 5.1; Fig. 5.4).

Current landscape distribution of three of the most common native trees, *Allophylus edulis*, *Myrcianthes cisplatensis*, and *Sebastiania commersoniana*, suggest that dispersal syndrome, adult palms, and fire frequency may all have a role in the encroachment process (Rolhauser et al. 2007). After 35 years of cattle exclusion

Fig. 5.4 Aspects of the *B. yatay* palm savannas in (**a**) cattle grazed and (**b**) protected areas at El Palmar National park. Cattle exclusion and fire control triggered invasion by both native and exotic tree species, especially in dense palm savanna stands (photos by A. Rolhauser)

and relative fire control, densities of *A. edulis* and *M. cisplatensis*, two bird-dispersed species, were found to decline with distance from the riparian forest edge. Since in 1970 trees were virtually absent from the palm savanna, the observed pattern is consistent with populations expanding as a front wave (Skellam 1951). For *M. cisplatensis*, the estimated distance of tree expansion in dense palm stands unburned

Table 5.1 Tree species recorded in the understorey of *Butia yatay* palm savannas at El Palmar National Park (Mesopotamic pampa)

Species name (family)	Distribution	Dispersal mode
Frequent in dense/sparse palm savannas		
Allophylus edulis (Sapindadeae)	P, E	Birds
Myrcianthes cisplatensis (Myrtaceae)	P, E	Birds
Sebastiania spp. (Euphorbiaceae)	P	Elastic dehiscence
Myrcia ramulosa (Myrtaceae)	P	Birds
Schinus longifolius (Anacardiaceae)	P, E	Birds
Sapium haemaospermum (Euphorbiaceae)	P	Birds
Melia azedarach (Meliaceae)	Exotic	Birds, bats
Frequent in dense palm savannas		
Hexachlamys edulis (Myrtaceae)	P	Gravity + mammals
Guettarda uruguensis (Rubiaceae)	P	Birds
Maytenus ilicifolia (Celastaceae)	P	Birds
Frequent in sparse palm savannas		
Acacia caven (Fabaceae)	E	Gravity + mammals
Prosopis affinis (Fabaceae)	E	Gravity + mammals
Occasional in palm savannas		
Acacia bonariensis (Fabaceae)	P, E	Gravity
Blephalocalyx salicifolius (Myrtaceae)	P	Birds
Celtis tala (Cannabaceae)	P, E	Birds
Erythrina crista-galli (Fabaceae)	P	Gravity + water
Nectandra angustifolia (Lauraceae)	P	Birds
Ocotea acutifolia (Lauraceae)	P	Birds
Pouteria salicifolia (Sapotaceae)	P	Gravity + water
Myrsine laetevirens (Myrcinaceae)	P	Birds
Scutia buxifolia (Rhamnaceae)	P, E	Birds
Fraxinus pennsylvanica (Oleaceae)	Exotic	Wind
Gleditsia triacantos (Fabaceae)	Exotic	Gravity + water + mammals
Ligustrum lucidum (Oleaceae)	Exotic	Birds
Magnolia grandiflora (Magnoliaceae)	Exotic	Birds
Prunus persica (Rosaceae)	Exotic	Gravity + animals

The occurrence of each species in other phytogeographic units and its dispersal mode are shown. *P* riparian habitats supporting forests of the subtropical Paranaense province; *E* Ñandubay forests of the Espinal province. *Exotics* refer to species not native from South America

over the previous 25 years (~2 km) was significantly longer than in either sparse palm stands or stands burnt within the previous 15 years (~1.5 km).

In contrast, the distribution of *S. commersoniana*, a self-dispersed species, exhibits no clear trend with distance from the forest edge. Density of this species is higher in dense than in sparse palm stands, where it is extremely rare. These patterns suggest that, under current conditions of cattle exclusion and fire suppression, palm trees would have facilitated tree encroachment. For bird-dispersed species, this may be partly generated by the palm trees acting as perches for dispersers. However, association of *S. commersoniana* with dense palm stands further suggests that other forms of facilitation or environmental control might be relevant. For *M. cisplatensis*, burning appears to have limited tree encroachment rates in these savannas (Rolhauser et al. 2007).

Field experiments revealed the local influences of *B. yatay* and established adults of *A. edulis* and *S. commersoniana* on seed arrival and seedling performance of the latter two riparian species (Rolhauser et al. 2011). Seed traps and seedlings of both species were placed in open grassland patches, and beneath palm trees, *A. edulis* adults and *S. commersoniana* adults. Palm tree and grassland microsites were used to evaluate the influence of resident palm trees on early encroachment. Seed traps under palm trees captured more seeds of the bird-dispersed *A. edulis* than those in herbaceous patches, whereas no seeds of the self-dispersed *S. commersoniana* were trapped. Survival of *A. edulis* seedlings was lower under palm trees than in grassland, mainly because seedlings were more likely to be damaged by trampling and burrowing animals. Survival of *S. commersoniana* seedlings did not differ between palm trees and grassland.

Patterns of seed arrival and seedling performance under established *A. edulis* and *S. commersoniana* trees were used to examine intra- and interspecific effects during advanced stages of encroachment (see Rolhauser et al. 2011). *A. edulis* seed occurred massively under conspecific trees, but seedling growth was strongly suppressed under conspecific adults. *S. commersoniana* seeds were rare and were trapped only under *Sebastiania* trees. Seedlings of this species also showed reduced growth rates under conspecifics, but increased survival beneath heterospecific, *A. edulis* trees.

Species-specific influences of direct and indirect biotic interactions appear to be instrumental in the encroachment process (Rolhauser et al. 2011). Whereas palm trees may not affect the early dynamics of *S. commersoniana*, they exert animal-mediated indirect effects that may either boost or slow early encroachment by *A. edulis*. Further, seedlings of both riparian species grew better in heterospecific than in conspecific adult microsites, a pattern consistent with the "escape hypothesis" (Howe and Smallwood 1982). Lastly, *A. edulis* trees might facilitate establishment of *S. commersoniana* provided seeds arrive in *A. edulis* patches. The ongoing encroachment of dense palm stands by these riparian trees (Rolhauser et al. 2007) suggests that, under current management regimes, the role of palm trees as perches for frugivorous birds outbalances their role as attractors of disturbing animals. In contrast, lack of a long dispersal mechanism appears to limit the expansion of *S. commersoniana* across these savannas.

5.5.1.2 Espinal Semiarid Savannas

The vegetation physiognomy of the southern Espinal (Caldén district, Fig. 5.2) varies from typical *P. caldenia*-grass savannas to dense *P. caldenia* woodlands in relatively humid lowlands and open treeless grasslands on drier sandy soils (Cabrera 1976; Lerner 2005). Current limits between these plant associations as well as between the southern Espinal and the Pampas have been notoriously modified by human activity. Since the mid-1800s, land clearing for agriculture has kept pushing the Espinal forest boundary in both west- and southward directions

(Cabrera 1976). Also, domestic grazing promoted the encroachment of semiarid savannas and grasslands by native trees, mainly *P. caldenia* and *G. decorticans* (León and Anderson 1983; Distel et al. 1996; Peláez et al. 1992).

Livestock generally facilitates woody encroachment in the southern Espinal. The introduction of cattle likely increased seed dispersal of *P. caldenia* and *G. decorticans* (León and Anderson 1983), as these were dispersed by native herbivores such as *Lama guanicoe* (guanaco) and *Odococeilus bezoarticus* (pampa deer). Ingestion of seed pots by cattle enhances seed viability by preventing predation by bruchids, and also increases germination via scarification (Lerner and Peinetti 1996). Grazing indirectly enhances tree seedling recruitment by creating safe microsites where grass competition is relaxed, and this effect likely increased under domestic stocking rates (Villalobos et al. 2005). Further, grazing impacts would have shifted when sheep were replaced by cattle as main domestic grazer. Sheep removal presumably decreased consumption of tree seedlings, thus accelerating tree encroachment on overgrazed areas (Dussart et al. 1998).

In addition, domestic grazing modified the natural fire regime (Dussart et al. 1998; Fernández and Busso 1999) with varied impacts on grass/woody species dynamics. Fire facilitates recruitment of *P. caldenia* seedlings by reducing grass biomass and increasing soil temperature and nutrient content (Villalobos et al. 2007). However, frequent burning may prevent seedling development into older life stages (Bond et al. 2005). The current disturbance regime of the Espinal, which is characterised by a low frequency of wildfires, increases the chances of tree seedling establishment by providing enough time for tree development between fire events.

5.5.1.3 Coastal Xerophytic Forests

The *Tala* district of the Espinal forms a narrow, discontinuous strip of xerophytic forests bordering the northern and eastern Pampa grasslands (Fig. 5.2). The limits between the two vegetation types are abrupt and determined by soil properties (Parodi 1940; Cabrera 1958). The extent of these coastal forests has been progressively reduced through human disturbance (Arturi and Goya 2005). Although abundant *C. tala* seeds are dispersed from mature trees, seedlings are typically scarce within the forest. Seedling densities up to 10 plants/m^2 can be found, but they mostly die by early winter. High winter mortality is related to low frost tolerance, which may account for the lack of *C. tala* establishment in open areas, away from the forest understorey (Arturi and Goya 2005).

Fruits of *C. tala* are readily consumed by birds, which disseminate seeds away from the forest. It is not uncommon to find isolated trees (<50 cm height) in nearby ungrazed grasslands, but there is no evidence of grassland encroachment by this native tree (Arturi and Goya 2005). Given that seeds would be easily dispersed onto adjacent grasslands, competition from resident grasses may be preventing *C. tala* encroachment (Arturi and Goya 2005). However, roadsides, channels and fencelines, and forestry plantations are focal points for the natural spread of *C. tala*. Disturbances removing the herbaceous cover, and planted woods where grasses are suppressed by canopy shading, both provide suitable microsites for tree seedling recruitment and survival.

Most *C. tala* forests are being heavily invaded by several exotic trees, including *L. lucidum, Ligustrum sinensis, M. azedarach, G. triacanthos, M. alba, Acer negundo, Fraxinus* sp., *Pinus* spp., and *Populus* spp. (Cagnoni et al. 1996). Patches of coastal forest may represent stepping stones for dispersal and subsequent invasion of adjacent grasslands by species like *L. lucidum* and *G. triacanthos*, which produce large seed crops and are dispersed by native fauna or cattle (Montaldo 1993; Ghersa and León 1999; see Table 5.1). Thus, a combination of anthropogenic disturbance and forested patches would promote grassland encroachment by either native or exotic tree species.

5.5.2 Invasion by Introduced Tree Species

5.5.2.1 Rolling Pampa: Exotic Trees in Agricultural Landscapes

The Rolling pampa is the main agricultural zone of Argentina and represents the most heavily transformed section of the Pampas (Ghersa and León 1999). At present, trees are virtually everywhere, and invasive species threaten to become problematic weeds too. By the end of the twentieth century, all fields had been converted into zero-tillage cropping, which involves growing glyphosate-resistant soybean transgenic cultivars (De la Fuente et al. 2006). This shift to no-till agriculture increased the likelihood of tree invasion, which was previously prevented by annual disturbance from tillage practices.

A recent survey of the area by Ghersa et al. (2002) recorded a total of 40 woody species, mostly trees, invading roadside and riparian corridors (Fig. 5.3b). The exotic trees *G. triacanthos, M. azedarach*, and *M. alba* showed the highest occurrence, and the former two species were also frequent components of weed communities in zero-tillage cropfields (Ghersa et al. 2002; De la Fuente et al. 2006). The current extent of woody encroachment indicates that the restoration of native grasslands in the Rolling pampa would be highly unlikely (see below) without targeted control of invasive tree species.

5.5.2.2 Inland Pampa: Tree Species Invasion in Old Fields

Factors limiting the establishment of exotic trees in grassland patches have been experimentally evaluated in a long-term study site in the eastern Inland pampa (Fig. 5.2). The area comprises planted woodlots and windbreaks with species like *Eucalyptus* sp., *L. lucidum, G. triacanthos, U. pumila*, and *Robinia pseudoacacia* (acacia blanca). As in other areas of the Pampas, trees have escaped from plantation and invade noncropped, ungrazed fields and peridomestic areas, as well as abandoned railway corridors with relictual tussock grasslands (Facelli and León 1986; Mazía et al. 2001; Chaneton et al. 2004).

Early work documented the invasion of old-field communities by *U. pumila* from nearby woodlots (Facelli and León 1986; Fig. 5.3c). The magnitude of invasion was inversely correlated with the grass cover and plot age. Removal of resident grasses in a 4-year-old plot enhanced tree seedling recruitment. Still, copious arrival of wind-dispersed seeds partly counteracted the negative effect of grasses, as tree seedlings also occurred in intact vegetation plots (Facelli and León 1986). In another study, we manipulated vegetation cover in a 15-year-old seral community dominated by pasture grasses (Mazía et al. 2001). To control for propagule pressure and test different tree genotypes, we added seed of three exotic (*G. triacanthos*, *U. pumila* and *L. lucidum*) and one native species (*P. caldenia*). The experiment was replicated in understorey and treefall-gap microhabitats in a *L. lucidum* plantation. We asked whether woodlots provide suitable conditions for tree establishment, and may thus work as "stepping stones" for invasive tree species.

Only *G. triacanthos* and *P. caldenia* were recruited in grassland. Removal of herbaceous cover increased seedling emergence and first year survival of both tree species. Yet, 5 years later, only *G. triacanthos* had established successfully in disturbed plots, whereas all *P. caldenia* seedlings failed to survive. Established trees were 4–5 m in height and produced abundant seed pots. Surprisingly, *L. lucidum* and *U. pumila* did not emerge from sown seeds, possibly due to low viability and/or seed predation (Mazía et al. 2001). These results confirmed that, once dispersal limitation is removed, resident plants remain a major barrier to tree invasion (Fig. 5.1). Further, how effective are local filters to invasion will depend on the specific identity of the tree coloniser, regarding its susceptibility to various biotic or abiotic stresses. Although *P. caldenia* occurs in the "marginal" species pool of the western Inland Pampa, this xeromorphic species apparently lacks the seed/seedling traits to regenerate in mesic grassland.

In the forest experiment, all four sown trees emerged in both gap and understorey conditions, but did not survive pass the first summer from emergence in understorey microsites. Survivorship of *G. triacanthos* and *P. caldenia* was high during 1.5 years in the forest gap. However, all seedlings eventually died with gap closure in the ensuing years (Mazía et al. 2001). Thus, while forest patches may not offer better conditions for tree establishment than old-field grasslands, they provide early recruitment opportunities that vary depending on species identity and microhabitat. The common presence of *L. lucidum* and *G. triacanthos* saplings in tree plantations suggests that increased seed arrival may enhance the chance of successful invasion of forest patches by these exotics.

Both anthropogenic and natural disturbances may favour exotic invasions (Hobbs and Huenneke 1992; D'Antonio et al. 1999). Studies in the Inland pampa show that disturbance effects on tree establishment can be context-dependent. In a large-scale, 4-year experiment, we examined the influence of two disturbance agents, burning and animal diggings, on *G. triacanthos* recruitment in a late-successional old field and a relict of native tussock grassland (Chaneton et al. 2004; Mazía et al. 2010). To adjust for propagule pressure, 40 seeds/m^2 were added in each of several sowing

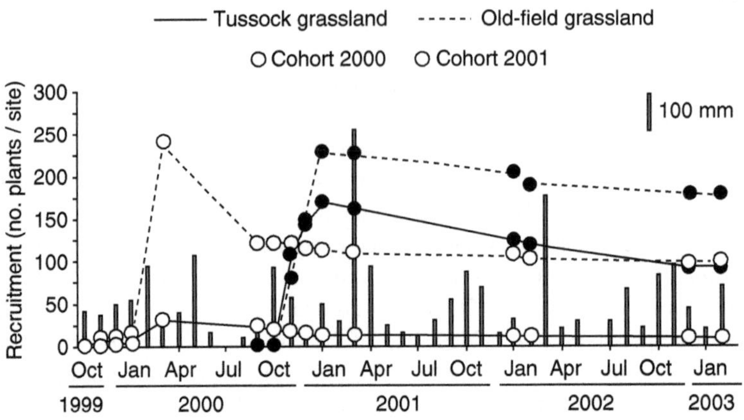

Fig. 5.5 Overall recruitment of *G. triacanthos* in old-field and relict tussock grasslands in the Inland pampa. Data show cumulative number of plants from two seedling cohorts (2000–2001) over 240 m^2. *Vertical bars* (*grey*) depict monthly rainfall (mm)

events to 12-m^2 plots, which were either burned or "burrowed" to simulate the digging activity of hairy armadillos (*Chaetrophractus villosus*) in a full factorial design (for details, see Mazía et al. 2010).

Emergence of *G. triacanthos* was fivefold higher in the old field than in the tussock grassland during the first, drier spring, but differences were reduced during the second, wetter spring (Fig. 5.5). Burning increased recruitment of the second-year cohort in both grasslands. Simulated armadillo disturbances prevented tree seedling emergence in the tussock grassland, especially in burned plots, and slightly increased recruitment in the old field (Fig. 5.6). Large- and small-scale disturbances thus promoted tree recruitment in the old field, whereas they exerted antagonistic effects in the relict tussock community (Mazía et al. 2010). Final establishment of *G. triacanthos* was twice as high in the old field as in the tussock grassland (Fig. 5.5). The greater resistance to tree invasion of the tussock grassland likely reflected the combined effects of grass competition and rodent granivory (Chaneton et al. 2004; Mazía et al. 2010). Indeed, seed loss to predators was higher in the tussock than in the old-field grassland (Fig. 5.7a), presumably because old fields offer a greater diversity of alternative food items to granivores. Hence, once propagule pressure is factored out, invasion by exotic trees becomes contingent on the recipient community, local disturbance regime, and annual rainfall patterns.

Recruitment of *P. caldenia* was similarly evaluated in the same grassland plots. Seedling emergence of *P. caldenia* from sown seed was abundant in both study grasslands. However, no seedling managed to survive beyond the first winter. Seed loss to granivores was greater in tussock than in old-field grassland (Fig. 5.7a). Rodent exclusion markedly increased *P. caldenia* emergence, while litter cover blocked seedling emergence within rodent exclosures (Fig. 5.7b). These findings show that different biotic factors may contribute to repel invasion by this native tree in both grasslands.

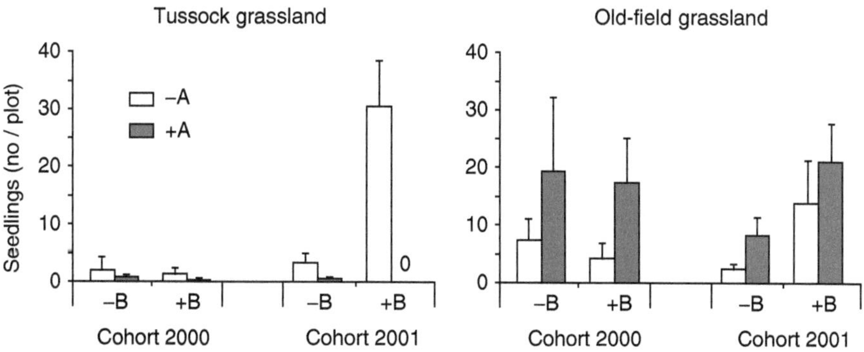

Fig. 5.6 Emergence of *G. triacanthos* under different disturbance regimes in old-field and relict tussock grasslands, Inland pampa. *Bars* show mean ± SE ($n=5$) for two seedling cohorts emerged in a dry (2000) and a wet (2001) year. *B* burning; *A* armadillo diggings. There were significant effects of B × A × cohort (tussock $F_{1,16}=9.46$, $P<0.01$) and B × cohort (old field $F_{1,12}=5.35$, $P<0.05$)

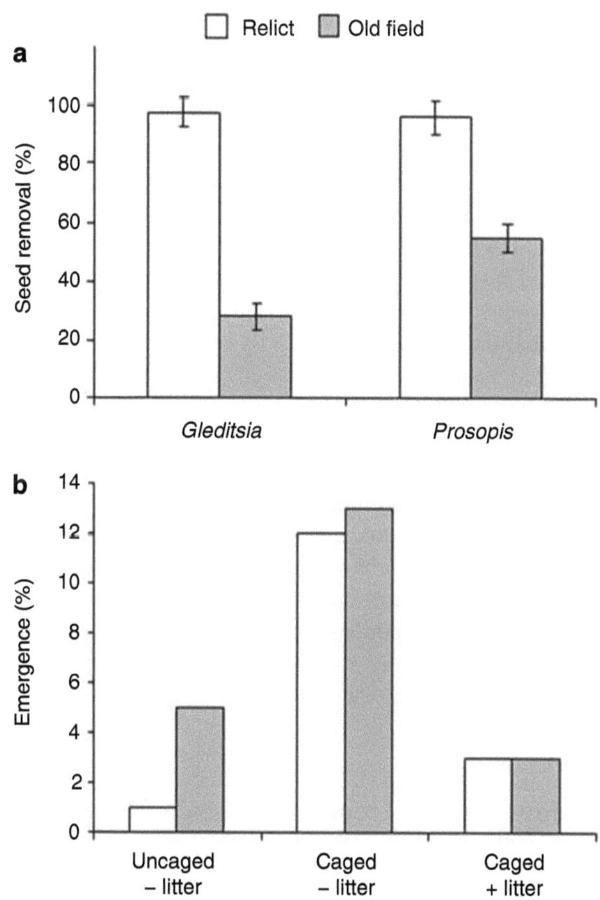

Fig. 5.7 (**a**) Removal of *G. triacanthos* and *Prosopis caldenia* seeds in old-field and relict tussock grasslands, Inland pampa. Site effect $F_{1,78}=135.7$, $P<0.0001$; site × species $F_{1,78}=7.1$, $P<0.01$. (**b**) Seedling emergence of *P. caldenia* as affected by rodent exclusion and litter cover. Relict, $\chi^2=30$, $P<0.0001$; Old field, $\chi^2=21.8$, $P<0.0001$ (both df = 2)

In contrast, we found that even undisturbed grasslands were not fully immune to invasion by *G. triacanthos* (Mazía et al. 2001, 2010; Chaneton et al. 2004).

5.5.2.3 Southern Pampa: Pine Invasions in Montane Grasslands

The western limit of the Southern pampas forms a dynamic ecotone with the semi-arid Caldenal savannas (Cabrera 1976; Fig. 5.2). Present environmental conditions suggest that *P. caldenia* might have dispersed towards the east, although fossil pollen records indicate that the southeastern limits of the Espinal remained rather stable throughout the Quaternary (Prieto 1996). It has then been suggested that biogeographical barriers could have prevented colonisation of the Southern pampas by native trees (Arturi and Goya 2005).

Exotic tree species were introduced to the area since the early nineteenth century. Although the landscape physiognomy was drastically altered, trees were not originally reported to invade open grasslands. Widespread tree invasions in the Southern pampas began with *Pinus* and *Eucalyptus* plantations in the 1970s (Zalba and Villamil 2002). Successful tree introductions were mediated by an intensive selection process of ecologically pre-adapted genotypes. Historical (dendrochronological) records indicate that both *Pinus halepensis* and *Pinus radiata* were the most invasive forestry species, and that their landscape spread has been assisted by large fire events (Zalba and Villamil 2002). Factors promoting pine invasions after fire include massive seed crops and seedling recruitment, which continued for 12 years in the case of *P. halepensis* (Zalba et al. 2008). Grazing by feral horses has also been found to facilitate the spread of *P. halepensis* by reducing the competitive ability of native grasses outside plantations (Villalobos et al. 2011).

As shown for other southern hemisphere regions (Simberloff et al. 2010), the novel presence of pirogenic species (*Pinus*) increased the flammability of the system, enhancing the likelihood of tree establishment by relaxing the intensity of grass competition. Most importantly, attempts to restore grasslands on invaded sites have been highly encouraging. Four years after disturbance from intensive pine removal, native grasses replaced pioneer forbs without the need to augment grass seed availability through sowing (Cuevas and Zalba 2011). The reversal of tree-induced vegetation changes in the Southern pampas contrasts with the lack of recovery of native plant assemblages during old-field succession in the Rolling and Inland pampas (Ghersa and León 1999; Tognetti et al. 2010).

5.6 Conclusion: Future Trends in Grassland/Forest Dynamics

Biotic interchanges between adjacent regions have been common throughout the history of the Earth, and still are to be expected as a result of species' range expansions (Elton 1958; Vermeij 2005). However, natural shifts in biogeographical limits take time to develop and rarely involve long-distance dispersal events. By contrast, human-assisted biotic invasions bridge large geographic scales, put in contact biotas with no co-evolutionary past, and occur at surprisingly fast rates (Williamson 1996; Vermeij 2005).

We have argued that in the humid Pampas pre-historic boundaries between natural grasslands and adjacent forest formations have remained largely stable, with little encroachment from native tree species. However, the last 300 years of human occupation have changed that pattern by creating a rural landscape mosaic with dynamic interfaces between cropland, forest stands, and grasslands (Vega et al. 2009). Anthropogenic activity has accelerated the addition of tree life forms to remnant grassy vegetation on a regional scale, by introducing woody genotypes with novel traits, by increasing propagule pressure and facilitating dispersal, and by altering disturbance regimes. These processes implied the removal of long-standing barriers to tree species establishment in open grassland (Fig. 5.1).

Furthermore, recent surveys suggest that the spread of exotic invasive trees might exert a cascading influence on the distribution of native woody species from adjacent forests. A screening of the central Pampas in 2×2 km cells using Google Earth revealed that trees occur in over 80 % of the cells. Interestingly, field observations indicate that many planted woodlots are being invaded by native forest species, such as *C. tala*, *A. caven*, and *S. australis*, and by widespread exotics, mainly *G. triacanthos*, *M. alba*, and *L. lucidum*. This pattern strongly suggests that tree patches increase the chance of directed immigration of other tree species via animal dispersers. The extent to which such spatial associations are only a matter of increased seed arrival or they reflect some kind of local facilitation to tree recruitment (Tecco et al. 2006; Rolhauser et al. 2011) is not as yet clear. It thus appears that several invasive tree species are past the initial lag-phase characteristic of invasion dynamics at landscape to regional scales (Williamson 1996), and that there might still be a substantial "invasion debt" (*sensu* Richardson and Rejmánek 2011) to be paid in the Pampas.

Historically, successional pathways in the study region have been driven by changes in grazing pressure, soil disturbance from cropping, and pasture ageing (Ghersa and León 1999; Chancton et al. 2002; Tognetti et al. 2010). Despite obvious alterations of the original grassland composition, the herbaceous life form remained the primary state of the system. Grassland fragmentation and the pervasive introduction of exotic invasive trees have given way to novel, woody-dominated community states developing in old fields, noncultivated areas, and grassland relicts. The overarching implication is that the Pampa grasslands may be doomed, as they would be inevitably converted into savannas or parklands if energy subsidies are not applied to control tree encroachment.

Acknowledgements We are most grateful to Professor Rolando León for inspiration and discussion, and Randall Myster for the invitation to contribute to this volume. We thank the National Parks Administration of Argentina for permission to work at El Palmar, and the staff of Estancia San Claudio for logistic assistance. Our research is partly funded by Agencia Nacional de Promoción Científica y Tecnológica (FONCYT), Consejo Nacional de Investigaciones Científicas (CONICET), and Universidad de Buenos Aires (UBACYT grant scheme).

References

Arturi MF, Goya JF (2005) Estructura, dinámica y manejo de los talares del NE de Buenos Aires. In: Arturi MF, Frangi JL, Goya JF (eds) Ecología y manejo de los bosques de Argentina. Universidad de La Plata, La Plata

Baldi G, Guerschman JP, Paruelo JM (2006) Characterizing fragmentation in temperate South America grasslands. Agric Ecosyst Environ 116:197–208

Barnosky AD, Lindsey EL (2010) Timing of Quaternary megafaunal extinction in South America in relation to human arrival and climate change. Quat Int 217:10–29

Bond WJ (2008) What limits trees in C_4 grasslands and savannas? Annu Rev Ecol Evol Syst 39:641–659

Bond WJ, Woodward FI, Kidgley GF (2005) The global distribution of ecosystems in a world without fire. New Phytol 165:525–538

Bredenkamp GJ, Spada F, Kazmierczak E (2002) On the origin of northern and southern hemisphere grasslands. Plant Ecol 163:209–229

Burkart A (1975) Evolution of grasses and grasslands in South America. Taxon 24:53–66

Burkart SE, León RJC, Conde MC, Perelman SB (2011) Plant species diversity in remnant grasslands on arable soils in the cropping Pampa. Plant Ecol 212:1009–1024

Cabrera AL (1958) Fitogeografía. In: Aparicio D (ed) La Argentina—Suma de geografía. Peuser, Buenos Aires, pp 101–207

Cabrera AL (1976) Regiones fitogeográficas Argentinas. In: Ferreira Sobral EF (ed) Enciclopedia Argentina de Agricultura y Jardinería. ACME, Buenos Aires, pp 1–85

Cabrera AL, Willink A (1980) Biogeografía de America Latina, Monografías. Secretaría General OEA, Washington, DC

Cagnoni M, Faggi AM, Ribichich A (1996) La vegetación de la Reserva "El Destino" (Partido de Magdalena, Provincia de Buenos Aires). Parodiana 9:25–44

Carnevali R (1994) Fitogeografía de la Provincia de Corrientes Gobierno de la Provincia de Corrientes. INTA, Corrientes

Chaneton EJ, Perelman SB, Omacini M, Leon RJC (2002) Grazing, environmental heterogeneity, and alien plant invasions in temperate Pampa grasslands. Biol Invasions 4:7–24

Chaneton EJ, Mazía CN, Machera M, Uchitel A, Ghersa CM (2004) Establishment of honey locust (*Gleditsia triacanthos*) on burned pampean grasslands. Weed Technol 18:1325–1329

Christy M (1892) Why are the prairies treeless? Proc R Geogr Soc Mon Rec 14:78–100

Ciccero P, Balabusic A (1994) Plan de Manejo Parque Nacional el Palmar. Administración de Parques Nacionales, Buenos Aires

Crawley MJ (1987) What makes a community invasible? In: Gray AJ, Crawley MJ, Edwards PJ (eds) Colonization, succession and stability. Blackwell, Oxford, pp 429–453

Cuevas YA, Zalba SM (2011) Recovery of native grasslands after removing invasive pines. Restor Ecol 18:711–719

D'Antonio CM, Dudley TL, Mack M (1999) Disturbance and biological invasions: direct effects and feedbacks. In: Walker LR (ed) Ecosystems of disturbed ground. Elsevier, Amsterdam, pp 413–452

Darwin C (1893) Journal of researches into the geology and natural history of the various countries visited by the H.M.S. Beagle. Henry Colburn, London

Davis MA, Wrage KJ, Reich PB (1998) Competition between tree seedlings and herbaceous vegetation: support for a theory of resource supply and demand. J Ecol 86:652–661

De la Fuente EB, Suárez SA, Ghersa CM (2006) Soybean weed community composition and richness between 1995 and 2003 in the Rolling pampas (Argentina). Agric Ecosyst Environ 115:229–236

De Patta Pillar V, Müller SC, de Souza Castilhos ZM, Ávila Jacques AV (2009) Campos sulinos—conservação e uso sustentável da biodiversidade. Ministério do Meio Ambiente, Brasilia

Dimitri MJ, Rial OR (1955) La protección de la naturaleza en la provincia de Entre Ríos. Natura 2:135–142

Distel RA, Pelaez DV, Bóo RM, Mayor MD, Délia ER (1996) Growth of *Prosopis caldenia* seedlings in the field as related to grazing history of the site and in a greenhouse as related to different levels of competition from *Stipa tenuis*. J Arid Environ 32:251–257
Dukes JS, Mooney HA (1999) Does global change increase the success of biological invaders? Trends Ecol Evol 14:135–139
Dussart E, Lerner P, Peinetti R (1998) Long-term dynamics of two populations of *Prosopis caldenia* Burkart. J Range Manage 51:685–691
Ellenberg VH (1962) Wald in der Pampa Argentiniens? Veröff Geobot Inst ETH Stiftung Rübel Zürich 37:39–56
Elton CS (1958) The ecology of invasions by plants and animals. Chapman & Hall, London
Facelli JM, León RJC (1986) El establecimiento espontáneo de árboles en La Pampa Un enfoque experimental. Phytocoenologia 14:263–274
Fernández OA, Busso CA (1999) Arid and semi-arid rangelands: two thirds of Argentina. In: Arnalds O, Archer S (eds) Case studies of rangeland desertification. Agricultural Research Institute Report No. 200, Reykjavik, pp 41–60
Funk JL, Cleland EE, Suding KN, Zavaleta ES (2008) Restoration through reassembly: plant traits and invasion resistance. Trends Ecol Evol 23:695–703
Garavaglia JC (1999a) Pastores y labradores de Buenos Aires. Una historia agraria de la campaña bonaerense 1700–1830. Ed. de la Flor, Buenos Aires
Garavaglia JC (1999b) Un siglo de estancias en la campaña de Buenos Aires: 1751–1853. Hisp Am Hist Rev 79:703–734
Ghersa CM, León RJC (1998) Ecología del paisaje pampeano: consideraciones para su manejo y conservación. In: Contin M (ed) El patrimonio paisajista: aspectos sociales y ambientales. LINTA-CIC, La Plata, Argentina. pp 61–80
Ghersa CM, León RJC (1999) Successional changes in agroecosystems of the Rolling pampa. In: Walker LR (ed) Ecosystems of disturbed ground, Ecosystems of the world 16. Elsevier, Amsterdam, pp 487–502
Ghersa CM, De la Fuente E, Suarez S, León RJC (2002) Woody species invasion in the rolling Pampa grasslands, Argentina. Agric Ecosyst Environ 88:271–278
Gibson DJ (2009) Grasses and grassland ecology. Oxford University Press, Oxford
Goveto L (2006) Historical fire occurrence in the savanna of El Palmar National Park: climatic and floristic evidences. MSc thesis, Universidad de Buenos Aires, Buenos Aires
Henning I (1988) The Pampas problem [Zum Pampa-Problem]. Erde 119:25–30
Hobbs RJ (2000) Land-use changes and invasions. In: Mooney HA, Hobbs RJ (eds) Invasive species in a changing world. Island Press, Washington, DC, pp 55–64
Hobbs RJ, Huenneke LF (1992) Disturbance, diversity, and invasion implications for conservation. Conserv Biol 6:324–337
Howe HF, Smallwood J (1982) Ecology of seed dispersal. Annu Rev Ecol Syst 13:201–228
Iriarte J (2006) Vegetation and climate change since 14,810 ^{14}C yr B.P. in southeastern Uruguay and implications for the rise of early Formative societies. Quat Res 65:20–32
Iriondo M (1999) Climatic changes in the South American plains: records of a continent-scale oscillation. Quat Int 57(58):93–112
Johnson CN (2009) Ecological consequences of Late Quaternary extinctions of megafauna. Proc R Soc Lond B 276:2509–2519
Keddy PA (1992) Assembly and response rules: two goals for predictive community ecology. J Veg Sci 3:157–164
Lamarque LJ, Delson S, Lortie CJ (2011) Tree invasions a comparative test of the dominant hypotheses and functional traits. Biol Invasions 13(9):1969–1989
León RJC, Anderson DL (1983) El límite occidental del pastizal pampeano. Tuexenia 3:67–83
Lerner P (2005) El Caldenal: dinámica de poblaciones de caldén y procesos de expansión de leñosas en pastizales. In: Arturi MF, Frangi JL, Goya JF (eds) Ecología y manejo de los bosques de Argentina. Universidad de La Plata, La Plata

Lerner P, Peinetti R (1996) Importance of predation and germination losses from the seed bank of caldén (*Prosopis caldenia*). J Range Manage 49:147–150

Levine JM, Adler PB, Yelenik SG (2004) A meta-analysis of biotic resistance to exotic plant invasions. Ecol Lett 7:975–989

Lockwood JL, Cassey P, Blackburn T (2005) The role of propagule pressure in explaining species invasions. Trends Ecol Evol 20:223–228

Mack RN (2003) Phylogenetic constraints, absent life forms, and preadapted alien plants: a prescription for biological invasions. Int J Plant Sci 164:S185–S196

Markgraf V, McGlone M, Hope G (1995) Neogene paleoenvironmental and paleoclimatic change in southern temperate ecosystems – a southern perspective. Trends Ecol Evol 10:143–147

Martinez-Crovetto R, Piccinini BG (1950) La vegetación de la República Argentina 1—Los palmares de *Butia yatay*. Rev Investig Agric 4:153–242

Mazía CN, Chaneton EJ, Ghersa CM, León RJC (2001) Limits to tree species invasion in pampean grassland and forest plant communities. Oecologia 128:594–602

Mazía CN, Chaneton EJ, Machera M, Uchitel A, Feler MV, Ghersa CM (2010) Antagonistic effects of large- and small-scale disturbances on exotic tree invasion in a native tussock grassland relict. Biol Invasions 12:3109–3122

Milton SJ, Wilson JRU, Richardson RM, Seymour CL, Dean WRJ, Iponga DM, Proches S (2007) Invasive alien plants infiltrate bird-mediated shrub nucleation processes in arid savanna. J Ecol 95:648–661

Montaldo N (1993) Dispersión por aves y éxito reproductivo de dos especies de *Ligustrum* en un relicto de selva subtropical en la Argentina. Rev Chil Hist Nat 66:75–85

Myers JA, Harms KE (2009) Seed arrival, ecological filters, and plant species richness: a meta-analysis. Ecol Lett 12:1250–1260

Nores M, Cerana MM, Serra DA (2005) Dispersal of forest birds and trees along the Uruguay River in southern South America. Divers Distrib 11:205–217

Parodi LR (1940) La distribución geográfica de los talares de la Provincia de Buenos Aires. Darwiniana 4:33–56

Parodi LR (1942) Por qué no existen bosques naturales en la llanura bonariense si los árboles crecen en ella cuando se los cultiva? Rev Cent Estud Agron 30:387–390

Parodi LR (1947) La estepa pampeana. In: Hauman L, Burkart AE, Parodi LR, Cabrera AL (eds) La vegetación de la Argentina. GAEA Geografía de la República Argentina, Buenos Aires, pp 143–207

Peláez DV, Bóo RM, Delía ER (1992) Emergence and seedling survival of calden in the semiarid region of Argentina. J Range Manage 45:564–568

Perelman SB, León RJC, Oesterheld M (2001) Cross-scale vegetation patterns of Flooding pampa grasslands. J Ecol 89:562–577

Prieto AR (1996) Late Quaternary vegetational and climatic changes in the Pampa grassland of Argentina. Quat Res 45:73–88

Prieto AR (2000) Vegetational history of the late glacial-Holocene transition in the grasslands of eastern Argentina. Palaeogeogr Palaeoclimatol Palaeoecol 157:167–188

Prieto AR, Blasi AM, De Francesco CG, Fernández C (2004) Environmental history since 11,000 14C year B.P. of the northeastern Pampas, Argentina, from alluvial sequences of the Luján river. Quat Res 62:146–161

Rejmánek M, Richardson DA (1996) What attributes make some plant species more invasive? Ecology 77:1655–1661

Ribichich AM (2002) El modelo clásico de la fitogeografía de Argentina: un análisis crítico. Interciencia 27:669–675

Ribichich AM, Protomastro J (1998) Woody vegetation structure of xeric forest stands under different edaphic conditions and disturbance histories in the Biosphere Reserve 'Parque Costero del Sur', Argentina. Plant Ecol 139:189–201

Richardson DA, Rejmánek M (2011) Trees and shrubs as invasive alien species—a global review. Divers Distrib 17:788–809

Ricklefs RE, Schluter D (1993) Species diversity in ecological communities. University of Chicago Press, Chicago

Riginos C (2009) Grass competition suppresses tree growth across multiple demographic stages. Ecology 90:335–340

Rolhauser AG, Biganzoli F, Pignataro G, Nordenstahl M, Batista WB (2007) Distribución de las poblaciones arbóreas en el paisaje del Parque Nacional El Palmar. In: Matteucci SD (ed) Panorama de la Ecología de Paisajes en Argentina y Países Sudamericanos. Ediciones INTA, Buenos Aires, pp 329–334

Rolhauser AG, Chaneton EJ, Batista WB (2011) Influence of conspecific and heterospecific adults on riparian tree species establishment during encroachment of a humid palm savanna. Oecologia 167:141–148

Sala OE (2001) Temperate grasslands. In: Chapin FS, Sala OE, Huber-Sannwald E (eds) Global biodiversity in a changing environment: scenarios for the 21st century. Springer, New York, pp 121–137

Sankaran M, Ratnam J, Hanan NP (2004) Tree-grass coexistence in savannas revisited—insights from an examination of assumptions and mechanisms invoked in existing models. Ecol Lett 7:480–490

Schmieder O (1927) The Pampa—a natural or culturally induced grassland? Univ Calif Publ Geogr 2(8):255–270

Simberloff D (2009) The role of propagule pressure in biological invasions. Annu Rev Ecol Evol Syst 40:81–102

Simberloff D, Nuñez MA, Ledgard NJ, Pauchard A, Richardson DM, Sarasola M, Van Wilgen BW, Zalba SM, Zenni RD, Bustamante R, Peña E, Ziller SR (2010) Spread and impact of introduced conifers in South America: lessons from other southern hemisphere regions. Austral Ecol 35:489–504

Skellam JG (1951) Random dispersal in theoretical populations. Biometrika 38:196–212

Smit C, Ouden JD, Müller-Scharer H (2006) Unpalatable plants facilitate tree sapling survival in wooded pastures. J Appl Ecol 43:305–312

Soriano A (1992) Río de la Plata grasslands. In: Coupland RT (ed) Natural grasslands: introduction and western hemisphere, Ecosystems of the world 8A. Elsevier, Amsterdam, pp 367–407

Tapson AJ (1962) Indian warfare on the Pampa during the colonial period. Hisp Am Hist Rev 42:1–28

Tecchi RA (1983) Contenido de silicofitolitos en suelos del sector sudoriental de la Pampa Ondulada. Ciencia del Suelo 1:75–82

Tecco PA, Gurvich DE, Díaz S, Pérez-Harguindeguy N, Cabido M (2006) Positive interaction between invasive plants: the influence of *Pyracantha angustifolia* on the recruitment of native and exotic woody species. Austral Ecol 31:293–300

Theoharides KA, Dukes JA (2007) Plant invasion across space and time: factors affecting nonindigenous species success during four stages of invasion. New Phytol 176:256–273

Tognetti PM, Chaneton EJ, Omacini M, Trebino HJ, León RJC (2010) Exotic vs. native plant dominance over 20 years of old-field succession on set-aside farmland in Argentina. Biol Conserv 143:2494–2503

Tonni EP, Cione AL, Figini AJ (1999) Predominance of arid climates indicated by mammals in the pampas of Argentina during the Late Pleistocene and Holocene. Palaeogeogr Palaeoclimatol Palaeoecol 147:257–281

Valiente-Banuet A, Rumebe AV, Verdú M, Callaway RM (2006) Modern Quaternary plant lineages promote diversity through facilitation of ancient Tertiary lineages. Proc Natl Acad Sci U S A 103:16812–16817

Vega E, Baldi G, Jobbágy EG, Paruelo JM (2009) Land use change patterns in the Rı´o de la Plata grasslands: the influence of phytogeographic and political boundaries. Agric Ecosyst Environ 134:287–292

Vermeij GJ (2005) Invasion as expectation: a historical fact of life. In: Sax DE, Stachowicz JJ, Gaines SD (eds) Species invasions: insights into ecology, evolution, and biogeography. Sinauer Associates, Sunderland, pp 315–340

Villalobos AE, Pelaez DV, Délia ER (2005) Factors related to establishment of *Prosopis caldenia* Burk seedlings in central rangelands of Argentina. Acta Oecol 27:99–106

Villalobos AE, Pelaez DV, Bóo RM, Mayor MD, Délia ER (2007) Effect of a postfire environment on the establishment of *Prosopis caldenia* seedlings in central semiarid Argentina. Austral Ecol 32:581–591

Villalobos AE, Zalba SM, Pelaez RV (2011) *Pinus halepensis* invasion in mountain pampean grassland: effects of feral horses grazing on seedling establishment. Environ Res 111:953–959

Walter H (1967) Das Pampaproblem in vergleichend-ökologischer Betrachtung und seine Lösung. Erdkunde 21:181–203

Webb SD (1978) A history of savanna vertebrates in the new world. Part II: South America and the great interchange. Annu Rev Ecol Syst 9:393–426

Whittaker RH (1975) Communities and ecosystems. Macmillan, New York

Williamson M (1996) Biological invasions. Chapman & Hall, London

Yazawa T (1989) The Pampa problem: a critical review. Geogr Rev Jpn Ser A 62:389–408

Zalba SM, Villamil CB (2002) Woody plant invasion in relictual grasslands. Biol Invasions 4:55–72

Zalba SM, Cuevas YA, Bóo RM (2008) Invasion of *Pinus halepensis* Mill following a wildfire in an Argentina grassland nature reserve. J Environ Manage 88:538–546

Part II
Tropical Forest-Grassland Ecotones: Old Fields, Savannas, and Llanos

Chapter 6
Post-agricultural Ecotones in Puerto Rico

Randall W. Myster

6.1 Introduction

The study of pastures and other ecotones recovering from agriculture—old fields—is important to world-wide social and environmental issues such as forest regeneration, forest ecosystem restoration, sustainability of agriculture, maintenance of biodiversity, and impacts of global climate change on forest dynamics. Furthermore these ecotones may serve as buffers between more pristine "primary" forest and urban/suburban areas (Brown and Lugo 1990a). Indeed because the amount of land area either converted to, or recovering from, agriculture is expected to only increase in the future everywhere on earth, studying the recovery of these areas will continue to be vital to human density.

Recovering fields in Puerto Rico, and elsewhere in the tropics, form an ecotone between tropical forests and various kinds of grasslands. These grasslands may include (1) native grasses that have established naturally, (2) native grasses that cattle have facilitated through various means, (3) exotic grasses that were either transported by accident or planted purposely, (4) remnants of previous planted crops that are grasses such as sugarcane (*Saccharum officinarum*) and banana (*Musa* spp.), and/or (5) grasses which associate with previously planted grass crops.

6.2 A Brief History of Cultivation in Puerto Rico

The Taínos were the indigenous peoples of the Bahamas, the greater Antilles, and the northern lesser Antilles. They lived in and used the coastal forests of Puerto Rico for cultivation of manioc (*Manihot esculenta*) and corn (*Zea mays*: Gomez and

R.W. Myster (✉)
Department of Biology, Oklahoma State University, Oklahoma City, OK 73107, USA
e-mail: rwmyster@gmail.com

R.W. Myster (ed.), *Ecotones Between Forest and Grassland*,
DOI 10.1007/978-1-4614-3797-0_6, © Springer Science+Business Media New York 2012

Ballesteros 1980). The Taínos cultivation techniques were probably similar to the "slash and burn" cultivation still common throughout Central and South America today (Myster 2007). That technique consists of first cutting down the forest trees to the ground and then burning them. Although this produces a short pulse of nutrients like phosphorus, it generally volatilizes both organic matter and nitrogen while raising soil temperature and, in the long term, makes soil poor in both nutrients and the seed bank. This technique may have also included a rotation of fields rather than crops, accompanied by long fallow periods, where the decision as to when to rotate fields and how long to leave them fallow depends on such factors as weed and insect loads in the fields, soil fertility, production, labor availability, and local dietary needs. Given the low soil fertility and other conditions, productivity in these fields would eventually decline below the effort needed to maintain them, at which time they would be abandoned and new forested areas cleared.

With the arrival of the Spanish—along with their oxen, cows, and horses—in 1493 AD, cultivation on the island changed dramatically. The Spanish added slave and animal power to the techniques of the Taíno and expanded cultivation to higher elevational forested areas (but guava [*Psidium guajava*] trees hindered initial efforts) to promote export, not the sustainable, low impact cultivation done by the Taínos. In particular, sugarcane (*S. officinarum*) was first planted in Puerto Rico in 1523 AD and considerable deforestation accompanied its development as a crop throughout the seventeenth and eighteenth centuries (Domínguez Cristóbal 2000). Since then "cane" has became the major agricultural crop in Puerto Rico covering, at one time, 14 % of the island's total land area (Pico 1969). Much of it was planted on alluvial soils on the coastal plains of Puerto Rico, and so these alluvial valleys were the first lands to be significantly deforested in Puerto Rico because they represented the best agricultural conditions (flat lands, deep fertile soils) available.

Sugarcane plantations were often converted to pasture—by letting in cattle which facilitated passive grass establishment and growth by trampling of vegetation, creation of hummocks, deposition of dung, and soil compaction, rather than by active grass planting or seeding (as was done with the exotic grass *Setaria sphacelata* in Ecuadorean pastures: Myster 2004a)—when they were no longer productive (Thomlinson et al. 1996). As the industrial revolution came to Puerto Rico, cultivation came to use less animal power and more mechanical field clearing techniques (e.g., using a bulldozer) as well as applications of fertilizers, pesticides, and herbicides, bringing about changes in both the physical and chemical properties of the soil. With the development of manufacturing on the island after World War II, sugarcane cultivation declined and was terminated in the 1980s leading to the abandonment of large areas to pastures and a variety of secondary forests (Foster et al. 1999; Ramos-Gonzalez 2001; Grau et al. 2003).

Coffee (*Coffea* spp.) was introduced in 1736 AD but its cultivation was always very secondary in scope to sugarcane, becoming a dominant crop only in higher elevational forests where sugarcane cultivation was not profitable (Dietz 1986). Coffee was planted and grown as other crops are—called "sun" coffee—but also underneath *Inga vera* or *Erthrina poeppigiana* trees—called "shade" coffee—presumingly for the nitrogen (N) enrichment of the soil that these trees can provide (Weaver and

Birdsey 1986; Marcano-Vega et al. 2002). Coffee was also largely abandoned after World War II (Rudel et al. 2000). Other crops included banana/plantain, tobacco, pineapple, corn, cotton, rice, and various fruit trees but they were cultivated at much smaller scales and often for local consumption. Tree plantations (e.g., *Leucaena leucocephala, Casuarina equisetifolia, Eucalyptus robusta, Pinus caribaea*) will not be considered here as agriculture. As for pastures and other post-agricultural ecotones, I only consider published studies performed on the island of Puerto Rico in this review (but see Myster 2004b, 2007; Zimmerman et al. 2007 for other Neotropical areas).

6.3 Cultivation and the Establishment of the Old Field Ecotone

The nature of the recovery of pastures and abandoned agricultural ecotones is heavily influenced by their disturbance regime, characterized by severity, size, spatial location, and frequency (Myster 2007). The severity of these areas is moderate, compared to landslides (Myster and Walker 1997) because after agricultural use there remains an intact soil profile, roots, and humus layer, but more severe than natural treefall gap formation (Everham et al. 1996) and hurricanes (Zimmerman et al. 1995; Pascarella et al. 2004) because all initial vegetation has been removed. Hurricanes, however, may have their greatest effect when they interact with other disturbances. For example, hurricanes may reduce basal area in older pastures but increase it in younger pastures (Flynn et al. 2010), and recovering fields closer to hurricanes may contain less *Cecropia schreberiana* (Zimmerman et al. 1995) and less *Spathodea campanulata* (Flynn et al. 2010) compared to those farther away

Fields in Puerto Rico usually do not exceed a few tens of hectares in size (Grau et al. 2003) but their crops often require specific locations on the island, as noted above. Frequency includes the number of times a field has been cultivated, the duration of each cropping period, the time between fallow periods, and the order of crop rotation. As the length of time an area is in crop increases, the length of successional recovery time may also lengthen. For example, a short cropping period can lead to large recruitment from the seed bank where trees regenerate quickly from remaining stump or root sprouts. But as the cropping period increases the seed bank and other plant remnants decay thus slowing tree invasion. Sustained cropping may also cause loss of soil carbon (C), a mixing of the soil horizons, and a higher degree of soil aeration.

Among other conditions that influence recovery are (1) lingering effects of the past crop—its "signature"—such as the crop itself remaining and/or facilitating the invasion and growth of other plants (Chinea and Helmer 2003), (2) the year and season of abandonment, (3) the amount of border the field shares with the forest or other sources of seed rain and the general fragmentation of the surrounding landscape, and (4) what happened at the time of abandonment, for example, was the field plowed under or left fallow (Myster 2007). The use of fire in Puerto Rico is infrequent (Grau et al. 2003).

6.4 Long-Term Permanent Plot Studies and Permanent Plot Studies That Incorporate Chronosequences

It has long been known that the establishment and regular sampling of permanent vegetative plots, set up at the time of the abandonment, provides the best information about the nature of vegetative recovery and dynamics of fields by showing the actual sequences of plant abundance and species change over time (Pickett 1982). The patterns revealed from these plot data are important in their own right, but they also help to guide field experimentation (Myster 1993, 2007) and provide parameters for modeling (Myster 2010b). Among the most important patterns are those that show structural and conceptual stages of post-agricultural succession. For example a first stage dominated by the field's past crop and its various signature effects, a second stage of patch dynamics among grasses, fast-growing herbs, past crop remnants and their plant associates, vines, asexual shrubs, and early "pioneer" tree invaders, a third stage of wholesale invasion of tree species of differing shade and other strategies, leading to a fourth stage, development of a closed-canopy "secondary" forest. Perhaps most importantly we may observe, by careful mapping of individual plants within these fields (Myster 2003b), the plant–plant replacement process which creates all subsequent population and community patterns (Myster 2012a). In these recovering fields, replacements are working everywhere and at all times.

A long-term permanent plot study was set up in a natural, unseeded pasture in Puerto Rico that had been grazed for decades before it was abandoned. This study pasture borders the Luquillo Experimental Forest (LEF) of northeastern Puerto Rico (18°20′N, 65°45′W), close to the town of Sabana. The LEF is classified as tropical montane wet forest (Walter 1979), dominated by tabonuco (*Dacryodes excelsa*), ausubo (*Manilkara bidentata*), and motillo (*Sloanea berteriana*), and is one of the sites maintained by the US long-term ecological research (LTER) program. The pasture itself is located at low elevation, receives between 2 and 5 m of rainfall/year, has an average temperature of 18 °C, and has soils that are fertile and volcaniclastic in origin (Thomlinson et al. 1996). Twenty-five 5 m × 2 m contiguous plots (the total plot was 25 m × 10 m) were laid out on the border with the LEF in 1996 (Myster 2003b) with the long side parallel to the forest in order to maximize any forest edge effects. There were not any remnant trees or sprouting tree roots in the plots at the beginning of the study. Starting in May of 1997 and continuing annually for 10 years, the plots were sampled for percent cover of each plant species. Percent cover—an indication of a species' ability to capture light and, therefore, to dominate these areas in the process of becoming forested communities (Myster 2003b)—was estimated visually in relation to each plot's area. The complete data set is archived as LTERDATB #97 at the web site http://www.luq.lternet.edu.

The first analysis of this data used only the first 5 years of sampling and showed that (1) grass dominated peaking at year 3 but still constituting 75 % of the total plant cover after 5 years, (2) forbs also peaked at year 3 and then declined to 40 % after 5 years, and (3) both ferns and woody species increased steadily to 45 and 15 % total cover respectively after 5 years (Myster 2003b). During these first 5 years

Table 6.1 All tree (T) and shrub (S) stems, at least 50 cm in height, sampled in the Sabana pasture permanent plot during the first 5 years of succession

Species	Year 1	Year 2	Year 3	Year 4	Year 5
Syzygium jambos (T)	18	80	76	70	65
Calophyllum calaba (T)	5	45	31	18	8
Clidemia hirta (S)	1	39	29	17	9
Tabebuia heterophylla (T)	4	11	10	12	5
Miconia impetiolaris (S)	4	24	22	21	20
Miconia prasina (S)	3	39	29	22	19
Citrus spp. (T)	2	3	3	3	3
Guarea guidonia (T)	1	4	3	5	2
Eugenia pseudopsidium (T)	2	7	6	8	7
Myrcia splendens (T)	0	2	3	1	1
Prestoea montana (T)	0	1	2	2	2
Inga laurina (T)	2	5	4	5	3
Ocotea leucoxylon (T)	0	0	1	2	1
Andira inermis (T)	0	2	5	8	6
Inga vera (T)	2	9	11	12	10
Casearia sylvestris (T)	1	3	1	4	4
Miconia racemosa (S)	3	12	13	7	5
Psychotria berteriana (S)	0	2	1	2	2
Piper hisperdium (S)	2	23	10	3	3

of succession, total percent cover ranged between 160 and 300 due to strata formation, percent species turnover declined to 25 in year 4, species richness plateaued at 20 and species evenness stayed around 0.75. Focusing only on the trees and shrubs, the woody plants were present in the first sampling (Table 6.1) with greatest density close to the forest edge. The dominant species were the trees *Syzygium jambos* (an exotic species), *Calophyllum calaba*, and *Tabebuia heterophylla* and the dominant shrubs *Clidemia hirta* and *Miconia* spp., all peaking in abundant individualistically (Pickett 1982) in different years. Tree species found after tree fall (Devoe 1989) and landslide (Myster and Walker 1997) in Puerto Rico and, consequently, thought to be common "pioneer" trees (e.g., *Cecropia* spp., *Shefflera* spp.) were not present in these data. Together these woody species showed (1) a maximum productivity of 400 g/m^2/year and a maximum total number of stems of 300, both in year 2, (2) an increasing mean height of 240 cm in year 5, and (3) a plateauing of total basal area of 950 cm^2 after year 2 (Myster 2003b). The spatial locations of each tree by basal diameter and species in year 5 were mapped and showed that trees invaded and grew from the forest edge, again in an individualistic manner.

Next the first 10 years of sampled plot data were used to implicate plant interactions by generating Spearman coefficients of rank associations (Table 6.2) because controlled field experiments performed elsewhere have confirmed that species that interact significantly also form significant associations with each other (Tirado and Pugnaire 2005). The pasture in Puerto Rico had a total of 307 significant pair-wise positive associations between species and 55 significant pair-wise negative associations

Table 6.2 Significant Spearman rank correlation coefficients among the top 25 most common plant species in the Sabana pasture over the first 10 years of succession

	B	C	D	E	F	G	H	I	J	K	L	M	N	O	P	Q	R	S	T	U	V	W	X	Y	Z
A	+−	+	−	++		+		++	−	+	+++			++			+		−					−	++−
B		++++++	++	+++++++	++−	+++			++	+++++	++	+	−	+++				+	+	+++	+−−−	++	++−	++++	++
C			+	+++		+	+	++	++++	++	+++	+	+++	++	+++							++	++++	++++	++
D				+	++	++			+	+	+		+	+−		++	+		+						
E					+−	+	++++	++		+++++	+++++++	++	+	++−	++	−−−	+	−	++	++	+++−	+−	++++	+++	++
F						++		+		+	−				−		+					+	+−		
G							+	++	++	+	++++	−		−−	+	+	−−	+++		+	−−	+	++	−−	+
H										+++++		+	+++	−−	++++			+		+	+	+		−	++
I										++	+	+		++	+	+++	+	+	+			+	+	+	+++++
J											++		++	++	+				+++++				++	++	++
K														++−	++	++	+−	+	+	++		+++	+−	++	+++
L												++	+	+	++	+			+++			−−−	++−−	+++	+
M														++++	++		+	++++	+			++	+		
N														++	+					+		++	++		
O																	−−	−	+		+−		−	−	+
P																			+			++	+	+++	+
Q																									
R																						++			
S																				+	−	+			
T																					+	++++		++	+
U																									
V																							++−	++	
W																									+
X																								++++	+
Y																									+

Significant positive associations are indicated by a "+" and significant negative associations are indicatedby a "–." There can be a maximum of ten significant associations between any two species, one for each sampling year. Plant species and families are indicated as follows: A—*Bromelia* spp.; B—*G.guidonia*; C—*O. leucoxylon*; D—*Citrus frutus*; E—*S. jambos*; F—*Desmodium* spp.; G—*Gleichenia bifida*; H—*I. laurina*; I—*Citrus limon*; J—*C. sylvestris*; K—*P. montana*; L—*C. calaba*; M—*M. prasina*;N—*E. pseudopsidium*; O—*T. heterophylla*; P—*Eugenia malaccensis*; Q—*P. hisperdium*; R—*A. inermis*; S—*Psychotria brachiata*; T—*M. racemosa*; U—*P. berteriana*; V—*Xanthosoma* spp.; W—*C. hirta*;X—*Panicum* spp.; Y—*M. splendens*; and Z—*Ocotea sintenisii*

Table 6.3 Estimates of above-ground biomass (in g/m²) for each of the ten most common tree species in each of the 10 years of the study

Species	Year 1	Year 2	Year 3	Year 4	Year 5	Year 6	Year 7	Year 8	Year 9	Year 10
S. jambos	35.1	202.4	77.7	110.5	102.2	89.9	82.2	76.6	85.7	77.9
P. berteriana	0.8	0.7	1.7	0.7	2.7	3.2	2.9	3.3	4.5	5.8
Piper glabrescens	2.1	8.3	7.6	6.1	12.2	9.8	12.5	9.4	13.6	14.4
M. impetiolaris	17.2	37.6	21.8	31.1	25.8	26.6	29.3	33.3	37.6	44.8
M. prasina	13.3	53.4	25.6	33.9	26.6	28.4	27.7	36.9	42.2	48.9
C. calaba	9.5	115.5	36.9	20.6	21.5	20.5	19.9	15.6	22.3	18.8
M. racemosa	0.7	3.3	3.4	3.2	2.3	3.3	4.5	4.2	3.1	3.9
Clibadium umbella	21.1	49.4	26.1	38.5	37.9	44.4	48.8	52.2	46.2	54.3
Piper hispidum	1.1	7.3	7.1	6.2	13.7	14.4	14.8	15.7	15.1	16.8
I. vera	2.2	24.4	2.8	4.1	7.4	9.8	12.2	13.8	15.5	17.7

over the 10 year sampling time period. Because most of the significant associations were positive, facilitation may be more common than competition in these stressful, early successional environments (Callaway and Walker 1997). Care must be taken, however, not to fall into the Nietzsche's "Cornaro" misinterpretation, because species that associate together may not be reacting to each other but to some other species or environment factor. Species that formed the most positive associations with other plants were the trees *S. jambos, Guarea guidonia, Ocotea leucoxylon,* and *Prestoea montana.* Species that formed the most negative associations with other plants were the grass *Panicum* spp., the fern *Gleichenia bifida*, and the tree *T. heterophylla*. The positive associations from Table 6.2 were also used to construct a clustering hierarchy of increasing species inclusiveness, in order to define post-agricultural communities (Myster 2012b).

Using the first 10 years of sampling, allometric above-ground biomass equations were developed for each of the ten most abundant tree and shrub species, which account for over 90 % of the total stems and over 90 % of the total basal area. Above-ground biomass was computed (Table 6.3) and the most productive species were *S. jambos, Miconia impetiolaris, Miconia prasina, C. calaba*, and *Clibadium umbella*. This productivity data was then plotted against turnover. The first 2 years of succession showed a slightly positive slope between productivity and turnover, but starting in year 3 there was a significant negative slope which became more negative with time. Turnover was also plotted against richness. There was a slightly negative slope between richness and turnover in the first year which became more negative with time.

The same plot design and sampling protocol was employed in a sun coffee plantation located on private land in 1996. The plots were sampled in 1997 and in 1998, but could not be continued because the farmer wanted the land back for other purposes. Plots showed that the coffee trees themselves persisted, having 30 % cover in year 1 and 23 % cover in year 2. Like the pasture plot, grass dominated (94 % in year 1, 90 % in year 2) and *Ichnanthus pallens* most common, with forbs (45, 50 %: *Phytolacca icosandra* and *Musa* spp. most common) and woody vines (64, 67 %: same species as in the pasture plots) also common. Ferns were at low cover (5 % in year 1, 6 % in year 2: same species as in the pasture plots) but at greater levels than in the pasture plots (Myster 2003b). Trees were much less common after coffee cultivation than in the pasture plots and the three most dominant trees and shrubs found in pasture were absent in coffee. However, *T. heterophylla*, *Miconia* spp., and fruit trees were just as common. *Andira inermis* and *G. guidonia* attained greater cover in coffee than in pasture but *Inga* spp. was not found. There were several species unique to the coffee plots (Myster 2003b) and *C. schreberiana* was again absent.

The same plot design and sampling protocol were again employed in four pastures of different starting ages located close to the LEF: CoRd (5 years old), Los Davilas (13 years old), CoRd 9948 (18 years old), and Puente Roto (23 years old). These four pastures were first sampled in 2000 and then sampled annually. All those data are archived as LTERDATB #100 at the LEF web site http://www.luq.lternet.edu. The first 10 years of sampling in the permanent plots of the Sabana pasture was then combined with the first 5 years of sampling in the other four pastures, started at different times. This hybrid study, combining permanent plot data with chronosequences, hoped to provide enough overlapping of sampling years and pastures to predict succession in these pastures better than using chronosequences by themselves. Total plant richness and total plant cover did form long continuous patterns among the samplings, but species composition was too unique to each pasture to be able to combine into a single timeline. Consequently this approach may work best for broad structural parameters, but not for predicting individual species abundances (Myster and Malahy 2008). Finally the three permanent plot data sets (Sabana pasture, coffee plantation, the four pastures that start in different years) were used for an analysis of dominance–diversity curves. All pastures and the coffee plantation showed a significant log-normal pattern, with only the Los Davilas pasture not flattening over successive samplings over time. Also, there was no lag time seen for any pasture compared to coffee plantation (Myster 2010a).

6.5 Chronosequence Studies Only

When the establishment of permanent plots in newly abandoned pastures or agricultural fields is not possible, many have turned to the use of chronosequences. There researchers sample fields of different ages—years since abandonment—and arrange that data chronologically, assuming that however these fields differ, their age is most important in determining their structure and function. Given the complexity of the disturbance regime and the range of factors that can influence how fields recover (as discussed previously), however, this is very unlikely. Indeed, this assumption has never been validated (Myster and Malahy 2008). Even so these studies may be useful in generating and focusing hypotheses (see Fig. 4.3 in Zimmerman et al. 2007) for future testing using permanent plots, or perhaps hybrid plots as outlined above. But care must be taken, in every such study, not to present these data as "trends," they are simply a collection of individual fields, sampled in various ways, which differ at least in year since abandonment.

6.5.1 Vegetation

Near the LEF, woody plant density, basal area, species number, and species diversity were low in 10-year pastures, but increased rapidly in 10- and 15-year-old pastures (Aide et al. 1995, 1996). Also pastures above 100 m in elevation had *M. prasina* as the dominant pioneer species, with a larger group of pioneers at lower elevations (Aide et al. 1996). Pastures of age 40 years have densities, basal areas, aboveground biomass, and species richness similar to old growth forest sites, but with a different species composition with several exotic species (Aide et al. 2000).

Similarly in other parts of Puerto Rico, pastures and abandoned coffee plantations 25–30 years since abandonment had basal area and species richness similar to surrounding forests, but not in species composition with several exotic species (Pascarella et al. 2000). In Cordillera Central of Puerto Rico, forest structure was similar in 30-year-old pastures, sun coffee plantations, and shade coffee plantations (Marcano-Vega et al. 2002). *Coffee arabica, I. vera*, and *E. poeppigiana* persisted in shade coffee, while pastures and sun coffee also contained species commonly found there. In the Karst region of Puerto Rico, abandoned pastures had greater woody species diversity and greater woody stem density than coffee plantations of similar age but basal area was similar between them with *S. campanulata* and *G. guidonia* as the dominant species in pastures and coffee plantation, respectively (Rivera and Aide 1998). The dryer forests of southwestern Puerto Rico were dominated by introduced species and after 45–60 years these recovering pastures and agricultural fields lagged behind both mature forests and similar post-agricultural areas in public land in basal area, tree density, and species richness (Colon et al. 2011). On recovering farmland in the Humacao municipality, soil type had effects on basal area, species diversity, and abundance of exotics—lower on alluvial compared to plutonic or volcanic substrates, higher areas of age 13–30 years had more exotics compared to lower areas, and bulldozing prior to abandonment reduced species diversity but increased percent exotics (Chinea 2002).

Three stands of recovering sugarcane plantations of ages 25–40 years old on moist alluvial sites had high basal area, intermediate tree density, high complexity index, and relatively low species richness compared to more mature forests. Of the 31 species, 11 species were introduced including *S. campanulata*, a wind-dispersed, nearly deciduous introduced species (Martínez and Lugo 2007).

6.5.2 Carbon

Aboveground carbon (C) accumulation showed significant log-linear trends with time for pastures and recovering agricultural areas in Puerto Rico (see Marín-Spiotta et al. 2007b and references therein). Silver et al. (2000) found that overall aboveground biomass had significantly faster biomass accumulation during the first 20 years of succession (6.17 Mg/ha/year) than the subsequent 60 years. Typically, forest regeneration on former croplands and pastures results in net C sequestration aboveground due to greater C stocks in longer-lived trees with the accumulation of woody biomass (Marín-Spiotta et al. 2007a, b). As shown in Silver et al. (2000), C accumulation on abandoned agricultural fields is greater than abandoned pastures, which is greater than previously cleared sites. In a wet secondary forest chronosequence on abandoned pastures in Puerto Rico, 80-year-old secondary forests had greater aboveground biomass than primary forests because of the dominance of woody species in the former, and a heavy dominance of palms in primary forests (Marín-Spiotta et al. 2007b). The specific species planted, and whether they are native or exotic, can have a strong effect on later composition of the stand. One advantage of some exotics is that they are fast growing and rapidly accumulate C.

Secondary forests attain high rates of litterfall relatively quickly, faster than above-ground biomass, within the first 25 years of succession and soon after plateau to a production rate of a primary forest (Ostertag et al. 2008), but litter chemistry and leaf decomposition rates were not different between young and old pastures. Indeed, litterfall biomass is likely to reflect development of biomass during succession. During secondary succession, coarse woody debris should be accumulating, but initially may be quite low as decomposition rates and slow inputs may initially diminish stocks after land clearing.

The type and intensity of prior land use may affect the direction and rate of soil organic matter (SOM) content during reforestation. Studies in Puerto Rico (Silver et al. 2004) have reported increases in SOM associated with an increase in plant biomass after pasture abandonment. While aboveground biomass typically increases with secondary forest age, soil C pools often show no relationship with time since agricultural or pasture abandonment (Marín-Spiotta et al. 2007a, b). One study, however, found that 26-year-old pastures had more SOM than areas recovering from crops (Sotomayor-Ramirez et al. 2009). In the Cayey Mountains of southeastern Puerto Rico, gains of new secondary forest C were compensated by losses of pasture-derived C from the soil, resulting in no net change with reforestation or during 80 years of secondary succession (Marin-Spiotta et al. 2006). Conversion of forest to pasture leads to less soil C and N loss than conversion of forest to crops, taking 40–50 years for C and 15–20 years for N to return to forest levels after abandonment (Brown and Lugo 1990b).

6.5.3 Soil Biota

Land conversion from forests to pasture or agricultural land, and the resulting loss of SOM, soil nutrients such as nitrogen (N), phosphorus (P), and potassium (K), and water holding capacity, often lead to considerable changes in soil macrofauna, fungi, and microbes which in turn greatly affect biogeochemical processes and cycling. Chief among these processes is decomposition comprised of (1) fragmentation, which is the breaking of larger pieces of organic matter into smaller ones and is largely a consequence of the feeding activity of soil macrofauna, and (2) chemical alteration, which is the change of chemical elements of dead organic matter into other chemical compounds and is primarily associated with the activity of fungi and bacteria.

Among the macrofauna, earthworms may be particularly important and agriculture enhances the introduction of exotic earthworm species (González et al. 2006, 2007) which is dominant over native species in pastures. With the conversion of pastures to secondary forests, (1) the richness of earthworm species increases due to the presence of native, anecic species (Zou and Gonzalex 1997) but (2) the exotic earthworms are still dominant in terms of biomass and density during early succession, which can accelerate leaf litter decomposition (Hendrix et al. 1999; Liu and Zou 2002). The density of the exotic species, *Pontoscolex corethrurus*, can decrease, however, over time (Sánchez-de León and Zou 2004). Earthworm density is highest in active pastures, decreases during their recovery to forest (25–40 years since abandonment), and is lowest in mature forest (at least 77 years old: Sánchez-de León et al. 2003). One pasture was sampled and it had a greater density of endogeic earthworms and no anecic earthworms, compared to a near-by forest (Zou et al. 2006). Among three sampled coffee plantations, the most stressed field had the most exotic earthworms, but there was no other clear taxonomic pattern (Ramos et al. 2010).

Together with bacteria, fungi are the main initial decomposers of dead plant and animal materials and account for 80–95 % of the total decomposer biomass and heterotrophic soil respiration. Two groups of fungi common to soils are the single-celled organisms called yeasts and the multicellular filamentous organisms, such as molds, mildews, smuts, and rusts. The four most common genera of molds in soils are *Penicillium, Mucor, Fusarium*, and *Aspergillus* (see Li and González 2007). Fungi are usually the most abundant and effective decomposers, although bacteria play a key role in N fixation. Even so, the primary route by which N enters this soil is through the breakdown of legume leaf litter inputs that are enriched with N, mainly by fungi. Soil microbes are actively involved in the cycling of P as well. Land conversion disrupts the structure and activity of the soil fungi and microbial community, affecting biogeochemical cycling. Total fungal and microbial biomass is significantly reduced following conversion from forest to pasture or agricultural land. Consequently in pastures, the total amount of immobilized P in litter significantly decreased because of less fungal biomass in the pasture litter. Conversion from forest to pasture also dramatically reduces the abundance of spores and the diversity of arbuscular mycorrhizal fungi.

In a 20-year-old pine *P. caribaea* (Morelet) plantation and a secondary forest originating from the same abandoned farmland in the Guzmán sector of the LEF, total fungal biomass was reduced 84.6 and 87.5 % and total bacterial biomass was reduced 62.1 and 56.9 %, respectively, in the plantation and the secondary forest. Total and active fungal biomass was significantly higher in the plantation than in the secondary forest in all treatments during both the wet and dry seasons. Total and active fungal biomass was significantly greater in the wet season than in the dry season everywhere. Also, microbial activity in the plantation was more sensitive to the litter input than the secondary forest. Soil CO_2 efflux had higher correlation with total fungal biomass but lower correlation with total bacterial biomass in the plantation than in the secondary forest, suggesting that the fungal communities may dominate soil heterotrophic respiration in the plantation, while bacterial communities may take the control in the secondary forest. Active fungal biomass demonstrated higher correlations with soil CO_2 efflux than their corresponding total fungal and bacterial biomass (Li and González 2007).

6.6 Field Experiments

Because these pastures and post-agricultural areas were once rainforest, and because they are recovering back into types of rainforest, the dynamics of the trees within them is the major vegetative concern. In the past sections, I reviewed what is known about the structure and function of that recovery. Now I examine how, or if, tree seeds and seedlings (Grubb 1977) overcome various mechanisms and tolerances in order to invade, establish, and grow in these fields (Myster 2004b). One of the most important early issues is the nature and extent of any woody vegetation still remaining in the field after clearing from forest. Trees can sprout from coppiced tree trunks or root stocks, for example, and then function as "nurse" trees facilitating seed dispersal and perhaps other ways such as shading and killing lower growing forbs and grasses. Forest trees or shrubs may also spread into the field asexually (e.g., *Miconia* spp. in Puerto Rican fields) from the woody remnants allowing trees to invade quickly from the forest edges, not dependent on the seed bank or the survival of newly dispersed seed. Because of the intensity of the mechanisms which remove tree seeds, these woody functions may be a significant part of facilitation in these fields.

6.6.1 Tree Seed Mechanisms and Tolerances

Tree seeds have been found to be rare in the soil seed bank of pastures and fields (Zimmerman et al. 2000; Cubina and Aide 2001; Myster 2004a, 2006). They have been somewhat more common in the seed rain, but decrease sharply with distance from the forest edge in young pastures (Cubina and Aide 2001). This is to be expected because early trees are more wind-dispersed than later-invading species and also any seed-dispersing animals avoid the open field, preferring areas next to forests and woody patches for perching and hiding, creating a positive feedback loop for tree dispersal which slows tree invasion out into the field. Woody tree density was only 0.01 seeds/m^2 after 6 months in a pasture (Zimmerman et al. 2000) and they were a small part of the approximately 20 plant seeds/m^2 sampled in young Puerto Rican fields after 3 months (Myster 2004a). They did increase, however, under shrubs growing naturally in the pasture (Myster 2006).

After tree seeds are dispersed, they face a host of tropical predators where even quick germination without dormancy may not save them. Detailed studies performed in Puerto Rico have always shown huge and immediate seed losses to predators including 50–75 % lost after 14 days in coffee plantation and pasture (Myster 2003a), 50–79 % lost after 14 days in a pasture (Myster 2003c), and 67–90 % lost after 14 days in various fields (Myster 2004a). Within those general results, however, are important sources of variation (SOV) that show how some seeds survive. Those SOV include species-specific predator preferences, litter effects and different starting seed densities, field-types, patch-types seasons, ages, and seed sizes. In particular results imply that the density of predators—ants and small mammals have been implicated (Myster 2003c)—and their behavior and availability are affected by the structure of plant cover.

The analysis of data from the Puerto Rican pastures shows SOV in these significant main effects in (1) species (*S. jambos* had 50 % losses, *Solanum torvum* had 65 % losses, *M. prasina* had 76 % losses, and *Piper aduncum* had 79 % losses), (2) density (78 % losses at 5 seeds/dish, 69 % losses at 10 seeds/dish, 51 % losses at 20 seeds/dish), and (3) patch type (grass patches had 69 % losses, *Miconia* spp. patches had 65 % losses). There were also SOV in these significant interaction effects (1) between species and density for *P. aduncum* and *S. torvum* (88 % losses at low density vs. 85 % losses at medium density vs. 56 % losses at high density, 83 % losses at low density vs. 70 % losses at medium density vs. 27 % losses at high density, respectively), and (2) between species and patch-type for *M. prasina* (93 % losses in grass patches vs. 55 % losses in *Miconia* spp. patches). Another experiment in the same pastures and in a Puerto Rican coffee plantation again showed large species differences (*G. guidonia* had 34 % losses, *Gonzalagunia spicata* had 55 % losses, *Miconia racemosa* had 66 % losses, and *Casearia arborea* had 72 % losses), with predators taking seeds at spatial scales of 2–11 cm (Myster 2003a). Puerto Rican banana plantations and sugarcane plantations also demonstrated significant differences in species (*G. spicata* had 80 % losses, *Palicourea riparia* had 63 % losses, *P. aduncum* had 74 % losses, and *Miconia pracina* had 86 % losses) (Myster 2003c).

Seeds lost to pathogenic diseases have always been much less than those lost to predation, in the ranges 20–45 % after 14 days in coffee plantation and pasture (Myster 2003a) and 5–10 % 21 days in various fields (Myster 2004a) but may be more important in places such as sugarcane plantations with large mats of decaying cane biomass. Seed pathogens showed no clear spatial preferences in the coffee plantation (Myster 2003a). Germination percentages were low for most species, 15–35 % after 14 days in a coffee plantation and pasture (Myster 2003a), 5–10 % in various fields (Myster 2004a), and 2–90 % in a pasture which was reduced by removal of vegetation (Zimmerman et al. 2000). Germination occurred at larger spatial scales than predation, 11–23 cm, in the coffee plantation (Myster 2003a). Greenhouse and shadehouse studies, however, suggest that these percentages could be greater in other fields and studies (Myster and Everham 1999; Zimmerman et al. 2000) and that many tree seeds could germinate if they can escape predation and pathogens.

6.6.2 Tree Seedling Mechanisms and Tolerances

Detailed studies performed in Puerto Rico have shown 30–50 % survival of tree seedlings after 6 months in various fields (Myster 2004a) and 2–90 % survival after 6 months in a pasture (Zimmerman et al. 2000) with *G. guidonia*, *C. calaba*, and *A. inermis* surviving best, *Psychotria brachiata* least. There were also significant SOV among pasture tree species in (1) survival: *P. aduncum* 21 %, *C. schreberiana* 34 %, *G. guidonia* 45 %, *Veronica patens* 47 %, *S. torvum* 51 %, and *M. prasina* 55 % and (2) growth parameters: *Cecropia* (85 cm in mean height, 1.18 cm mean basal diameter, 3.1 leaves on average, 350 cm^2 mean leaf area), *Solanum* (90 cm, 0.76 cm, 5.5, 150 cm^2), *Piper* (66 cm, 0.82 cm, 5.9, 75 cm^2), *Veronica* (95 cm, 0.92, 7.2, 104 cm^2), and *Miconia* (72 cm, 0.75, 4.9, 122 cm^2). In general, seedlings had fatter stems and more leaves in the forest/pasture border and taller stems 5–10 m from the forest edge. Competition for light is implicated as a major mechanism affecting seedling establishment in fields and pastures, because forest trees compete primarily for light, but competition with grasses for water and soil nutrients may be more important in early succession. Herbivory levels (<2 % of leaf area on average) and pathogens levels (even less: Myster 2004a) are perhaps too low even to affect competition (Myster 1993).

6.6.3 Tree Mechanisms and Tolerances

Even though studies at the seed/seedlings stages are most important in understanding the present and future dynamics of these areas (Grubb 1977), studies with larger trees may also be needed. Such manipulations could include removing trees from field plots and manipulating herbivory and pathogenic levels. Indeed, association analysis (Table 6.2) along with these kinds of experiments may be able to address the relative roles of competition and facilitation in these fields (Callaway and Walker 1997). Competition may become more common as succession proceeds (Myster 1993, 2007) with facilitation occurring mainly in early succession when abandonment creates harsh, stressful conditions that are ameliorated by neighbors. As fields age, however, facilitation may become rarer as plants become bigger and competition increases for fewer resources.

6.7 Restoration Strategies

Because of the ubiquity of recovering pastures in Puerto Rico and the common practice of converting crop fields to pastures after abandonment, pasture restoration should take center stage in post-agricultural restoration in Puerto Rico. The purposes of pasture restoration involve attaining vegetation structure and functional characteristics as close as possible to the original rainforest, which in turn has the compound objectives of increasing biodiversity and recovering soils.

Consequently, restoration strategies should seek to overcome the negative effects, or to amplify the positive effects, of mechanisms and tolerances controlling pasture succession. Factors limiting tree regeneration in abandoned pastures can include scarcity of nutrients, soil compaction, lack of or excess soil humidity, high solar radiation, and intra- and inter-specific competition. To restore abandoned pastures, strategies have to counter these factors. Active measures are needed to encourage and favor forest regeneration including soil treatments, control of invasive vegetation, and planting appropriate tree species to serve as perches for dispersal agents and to ameliorate soil and microenvironmental conditions. Strategies to encourage natural regeneration of degraded pastures may also involve (1) care and management of remnant trees in pastures, (2) planting individual trees, or (3) groups of trees in blocks and windbreaks (Montaguini 2007).

First, there may be isolated trees in the pasture that are remnants of the original forest. Often the microenvironment below tree crowns is favorable for the establishment of species that cannot grow in pastures. For example, they may facilitate tree seed dispersal by birds, and create shade and accumulate litter to diminish grass and herbaceous growth. If those individuals are seedlings or saplings, then addition of artificial structures can increase their survival by providing protection against excess insulation, drought, or frost. Second, restoring abandoned pastures can involve planting trees individually, as seedlings, saplings, or mature trees, which may contribute to the recovery of environmental conditions favorable to tree regeneration processes, such as *Inga* spp. trees increasing soil nitrate concentration and organic matter. Native tree species are preferred over exotics, except in cases where the performance of native species is still not well known. It is often best to plant fast-growing trees. However, it may be necessary to first removing the invasive vegetation using herbicides, fire or mechanical weeding, and may involve tillage and root removal. Exotic tree species can also play an important role in the restoration of degraded pastures, especially when conditions of extreme degradation persist, such as eroded, denuded soil. Exotic species that can thrive in highly adverse conditions can be planted initially and later serve as facilitators for the introduction of native species (Lugo and Helmer 2004). However, one must guard against them escaping and becoming invasive.

Third, trees can be planted in monocultural groups—plantations—and multiply individual tree improvements of soil and microclimatic conditions, also serving as perches for birds, bats, and other seed dispersers. In Puerto Rico, the understory of plantations of the exotic *Albizia lebbek*, contained 22 species of trees and shrubs, compared to only one species in control plots (Parrotta 1992). Mixed plantations, however, may offer a more favorable environment for natural regeneration than pure plantations due to their multi-strata architecture. Mixed plantations may have a higher variety of microhabitats for seed dispersers and potentially create a greater variety of ecological niches allowing for the establishment of diverse regeneration, for example creating different conditions of shade and litter accumulation. Finally, in pastures that are far from sources of propagules, windbreaks may be important reservoirs of native tree species. Windbreaks can increase forest seed recruitment by serving as habitat and/or movement corridors for seed-dispersing birds. Indeed windbreaks within the landscape may increase tree recruitment if (1) connected to a forest, (2) connecting forest patches, and (3) including native, fruit-producing trees. Finally some degraded pasture soils can be recovered through application of fertilizers or the planting of trees that have a positive influence on soil properties.

6.8 Modeling and Future Research

The focus on trees should continue into modeling, which is just as important as field experimentation. Indeed, these two kinds of investigation need to work together where field experiments can supply parameters for modeling and modeling can direct experimental design. The fluid and continuous process of plant–plant replacement (Busing 1996) is at the core of old field community dynamics, generating all population and community patterns found there. Consequently, modeling in these fields and pastures should focus on how trees replace trees and other plants over time and space.

Because plants are sessile organisms, and thus space is intimately related to how they grow and reproduce, these models must contain for each individual (Gleason 1926) plant both a "neighborhood-space"—outside its plant body where it interacts with the environment through mechanisms and tolerances (sensu Turkington and Harper 1979)—and a "phyto-space" which is made up of both its necromass and its biomass (Myster 2012a). Models should also focus mainly on replacements at the seed and seedling stage of the plant life cycle (Grubb 1977). Tree neighborhood-spaces and phyto-spaces would become occupied, available, and then often reoccupied over time as replacements proceed. These replacements are not one-by-one, however, because most replacements actually involve (1) a single individual plant being replaced by several other plants, (2) several plants being replaced by a single individual plant (common during old field succession), and/or (3) a plant or plants replaced by the growth of the still living plants which surround it (e.g., the thinning phase of forest stand development). Plants respond to a *mechanism* when they interact directly with another living thing (e.g., dispersal, predation, pathogenic disease, herbivory) but *tolerate* a nonliving part of the environment by responding to it either directly (e.g., germination) or indirectly through another living thing (e.g., competition).

Many of these spaces, and thus the complexity of the plant–plant replacement process, can be seen and mapped by analysis of long-term permanent vegetation plots (Myster 2003b). In addition, field experiments can help us understand the mechanisms and tolerances which cause plant responses and thus create these neighborhood and phyto-spaces. Eventually every plant will die. At which point, the neighborhood space may become available to other plants immediately, but for the phyto-space this will depend on local decomposition rates.

Taken together then I am suggesting that the best way to investigate old field dynamics is to focus on: (1) how mechanisms and tolerances affect tree seeds and seedlings producing responses, (2) how spaces both within and without these tree individuals change as a result, and (3) how those changes lead to trees replacing (or not replacing) grasses, forbs, and other trees. Based on the field experiments and analysis from permanent plot sampling already completed and reviewed above, dispersal, seed predation, facilitation, and competition have major roles in controlling tree invasion, establishment, and growth in recovering pastures and fields in Puerto Rico. Facilitation may, however, take many forms. For example, trees can facilitate seed dispersal by supplying woody perches for birds and cover for ground animals. Also patches that are less preferred by predators may facilitate establishment for those seeds found there.

In addition to this modeling, I suggest these field samplings and experiments:

1. Studies in other parts of the island away from the LEF should be continued.
2. Permanent plots need to be set up and sampled annually in sugarcane, coffee, and other pastures, even if they are already some years old. In that case hybrid studies, like reviewed previously, can be done.
3. Sampling of ecosystem parameters, concerned with energy flow (e.g., light, C, temperature) and biogeochemical cycling (e.g., NPK, water, CO_2), needs to be done in the same way as the vegetation.
4. Along with the sampling and analysis of that data, field experiments need to be expanded with a focus on dispersal, seed predation, facilitation, and competition.

Plot data and field experiments can then be used for extensive modeling of plant–plant replacements. Finally, many more restoration projects need to be done using many of the techniques discussed.

Acknowledgments I thank Eda Melendez and her staff for their help with the LTER data sets, Maria Aponte for field assistance, and N.V.L. Brokaw for logistical support. I received support from grant DEB-0218039 from the National Science Foundation to the Institute of Tropical Ecosystem Studies, University of Puerto Rico, and the USDA Forest Service, International Institute of Tropical Forestry as part of the Long-Term Ecological Research Program in the Luquillo Experimental Forest. Additional support was provided by the Forest Service (U.S. Department of Agriculture) and the University of Puerto Rico.

References

Aide TM, Zimmerman JK, Herrera L, Rosario M, Serrano MS (1995) Forest recovery in abandoned tropical pastures in Puerto Rico. For Ecol Manage 77:77–86

Aide TM, Zimmerman JK, Rosario M, Marcano-Vega H (1996) Forest recovery in abandoned cattle pastures along an elevational gradient in Northeastern Puerto Rico. Biotropica 28:537–548

Aide TM, Zimmerman JK, Pascarella JB, Rivera L, Marcano-Vega H (2000) Forest regeneration in a chronosequence of tropical abandoned pastures: implications for restoration. Restor Ecol 8:328–338

Brown S, Lugo AE (1990a) Tropical secondary forests. J Trop Ecol 6:1–32

Brown S, Lugo AE (1990b) Effects of forest clearing and succession on the carbon and nitrogen content of soils in Puerto Rico and US Virgin islands. Plant Soil 124:53–64

Busing RT (1996) Estimation of tree replacement patterns in an Appalachian *Picea-Abies* forest. J Veg Sci 7:685–694
Callaway RM, Walker LR (1997) Competition and facilitation: a synthetic approach to interactions in plant communities. Ecology 78:1958–1965
Chinea JD (2002) Tropical forest succession on abandoned farms in the Humacao municipality of eastern Puerto Rico. For Ecol Manage 167:195–207
Chinea JD, Helmer EH (2003) Diversity and composition of tropical secondary forests recovering from large-scale clearings: results from the 1990 inventory in Puerto Rico. For Ecol Manage 180:227–240
Colon SM, Lugo AE, Ramos-Gonzalez OM (2011) Novel dry forests in Southwestern Puerto Rico. For Ecol Manage 262:170–177
Cubina A, Aide TM (2001) The effect of distance from forest edge on seed rain and soil seed bank in a tropical pasture. Biotropica 33:260–267
Devoe NN (1989) Differential seedling and regeneration in openings and beneath closed canopy in sub-tropical wet forest. Ph.D. dissertation, Yale University, New Haven
Dietz JL (1986) Economic history of Puerto Rico: institutional change and capitalist development. Princeton University Press, Princeton
Domínguez Cristóbal CM (2000) Panorama histórico forestal de Puerto Rico. Editorial de la Universidad de Puerto Rico, Río Piedras
Everham EM III, Myster RW, VanDeGenachte E (1996) Effects of light, moisture, temperature, and litter on the regeneration of five tree species in the tropical montane wet forest of Puerto Rico. Am J Bot 83:1063–1068
Flynn DFB, Uriarte M, Crk T, Pascarella JB, Zimmerman JK, Aide TM, Ortiz MAC (2010) Hurricane disturbance alters secondary forest recovery in Puerto Rico. Biotropica 42:149–157
Foster DR, Fluet M, Boose ER (1999) Human or natural disturbance: landscape-scale dynamics of the tropical forests of Puerto Rico. Ecol Appl 9:555–572
Gleason HA (1926) The individualistic concept of the plant association. Bull Torrey Bot Club 53:1–20
Gomez L, Ballesteros M (1980) Vida Y Cultura Precolombina de Puerto Rico. Editorial Cultural, Rio Piedras
González G, Huang CY, Zou X, Rodríguez C (2006) Earthworm invasions in the tropics. Biol Invasions 8(6):1247–1256
González G, Huang CY, Chuang SC (2007) Earthworms and post-agricultural succession. In: Myster RW (ed) Post-agricultural succession in the Neotropics. Springer, Berlin, pp 114–138
Grau HR, Aide TM, Zimmerman JK, Thomlinson JR, Helmer E, Zou X (2003) The ecological consequences of socioeconomic and land-use changes in postagriculture Puerto Rico. Bioscience 53:1–10
Grubb PJ (1977) The maintenance of species-richness in plant communities: the important of the regeneration niche. Biol Rev Camb Phil Soc 52:107–145
Hendrix PF, Lachnicht SL, Callaham MA, Zou X (1999) Stable isotopic studies of earthworm feeding ecology in tropical ecosystems of Puerto Rico. Rapid Commun Mass Spectrom 13:1295–1299
Li Y, González G (2007) Soil fungi and macrofauna in the Neotropics. In: Myster RW (ed) Post-agricultural succession in the Neotropics. Springer, Berlin, pp 93–113
Liu ZG, Zou X (2002) Exotic earthworms accelerate plant litter decomposition in a Puerto Rican pasture and a wet forest. Ecol Appl 12:1406–1417
Lugo AE, Helmer E (2004) Emerging forests on abandoned land: Puerto Rico's new forests. For Ecol Manage 190:145–161
Marcano-Vega H, Aide TM, Baez D (2002) Forest regeneration in abandoned coffee plantations and pastures in the Cordillera central of Puerto Rico. Plant Ecol 161:75–87
Marin-Spiotta E, Silver WL, Swanston CW, Torn MS, Burton SD (2006) Mineral control of soil carbon storage with reforestation of abandoned pastures. American Geophysical Union, Fall Meeting, abstract #B41D-0221

Marín-Spiotta E, Ostertag R, Silver WL (2007a) Long-term patterns in tropical reforestation: plant community composition and aboveground biomass accumulation. Ecol Appl 17:828–839
Marín-Spiotta E, Cusack DF, Ostertag R, Silver WL (2007b) Trends in above and belowground carbon with forest regrowth after agricultural abandonment in the Neotropics. In: Myster RW (ed) Post-agricultural succession in the Neotropics. Springer, Berlin, pp 22–72
Martínez OA, Lugo AE (2007) Post sugar cane succession in moist alluvial sites in Puerto Rico. In: Myster RW (ed) Post-agricultural succession in the Neotropics. Springer, Berlin, pp 73–92
Montaguini F (2007) Management for sustainability and restoration of degraded pastures in the Neotropics. In: Myster RW (ed) Post-agricultural succession in the Neotropics. Springer, Berlin, pp 265–295
Myster RW (1993) Tree invasion and establishment in old fields at Hutcheson Memorial Forest. Bot Rev 59:251–272
Myster RW (2003a) Seed regeneration mechanisms over fine spatial scales on recovering coffee plantation and pasture in Puerto Rico. Plant Ecol 166(2):199–205
Myster RW (2003b) Vegetation dynamics of a permanent pasture plot in Puerto Rico. Biotropica 35:422–428
Myster RW (2003c) Effects of species, density, patch-type and season on post-dispersal seed predation in a Puerto Rican pasture. Biotropica 35:542–546
Myster RW (2004a) Regeneration filters in post-agricultural fields of Puerto Rico and Ecuador. Plant Ecol 172:199–209
Myster RW (2004b) Post-agricultural invasion, establishment, and growth of neotropical trees. Bot Rev 70:381–402
Myster RW (2006) Shrub vs. grass patch effects on the seed rain and the seed bank of a five-year pasture in Puerto Rico. Ecotropica 12:12–19
Myster RW (2007) Introduction. In: Myster RW (ed) Post-agricultural succession in the Neotropics. Springer, Berlin, pp 3–21
Myster RW (2010a) Testing dominance-diversity hypotheses using data from abandoned plantations and pastures in Puerto Rico and Ecuador. J Trop Ecol 26:247–250
Myster RW (2010b) A comparison of tree replacement models in oldfields at Hutchenson Memorial Forest. J Torrey Bot Soc 137:113–119
Myster RW (2012a) Plants replacing plants: the future of community modeling and research. Bot Rev 78(1):2–9
Myster RW (2012b) A refined methodology for defining plant communities using data after agriculture in the Neotropics. ScientificWorld Journal, vol. 2012, Article ID 365409, 9 pages. doi:10.1100/2012/365409
Myster RW, Everham EM III (1999) Germination cues across the disturbance regime in the Puerto Rican rainforest. Trop Ecol 40:89–98
Myster RW, Malahy MP (2008) Is there a middle-way between permanent plots and chronosequences? Can J For Res 38:3133–3138
Myster RW, Walker LR (1997) Plant successional pathways on Puerto Rican landslides. J Trop Ecol 13:165–173
Ostertag R, Maria-Spiotta E, Silver W, Schulten J (2008) Litterfall and decomposition in relation to soil carbon pools along a secondary forest chronosequence in Puerto Rico. Ecosystems 11:701–714
Parrotta JA (1992) The role of plantation forests in rehabilitating degraded tropical ecosystems. Agric Ecosyst Environ 41(2):115–133
Pascarella JB, Aide TM, Serrano MI, Zimmerman JK (2000) Land-use history and forest regeneration in the Cayey mountains, Puerto Rico. Ecosystems 3:217–228
Pascarella JB, Aide TM, Zimmerman JK (2004) Short-term response of secondary forests to hurricane disturbance in Puerto Rico, USA. For Ecol Manage 199:379–393
Pickett STA (1982) Population patterns through twenty years of old field succession. Vegetatio 49:45–59
Pico R (1969) Nueva geografia de Puerto Rico: fisica, economia Y social. University of Puerto Rico, Editorial Universitaria, San Juan

Ramos C, Alfaro M, Borges S, Amador JA (2010) Earthworm taxonomic structure of coffee plantations at three soil associations in Puerto Rico (Oligochaeta:Glossoscolecidae, Megascolecidac). Zool Middle East 2:125–131

Ramos-Gonzalez OM (2001) Assessing vegetation and land cover changes in Northeastern Puerto Rico: 1978–1995. Caribb J Sci 37:95–106

Rivera LW, Aide TM (1998) Forest recovery in the Karst region of Puerto Rico. For Ecol Manage 108:63–75

Rudel TK, Perez-Lugo M, Zichel H (2000) When fields revert to forest: development and spontaneous reforestation in post-war Puerto Rico. Prof Geogr 52:386–397

Sánchez-de León Y, Zou X (2004) Plant influences on native and exotic earthworms during secondary succession in old tropical pastures. Pedobiologia 48:215–226

Sánchez-de León Y, Zou X, Borges S, Ruan H (2003) Recovery of native earthworms in abandoned tropical pastures. Conserv Biol 17:999–1006

Silver WL, Ostertag R, Lugo AE (2000) The potential for carbon sequestration through reforestation of abandoned tropical agricultural and pasture lands. Restor Ecol 8:1–14

Silver WL, Kueppers LM, Lugo AE, Ostertag R, Matzek V (2004) Carbon sequestration and plant community dynamics following reforestation of tropical pasture. Ecol Appl 14:1115–1127

Sotomayor-Ramirez D, Espinoza Y, Acosta-Martinez V (2009) Land use effects on microbial biomass, beta-glucosidase and beta-glucosaminidase activities and availability, storage and age of organic C in soil. Biol Fertil Soils 45:487–497

Thomlinson JR, Serrano MI, Lopez MT, Aide TM, Zimmerman JK (1996) Land-use dynamics in a post-agricultural Puerto Rican landscape (1936–1988). Biotropica 28:525–536

Tirado R, Pugnaire FI (2005) Community structure and positive interactions in constraining environments. Oikos 111:437–444

Turkington R, Harper JL (1979) The growth, distribution and neighbour relationships of *Trifolium repens* in a permanent pasture: I. Ordination, pattern and contact. J Ecol 67:201–218

Walter H (1979) Vegetation of the earth and the ecological systems of the geo-biosphere. Springer, New York

Weaver PL, Birdsey RA (1986) Tree succession and management opportunities in coffee shade stands. Turrialba 36:47–58

Zimmerman JK, Aide TM, Rosario M, Serrano M, Herrera L (1995) Effects of land management and a recent hurricane on forest structure and composition in the Luquillo experimental forest, Puerto Rico. For Ecol Manage 77:65–76

Zimmerman JK, Pascarella JB, Aide TW (2000) Barriers to forest regeneration in an abandoned pasture in Puerto Rico. Restor Ecol 8:350–360

Zimmerman JK, Aide TM, Lugo AE (2007) Implications of landuse history for natural forest regeneration and restoration strategies in Puerto Rico. In: Cramer VA, Hobbs RJ (eds) Oldfields: dynamics and restoration of abandoned farmland. Island Press, Washington, DC, pp 51–74

Zou X, Gonzalex G (1997) Changes in earthworm density and community structure during secondary succession in abandoned tropical pastures. Soil Biol Biochem 29:627–629

Zou X, Arandes-Perez B, Capo-Ramos D, Gutierrez-Santiage J, Rosario-Santiage M (2006) Comparisons of earthworm community structures between active pasture and an adjacent tropical wet forest in Puerto Rico. Carribb J Sci 42:311–314

Chapter 7
South Brazilian Forest-Grassland Ecotones: Dynamics Affected by Climate, Disturbance, and Woody Species Traits

Sandra C. Müller, Gerhard E. Overbeck, Carolina C. Blanco, Juliano M. de Oliveira, and Valério D. Pillar

7.1 Introduction

The southernmost part of Brazil is characterized by extended grasslands—the southern Campos. In some parts of the region, these grasslands occur in close contact with different forest types of the Atlantic forest biome or riparian forests, forming mosaics. Otherwise, they form large and continuous areas toward south and southwest, where they reach the Pampa grassland of Uruguay and Argentina. Throughout this region, forest expansion and/or shrub encroachment can be observed and have been related both to site conditions (abiotic variables and disturbance regimes) and climatic changes. We know today, based on studies of pollen records from peat profiles, that expansion of woody vegetation throughout the past centuries was driven by changes in the climatic conditions towards a climate increasingly favorable for forest development, and evidence exists that this process is halted or at least slowed down as consequence of land use and disturbances. Current vegetation dynamics in forest-grassland ecotones thus likely are linked both to climatic changes (on the regional scale) and vegetation management and/or disturbance regime (on the local scale).

S.C. Müller (✉) • C.C. Blanco • V.D. Pillar
Departamento de Ecologia, Instituto de Biociências, Universidade Federal do Rio Grande do Sul, Av. Bento Gonçalves 9500, Bloco IV, Prédio 43422, Caixa Postal 15007, 91501-970 Porto Alegre, RS, Brazil
e-mail: sandra.muller@ufrgs.br

G.E. Overbeck
Departamento de Botânica, Instituto de Biociências, Universidade Federal do Rio Grande do Sul, Av. Bento Gonçalves 9500, Bloco IV, Prédio 43433, 91501-970 Porto Alegre, RS, Brazil

J.M. de Oliveira
Programa de Pós-Graduação em Biologia, Universidade do Vale do Rio dos Sinos—UNISINOS, 93022-000 São Leopoldo, RS, Brazil

R.W. Myster (ed.), *Ecotones Between Forest and Grassland*,
DOI 10.1007/978-1-4614-3797-0_7,

In this chapter, we offer a synthesis of the current understanding of forest-grassland dynamics in southern Brazil. First, we present principal features of the region's vegetation and briefly discuss some features of the grasslands (known as Campos, in Portuguese or Spanish) compared to the Brazilian savannas (Cerrado). Then we address principal questions on grassland-forest dynamics in the region: Which are the main factors that maintain forest-grassland ecotones under current climatic conditions that allow for forest establishment? Which types of woody species—in terms of functional characteristics—effectively manage to establish in grasslands? In which way is this related to land use? Is there any evidence that the current spatial patterns of vegetation types will change as a consequence of expected climate change? The findings that we present to answer these questions mostly come from studies conducted in the southernmost state, Rio Grande do Sul (RS); other parts of southern Brazil—the states of Paraná and Santa Catarina—have been poorly studied regarding these questions. At the end of this chapter, we present prospects for further research.

7.2 Vegetation Patterns in Southern Brazil

The southern part of Brazil is situated at a transitional zone between tropical vegetation types to the north, and temperate vegetation to the South (Overbeck et al. 2007). Current climate in the southernmost state, Rio Grande do Sul, is Köppen's Cfa, with some parts with higher altitude under Köppen's Cfb (Kuintcher and Buriol 2001). Mean annual precipitation varies from 1,200 mm in the southern part to above 2,000 mm in the northeast; mean annual temperatures range from 15 to 20 °C, with mean temperatures in the coldest month ranging from 11 to 14 °C (Nimer 1990).

The natural vegetation is a mosaic of grassland, shrubland, and different forest types (Rambo 1956; Teixeira et al. 1986; Leite and Klein 1990). While grasslands prevail in most parts of southern RS, in the Pampa biome, the northern half of RS shows more pronounced mosaics of forest and grassland, which are part of the Atlantic Forest biome (IBGE 2004). This biome includes tropical rainforest (Atlantic forest *sensu stricto*), mixed ombrophilous forest (*Araucaria* forest), and seasonal forests (both deciduous and semideciduous) (Oliveira-Filho and Fontes 2000). Atlantic rainforest *sensu stricto* occupies the slopes and valleys in the east of the South Brazilian Plateau in northeastern RS and the coastal plain and highland slopes of the adjacent states of Santa Catarina (SC) and Paraná (PR). *Araucaria* forest, characterized by the conspicuous presence of the Brazilian Pine, *Araucaria angustifolia*, in the upper stratum, occurs on highlands above 500 m altitude and as patches in less altitudes areas of the Serra do Sudeste (Carlucci et al. 2011). Seasonal forests can be found in western SC and PR, northwestern RS and in river basins in the central lowlands of RS, as well as gallery forests in the Pampa biome. Both *Araucaria* forest and seasonal forests form mosaics with natural Campos.

Therefore, natural Campos are differentiated by both a phytogeographical perspective (Cabrera and Willink 1973) and in the current biome classification in

Brazil (IBGE 2004) that separates grasslands in the Pampa biome (occurring in larger extended areas with forests mostly restricted along rivers and to the granitic shield) from those forming more pronounced mosaics in the Atlantic forest biome. Edaphic and climatic variations contribute to physiognomic and floristic differences (Waechter 2002), e.g., with the presence of shrublands and semideciduous forests in the so-called Serra do Sudeste, on the granitic shield in southern RS, or the Acacia-Prosopis parkland, which can be considered a transition to the Chaco and Espinal formations further to the west. Composition of the species-rich grasslands likewise reflects the geographical situation between temperate and tropical climate, as C_3 and C_4 species co-occur, despite the dominance of the C_4 grasses in terms of biomass (Boldrini 2009).

The presence of extensive grasslands under a climate that apparently is adequate for the development of forests had puzzled the first European explorers and naturalists who traveled in the region and has been debated by ecologists for decades (Lindman 1906; Walter 1967; Eriksen 1978; Box 1986). Phytogeographical evidence, palaeoclimatic records, and data on origin of soil organic matter suggest that the grasslands are relicts of cooler and dryer glacial and postglacial phases that preceded the present-day climate, and are not the result of modern forest destruction by man (Rambo 1957, 1961; Klein 1975; Martin et al. 1993; Behling 2002). Palaeoecological data from sites presently covered with *Araucaria* forests confirm forest expansion from refugia in deep valleys and riverine areas over Campos is a recent process, which had already been suggested by Hueck (1966) who studied form and shape of gallery forests. Stratigraphic and palynological investigations in mires with peat layers (Behling 2002; Behling et al. 2004, 2007) showed that the climate was colder and drier than today until about 10,000 years before present (BP) and warmer, drier from 10,000 to about 4,000 years: grasslands dominated under these conditions. The process of forest expansion started in the Late Holocene, around 4,000 years BP, from then on under a cool and moist climate, and accelerated around 1,000 years BP, probably in response to the emergence of an even moister climate. Fire and grazing by domestic animals introduced by European settlers are considered to be the principal factors impeding expansion of forest over grassland vegetation in the past centuries and under current climatic conditions (Pillar and Quadros 1997).

7.3 Open Vegetation Types in Southern Brazil: Grassland, Not Savanna

Shrubs and some trees are conspicuous elements of the grassland vegetation in southern Brazil. At the beginning of the twentieth century, the Swedish naturalist Lindman (1906) noted that in the state of RS it was virtually impossible to find a grassland patch without presence of woody species. These woody species can be divided into two groups: pioneer species from the different forest communities that manage to establish within the herbaceous matrix, and grassland shrubs, principally members of the Asteraceae family, that can be found both sparsely distributed or in dense patches.

The presence of a continuous herbaceous layer and a woody component as well as the mosaic-like vegetation patterns of forest and grassland in many parts of the region have lead to the application of the term savanna to grasslands in southern Brazil, for example in the RADAMBRASIL classification of vegetation types (Teixeira et al. 1986; Leite and Klein 1990; IBGE 2004). The vegetation structure of the South Brazilian Campos, especially when abandoned or under the influence of fire and thus with dominance of high caespitose grasses (Overbeck 2005), certainly can resemble that of open savanna physiognomies, such as in the Cerrado (Brazilian savanna). However, the term savanna includes a climatic component. Walter (1971) defines savanna as "natural, homogenous zonal vegetation of the tropical summer-rain zone showing a closed grass cover and scattered, individual woody plants, either shrubs or trees." This climatic component of the definition of savannas clearly does not apply to the grasslands in the southern Brazil and the use of the term savanna is inappropriate.

The different vegetation physiognomies of Cerrado form a gradient of tree cover, from pure grassland (in Portuguese: "campo limpo"), grassland with scattered woody species ("campo sujo" or "campo cerrado," with a more pronouncedly expressed tree component in the latter case), savanna with a distinct two-layer structure and tree cover up to 30 % ("cerrado sensu stricto") to an open forest physiognomy ("cerradão") (Eiten 1972; Furley 1999; Cruz Ruggiero et al. 2002; Oliveira-Filho and Ratter 2002). Such a gradient does not exist in the southern Campos grassland. Cover of woody species may differ considerably in southern Campos, mostly depending on land management (grazing intensity and/or fire), but the woody component of the grassland itself almost exclusively consists of shrubs, some of them not even exceeding the grass layer, not of trees, as in the Cerrado, where total number of tree and shrub species in the savanna vegetation is roughly 800 (Oliveira-Filho and Ratter 2002). Those trees that are present in the South Brazilian Campos are pioneer species from forests that are not fully adapted to fire and the occurrence of them is linked to the dispersal from nearby forest patches. This contrasts with the situation in the Cerrado biome, where density of savanna trees increases with time since the last fire both in open and closed vegetation physiognomies and where savanna trees show a number of adaptations to frequent fires, such as thick bark that protected meristematic tissues, nutrient and carbohydrate reserves or high resprouting capacity (Hoffmann and Moreira 2002). Additionally to this increase in savanna species over intervals between fires, an increase of forest species, not adapted to fire, is possible at sites in proximity to forest when fire is excluded (Geiger et al. 2011)—a process similar to the situation in southern Brazil (see discussion below). Just as in the Cerrado, the cover of woody species (shrubs, in the case of Campos) may differ greatly among different Campos areas, but this does not allow for the definition of such a clear physiognomic gradient as in the Cerrado, even though the terms "campo limpo" and "campo sujo" have been used as well.

This brief account on the vegetation history in southern Brazil and on the differences in the physiognomy between southern Campos and the Cerrado suggest that we may expect rapid and strong processes of forest expansion in southern Brazil, at least if disturbance regimes or land management do not impede the recruitment and establishment of forest species and if soil conditions do not prevent forest vegetation. In the following section, we wish to briefly discuss the relative importance of the disturbance regime and of climate on vegetation processes at the forest-grassland interface and on processes of shrub encroachment in grasslands.

7.4 Factors Responsible for the Maintenance of Forest-Grassland Ecotones and Evidences of Forest Expansion in Southern Brazil

The dynamics of the often very sharp transition between forest and grassland, and the mechanisms of forest expansion processes that can be observed at some locations have been subject of a number of studies in the past 10 years (Oliveira and Pillar 2004; Duarte et al. 2006a, b; Müller et al. 2007; Schüler-da-Silva 2009; Santos et al. 2011). No evidence for restriction of forest vegetation by soil properties has been found (Lindman 1906; Rambo 1956; Garcia Martinez 2005). Shifts in dominance between forests and grasslands have been occurring since at least the Late Quaternary (Behling et al. 2004, 2007). Evidence from fossil pollen records (Behling 2002) and soil carbon isotopes (Duemig et al. 2008; Silva and Anand 2011) shows that forests have been expanding very slowly since at least the Late Holocene (<100 m in the past few millennia; Silva and Anand 2011), which enabled persistence of the mosaics under the modern humid climatic conditions, which favor forests. In addition, results from the application of the adaptive Dynamic Global Vegetation Model (DGVM) developed by Scheiter & Higgins (2009) to the region showed that this relative stable coexistence of forests and grasslands in southern Brazil even under the current mesic climate is maintained by strong feedbacks of grass biomass on fire frequency, suggesting that grasslands have persisted as an alternative stable state (Blanco 2011). Nowadays, transitions from forest to grassland are unlikely to occur without anthropogenic deforestation, and hence only two scenarios are conceivable: forest-grassland coexistence or forest dominance.

Grazing and fire are considered to be the main factors that impede forest expansion over grassland under current climatic conditions (Pillar and Quadros 1997). Grasses and grasslands are well adapted to defoliation by both herbivores and fire: grasses generally do not have chemical defences, while intercalary meristems at base of leaves ensure continued growth of the plant after top parts of leaves were removed, either by grazing or fire (Anderson et al. 1999). On the other hand, the establishment and posterior survival of woody species in grazed grasslands are constrained by continuous grazing and trampling, with the magnitude of the effect depending on grazing pressure. The grasslands of southern Brazil have been under grazing influence for a long time. Large mammals were present until 8,000 years BP, when species of Equidae (*Equus*), Camelidae and Cervideae became extinct in the fossil record (Schüle 1990; Kern 1997) Later, in the seventeenth century, Jesuit missionaries introduced cattle and horses. Until today, beef cattle production on Campos remains as an important land use of the region, with currently 31 % of the territory of RS under grazing (Cordeiro and Hasenack 2009). Although grazing by large mammals thus has been absent for a period of more than 7,000 years, before the introduction of cattle, no pronounced forest expansion took place, as apparently climatic conditions still were restrictive (see above). Recent studies indicate that grazing (and fire) suppression increases shrub and tree cover (Boldrini and Eggers 1996; Pillar and Quadros 1997; Oliveira and Pillar 2004).

Fire likely has been—from the extinction of native herbivores until the introduction of cattle—a key factor to explain the maintenance of current forest-grassland boundaries in southern Brazil, and still is in some parts of the region. Data from peat profiles with countings of charcoal particles indicate that fire became frequent around 7,000 years BP at one site (Behling et al. 2004) and 9,000 years BP at another site further west (Behling et al. 2005). This may be consequence of human influence coupled with climatically related increased fuel accumulation in the grassland. The seasonal forest and the Atlantic rainforest in southern Brazil are not flammable, that is, grassland fire will not go beyond the forest edge, while in the *Araucaria* forest it can occur but is rare. In contrast, the Campos, even where dominated by shrubs, are highly flammable if sufficient grass senescent biomass has accumulated (Overbeck 2005)—which can be the case after as little as 1 year. Regular fires inevitably cause the abrupt boundaries between grassland and forest communities, which has been observed in southern Brazil (Oliveira and Pillar 2004; Adelmann et al. 2011).

Our understanding of forest expansion processes and of factors that may maintain the ecotone patterns is primarily based on studies carried out in two conservation units on the highlands of South Brazilian Plateau, both protected from fire and cattle grazing, and an area in the northernmost portion of the granitic shield in the state of RS, with recurrent fire events but without grazing. The study sites on the highlands, Pró-Mata Center for Research and Nature Conservation (in the following: Pró-Mata) and Aracuri Ecological Station (Aracuri), are characterized by mosaics of Campos grasslands and *Araucaria* forests. Aracuri and Pró-Mata were excluded from grazing and fire in 1974 and 1994, respectively. Before serving for biodiversity conservation, the sites were used for cattle grazing and management included burning of the grassland every two or three winter seasons in order to remove accumulated senescent biomass. Forests were used for selective logging of *A. angustifolia* and other timber species. The site on the granitic shield is a hill (Morro Santana; "morro" is Portuguese for "hill") inserted in the urban matrix of the capital of RS, Porto Alegre, and represents the mosaic of Campos and Seasonal forest (although with presence of a high number of floristic elements from the Atlantic rainforest). Currently, the grasslands on Morro Santana are subject to frequent man-made fires. In both of these regions (Plateau highlands and granitic shield), two different types of forest expansion can be identified: formation of forest patches within the grassland matrix through nucleation and gradual advance of the forest border over grassland.

On Morro Santana, the comparison of aerial photographs spanning over a period of 40 years (from 1941 to 1985) has shown that the forest-grassland border has been remarkably stable under the current fire regime (anthropogenic fires with a frequency of roughly 3 years on average (Adelmann and Overbeck 2009)): little changes in cover of the main vegetation types have occurred. However, the past 20 years had a somewhat faster increase in forest cover (increment of 8.9 % in forest cover from 1985 to 2002), than those observed in the previous period (increment of 3.6 % in forest cover from 1941 to 1985) (Blanco 2011). While fire clearly can be considered responsible for the persistence of forest-grassland ecotones under current climatic conditions, the establishment of forest tree species within adjacent grasslands in clearly visible, as discussed further on.

Analyses of aerial photographs taken before (1974) and 5 years after fire and grazing suppression (1999) at Pró-Mata showed pronounced encroachment of woody species grasslands at sites close to forest borders, therefore suggesting a succession from grasslands dominated by tall tussock grasses to grasslands with shrubs and, finally, even though it is not known how long this process will take, forests (Oliveira and Pillar 2004) (Fig. 7.1a). Still at Pró-Mata, seed rain of forest tree species across the forest-grassland ecotones was higher in number and richer in terms of species in grasslands with presence of tall shrubs (dominant shrub species: *Baccharis uncinella*) and under isolated *Araucaria* trees within the grassland matrix than in tall tussock grassland without prominent woody species (Fig. 7.1b). Thus, the attractiveness of isolated trees or of shrub islands to bird dispersers contributes to the gradual increase of shrubs and trees close to the forest-grassland edges, and to forest nucleation under nurse *Araucaria* trees (Santos et al. 2011). At the Aracuri Ecological Station, after the exclusion of grazing and fire (1974), the colonization rate of *Araucaria* trees in grasslands was inversely related to the distance of the forest edge (Fig. 7.1c), clearly indicating a gradual advance from the forest border (Schüler-da-Silva 2009). In addition, it is interesting to note the higher frequency of young individuals (<20 years) colonizing the most distance grassland sites, evidencing the faster forest expansion beyond the forest border after exclusion of disturbances following the creation of the Ecological Station in 1974.

At the forest-grassland boundaries, young trees may succeed in establishing in the grassland matrix near the forest edge, but the probability of tolerating fires before reaching a certain height seems to be low. Young individuals of some forest pioneer tree species may resprout (see later), even though slowly, from the stem base or present clonal growth by producing root suckers (Hermann 2007). Likely, tree survival will be a function of stem height and fire intensity (as proposed in the model by Higgins et al. 2000 for African savannas), but empirical data on survival capacity of trees of different age or size under fire is missing for southern Brazil. It has been shown for grasslands both in the granitic hills as well as on the South Brazilian plateau that presence of isolated forest patches can also be associated to rock outcrops (Müller et al. 2007; Carlucci et al. 2011), likely due to two effects: On the one hand, rock outcrops and initial pioneer vegetation serve as perching structures and thus attract animal-dispersed seeds (Duarte et al. 2006b). On the other hand, at rock outcrops, it has been suggested that the grass layer and thus flammability is strongly reduced, which allows for establishment of trees at these safe sites (Pillar 2003). In addition, a recent study evaluating patterns of forest expansion on the highlands in South Brazil have found that once established, forest trees colonizing grassland

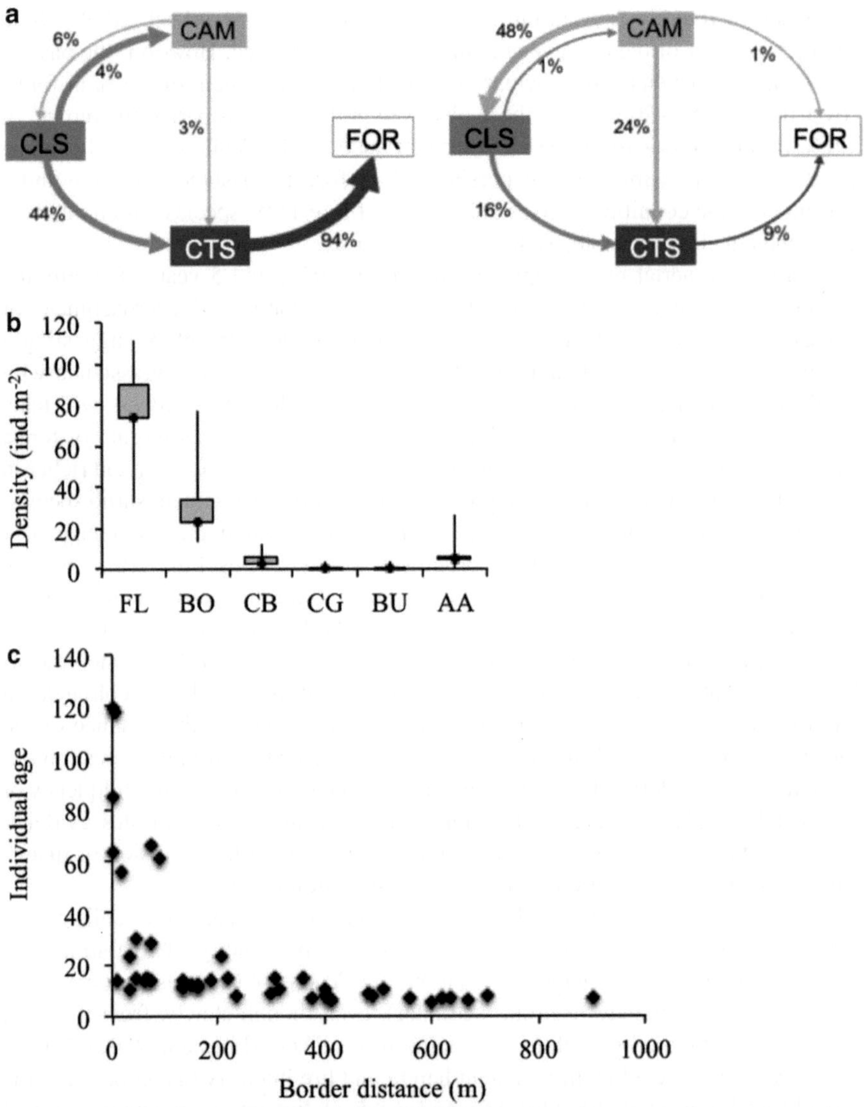

Fig. 7.1 Evidences of forest expansion: (**a**) Direction of vegetation change from 1974 to 1999 in Pro-Mata ecotones: at *left* the numbers refer to percentages of change relative to the area covered by each vegetation type in 1974, and at *right*, they refer to change relative to the total altered area that showed different vegetation types comparing 1974 and 1999 (campos—CAM, campos with low shrubs—CLS, campos with tall shrubs—CTS, forest—FOR) (data from Oliveira 2003; Oliveira and Pillar 2004). (**b**) Patterns of seed rain (density of seeds from woody species per square meter) under six vegetation physiognomies at forest-grassland ecotones at Pro-Mata: forest (FL), boundary (BO), campos with *Baccharis uncinella* (CB), tall tussock campos without shrubs (CG), natural perches of *B. uncinella* (BU) or of *Araucaria angustifolia* (AA) (data from Santos et al. 2011); (**c**) Relationship between the distance of the forest border and age of *A. angustifolia* individuals that colonized the grassland (data from Schüler-da-Silva 2009)

sites may enhance nutrient uptake due to symbiotic interactions (Silva and Anand 2011). This corroborates the nurse effect of isolated trees (Duarte et al. 2006b) on grasslands favoring further local wood encroachment, a positive feedback to forest expansion (Silva and Anand 2011) and another source of spatial heterogeneity generated by isolated forest patches growing in grasslands.

Nonetheless, it is well know that variation in site conditions, such as soil patchiness, topographic factors (declivity, exposition) as well as unpredictable seasonal drought, can promote spatial or temporal stress for the establishment and recruitment of tree individuals, especially for rainforest species (Knoop and Walker 1985; Kauffman et al. 1994). Differences in these conditions can be responsible for differences in the colonization of forest species in adjacent grassland both in space and time, leading to different structural patterns of forest-grassland boundaries. Further, the effects may be even stronger due to the interaction with disturbance. The number of studies in southern Brazil still is too small to allow a more detailed discussion of these factors.

7.5 Which Functional Traits Allow for Establishment and Persistence of Woody Species in Burned Grassland?

The woody species that occur in grasslands adjacent to forest patches in southern Brazil can be divided into main types: grassland shrubs and forest trees. Grassland shrubs are principally species from the Asteraceae family, especially from the genus *Baccharis*, but also from Fabaceae, Euphorbiaceae, and Malvaceae, among others (Boldrini 2009). In contrast, forest tree species establishing in the grassland matrix are species that commonly occur in the forests of the region. The species that are important colonizers in grasslands and are frequently seen as pioneers in grassland are *A. angustifolia* (Araucariaceae), *Myrsine coriaceae, Myrsine umbellata* (Primulaceae), *Symplocos uniflora* (Symplocaceae), *Eugenia uniflora, Blepharocalyx salicifolius* (Myrtaceae), and species from the genus *Schinus* and *Lithraea* (Anacardiaceae) (Oliveira 2003; Duarte et al. 2006b; Dadalt 2007, 2010; Müller et al. 2007; Schüler-da-Silva 2009). Curiously, not only typical pioneer species are advancing to adjacent grasslands: *Myrciaria cuspidata* (Myrtaceae) and *Matayba elaeagnoides* (Sapindaceae), for example, both are species that are preferentially found within forests (Müller et al. 2003, 2007). Except for *A. angustifolia*, those colonizers species are also found further in Uruguay, characterizing shrublands and woodlands, which considerably expanded over grasslands between nineteenth and twentieth centuries (Gautreau and Lezama 2009; Gautreau 2010).

One of the most important functional trait of woody species that persist in grasslands subject to fire is the ability to resprout after aboveground parts damage due disturbance events (Müller et al. 2007), as has been pointed out for Mediterranean, Australian and African ecosystems (Pausas 1999; Bond and Midgley 2001; Pausas et al. 2004; Esther et al. 2010). For grassland woody species that do not resprout after fire, however, the colonization of burned areas by seed dispersal or from the soil seed bank, as well as fast establishment (recruitment and survival of seedlings) seems to be successful alternative strategies, while for forest trees the current time intervals between fires may be too short.

It can be assumed that the proportion of resprouting species in the community increases with an increasing disturbance frequency and that seeder species must have higher relative growth rates and set seeds earlier than the resprouters (Bellingham and Sparrow 2000). In the forest-grassland study on Morro Santana, individuals from 76 woody species were found in the grassland areas (including border plots, i.e., plots still with a relatively dense grass stratum but adjacent to the forest), including both typical grassland shrubs and forest trees (Müller 2005, 2007). From these species, 65.8 % were resprouters, that is species with the ability to resprout from basal or underground structures after fire events, as can easily be observed directly in the field. More than 90 % of the individuals were grassland woody species (average of 1.95 individuals per square meter; ind./m^2). Forest trees were represented by 36 species, but with a very low number of individuals (average of 1.09 ind./m^2 in border plots—0–4.5 m from forest edge—and 0.13 ind./m^2 in grassland plots unburned for 4 years).

Figure 7.2 shows a spatial pattern of grassland and forest woody species that occur along a forest-grassland border on Morro Santana, in an area where grassland had not been burned for at least 4 years (in addition, the forest is too moist to burn and fires always stop at the boundary in this site) (Müller 2005). At this study site, the proportion of resprouter species is much greater than that of nonresprouters. When comparing adjacent grassland areas with fire return intervals of 4 and 1 year (Fig. 7.2a, b), it becomes clear that the density of resprouting grassland woody species increases considerably in more recently burned plots, indicating high resilience of these species to frequent fire disturbance. Current fire return interval in the grassland lies between 2 and 4 years (Overbeck et al. 2005; Müller et al. 2007; Fidelis et al. 2010a). Once established in interstitial places between tussock grasses, resprouting woody species individuals should persist regularly burns as species with the persistence niche strategy (Bond and Midgley 2001).

In frequently burned grassland ecosystems, most woody species depending on sexual reproduction (especially forest trees) need to develop to a height that allows them to survive fires (i.e., usually above the lethal fire zone, in our case, the grass layer), within the period between burns or despite the burns (Bond and van Wilgen 1996). An alternative successful establishment for such species can be to germinate in "safe sites," that is, at sites protected from at least very intense fires. Termite mounds (Furley 1999; Guillet et al. 2001), thickets or adult trees (Hochberg et al. 1994), and rock outcrops (Clarke 2002a, b; Pillar 2003; Müller and Forneck 2004; Carlucci et al. 2011) have been considered safe sites for woody plants in a matrix of frequently burned vegetation, due to the lower density and growth of grasses, decreasing thus, flammability, as discussed before (see photos in Fig. 7.3). Once safely established, adult individuals may serve as facilitators, being attractors to seed dispersers and/or improving seedling establishment by diminishing local water deficit (Hoffmann 1996; Li and Wilson 1998; Holl 2002; Duarte et al. 2006b), which can lead to the process (discussed above) of forest expansion by nucleation.

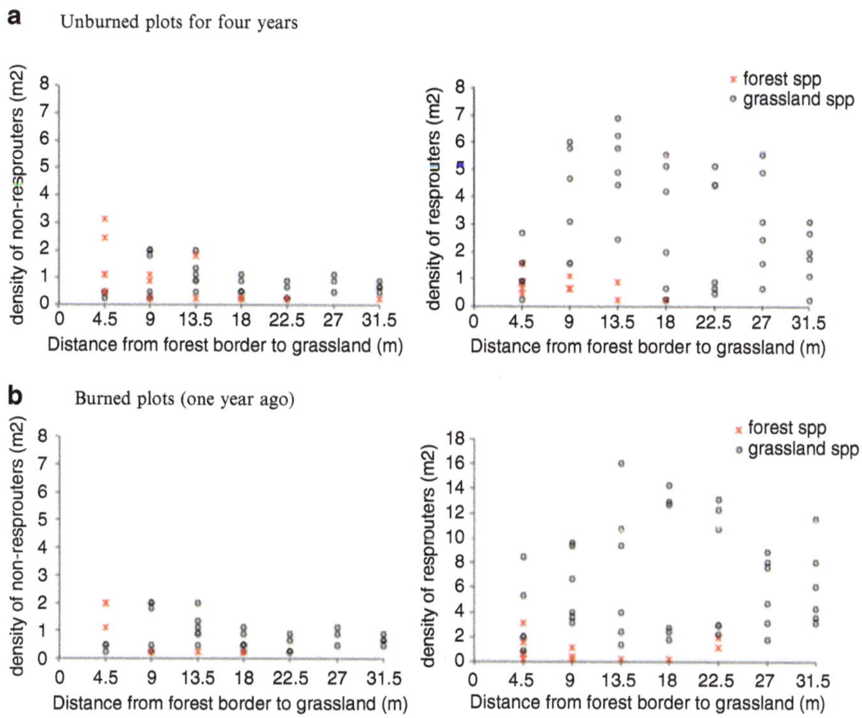

Fig. 7.2 Spatial patterns of resprouter and non-resprouter grassland woody species and forest tree species established on grassland adjacent to forests, i.e., along a forest-grassland border. Data are based on results of Müller (2005), on Morro Santana, Porto Alegre, RS. All woody plants with at least 10 cm height were considered. The graphs show in (**a**) plots in grassland unburned for at least 4 years, while in (**b**) plots of nearby areas that were experimentally burned 1 year before, considering the time of survey. The *vertical scale* of the last graph differs from the others

Furthermore, sexual reproduction is a strategy usually observed in grassland shrub species, mainly for anemochoric species of Asteraceae that may be resprouters or not, and usually produce great amounts of seeds. In all studied sites that we discuss in this chapter to illustrate the processes at forest-grassland ecotones in southern Brazil (Pró-Mata, Aracuri and Morro Santana), it is possible to observe an increase in density of seeder Asteraceae species (i.e., nonsprouters) in grassland where fire and grazing have been suppressed (see photos in Fig. 7.3). *Heterothalamus psiadioides, Baccharis dracunculifolia and B. uncinella* are some of the most abundant species with such a strategy in those areas (Oliveira 2003; Müller et al. 2007; Schüler-da-Silva 2009; Fidelis et al. 2010b; Santos et al. 2011); they are normally taller than the cogeneric resprouter species. On Morro Santana, *H. psiadioides* reached 10 ind./m^2 by seed germination after 1 year after a fire event (Müller 2005); before the fire event, density of *H. psiadioides* at the same survey plots had been only 0.53 ind./m^2. These species thus seem to be profit from after-fire conditions and high fire frequency. This dependence on postfire conditions possibly is a

Fig. 7.3 Examples of woody species established in grasslands of the Morro Santana forest-grassland ecotone, Porto Alegre, RS, Brazil. (**a**) Grassland area dominated by *Heterothalamus psiadioides*—a seeder shrub; (**b**) sprouts of *Baccharis cognata* after the burn; (**c**) individuals of *Myrciaria cuspidata* and *Maytenus casseniformis* at rocky outcrops; (**d**) seedlings of *Myrsine umbellata* established very close to the forest border

mechanism to avoid population senescence without new establishment (DeSimone and Zedler 2001). Shrub seedling density has been shown to be related to postfire conditions in chaparral and savanna-type vegetations, with some species depending on fire to persist in the community (Smith and Taiton 1985; Tyler 1995; DeSimone and Zedler 2001).

In addition to sexual reproduction and resprouting ability, vegetative reproduction from belowground structures is an important strategy in ecosystems under disturbance (Bond and van Wilgen 1996; Hoffmann 1998; Bellingham and Sparrow 2000; Drewa et al. 2002). In southern Brazil, no studies assessing the importance of vegetative reproduction on grassland shrub species neither on forest species that colonize grasslands have been conducted so far. There is only one study that identifies bud bank features along a gradient of grassland disturbance, indicating different types of underground structures and how many buds were present on them (Fidelis 2008), but the main focus of this study was on grasses and forbs, not on woody species (shrubs or forest trees). Field observations indicate that some species, like *M. coriaceae*, one of the most abundant forest tree species in excluded grassland areas nearby forests (Müller 2005; Schüler-da-Silva 2009), can show clonal growth by producing root suckers. The importance of undergrowth structures to resprouting ability and/or vegetative reproduction should be high for woody species in forest-grassland ecotones, as shown for Cerrado species (Hoffmann 1998), but there is no sufficient information for the species of southern Brazilian Campos. A continuum of responses concerning seeder and resprouter species especially under conditions of less intense disturbances, which are often associated with spatial heterogeneity of both abiotic variables and vegetation structure, can be observed in different species groups and vegetation types worldwide (Lloret et al. 1999; Guo 2001; Vesk and Westoby 2004).

Plant functional traits that are used to predict the plant responses to different environmental conditions or the effects of the plants on ecosystem functions are frequently related to leaf and stem features, biomass, maximal plant height, and seed mass (Westoby 1998; Díaz et al. 1999; Weiher et al. 1999; Westoby et al. 2002; Cornelissen et al. 2003). In the study on Morro Santana, in addition to the already mentioned resprouting ability, the traits habit (maximal species height), mean plant height, stem base (single or multistemmed), leaf area and leaf texture of woody plants were also found to be relevant for defining plant functional types associated to distance from the forest limit and time since fire (Müller et al. 2007). Such traits were measured in homogeneous populations according to plant phase (classes of height) and stem base within the same species. This approach helps revealing consistent vegetation patterns, since trait-stage shifts within a species from seedlings to adults (Cornelissen et al. 2003), and plants of the same species may have different responses to environmental factors when growing in different habitats (Bellingham and Sparrow 2000; Vesk and Westoby 2004). For example, variability of crown architecture has been considered a response to disturbance in a hierarchical framework along a continuum of disturbance severity or site productivity (e.g., moisture and/or fertility) (Bellingham and Sparrow 2000). Forest species growing as a monopodial tree on moist/fertile habitat (such as inside the forest), can present a multistemmed architecture if under adverse site conditions or periodic disturbance (such as in grasslands), as observed for species in Morro Santana hill (Müller 2005).

7.6 Interactions Between Fire and Climate Change on the Patterns of Woody Species in Forest-Grassland Ecotones

Present structural patterns of woody species in the grassland vegetation of southern Brazilian forest-grassland ecotones, in areas excluded from grazing, seem to be defined by current fire regime and spatial heterogeneity (distance from forest boundaries, aspect and presence of safe sites, e.g., rock outcrops) (Pillar 2003; Müller et al. 2007; Carlucci et al. 2011). Grassland shrubs species, either seeders or resprouters, predominate compared to forest tree species due to their successful strategies to survive or recruit despite fire events and to their fast life cycle. Additionally, many adult forest species respond to fire events with resprouting ability as well, but for the initial establishment phase they seem to be directly limited to safe sites (less flammable areas) or, between fire-intervals, those individuals that are able to germinate and grow under grassland conditions need to growth higher than the lethal zone (grass stratum) to survive recurrent burns by resprouting ability (Bond and van Wilgen 1996; Hoffmann et al. 2003). Most of the forest species can thus be considered as fire-sensitive species at least in their early development.

An increase in density of woody species by shrub encroachment and invasion of forest species has been observed worldwide (Archer 1990; Van Auken 2000; Roques et al. 2001; Cabral et al. 2003; Puyravaud et al. 2003), changing the vegetation physiognomy by vegetation shifts at boundaries and by increased presence of dense bush stands and woody thickets within the grassland matrix. These increases in woody vegetation have been related to climatic changes, that is to alterations in precipitation, temperature and CO_2 concentration, and may increase as the result of a complex trade-off with grass production and disturbance management (Bond and Midgley 2000; Roques et al. 2001; Sternberg 2001; Langevelde et al. 2003).

For southern Brazil, detected trends of increasing temperature (Marengo and Camargo 2008) and rainfall (Haylock et al. 2006) over the last decades (1960–2000, 2002) have been reported. The IPCC (2007) SRES scenarios project atmospheric CO_2 levels between 540 and 970 ppm (parts per million) in 2100, depending on the scenario and model used (and with some uncertainties), which will enhance even more the ongoing increases in local temperature and rainfall in this region. In a comparison of results from different global climate models for southern Brazil, Marengo (2007) showed possible temperature rises between 2 and 3 °C for climate scenario B2 and between 2 and 5 °C for A1 (mean temperatures), as well as changes in precipitation ranging from –0.5 to +0.5 mm/day, respectively, compared to 1961–1990 values. Local predictions from a bidimensional Dynamic Global Vegetation Model (2D-aDGVM), considering projected climate change from the IPCC (2007) A1B emission scenario (mean changes in temperature and precipitation from 2000 to 2100: mean annual temperature +1.8°; mean precipitation +300 mm/year), showed that potential vegetation growth (increments on tree height and grass biomass with no grass–tree interactions) and fecundity may be enhanced by elevated temperature and atmospheric CO_2 levels from preindustrial to future climate scenarios, and irrespective of plant type physiology (C_3-tree and C_4-grass) (Fig. 7.4) (Blanco 2011). However, according to the model, when considering grass–tree interactions, the predicted advantage on the rate of stem growth of C_3-trees compared to the increment in C_4-grass fine fuel biomass caused a reduction in the proportion of trees killed by fire in burned cells under rising CO_2 levels (not shown), and hence a slight increase in the rate of forest expansion, even under frequent fires, mainly for the next decades (Blanco 2011). This suggests that the photosynthetic advantage of C_3-trees over C_4-grasses in the presence of fire under the fertilization effect of rising CO_2 levels, which has been proposed as an explanation of shrub encroachment phenomena in grasslands and savannas worldwide (Bond and Midgley 2000), could also potentially affect the relative stability of forest-grassland boundaries observed until present and maintained under frequent fires, leading to a gradual advance of forest over grassland.

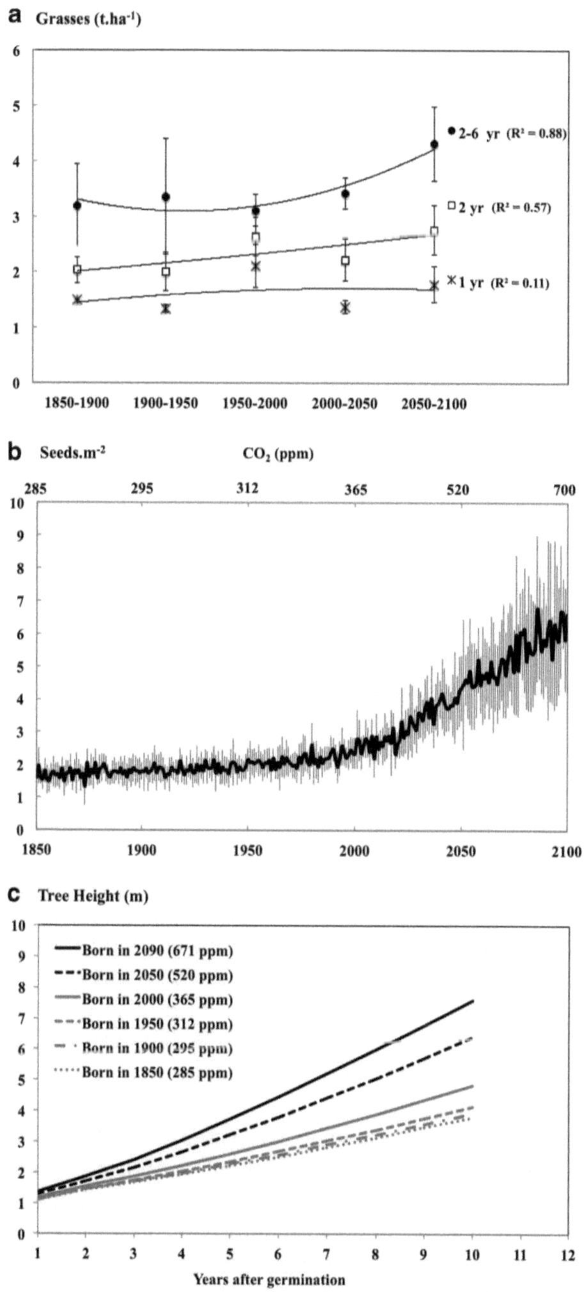

Fig. 7.4 Potential values (no grass–tree interactions) of total grass production (aboveground and belowground biomass) relative to different fire return intervals (**a**), seed production in a forest stand in the absence of fire (**b**) and tree height increment of isolated trees in the absence of fire (**c**) under changing climatic conditions in the Morro Santana site (30 °04′32″S; 51 °06′05″W, Porto Alegre, southern Brazil), using IPCC SRES A1B projected climate data (see text). Depicted values are the mean and the standard deviations after ten replicates of each simulation case(random initializing seed varying from 1 to 10). In (**c**), standard deviations are not shown because the values were <0.001 (data from Blanco 2011)

7.7 Forest-Grassland Ecotones in Southern Brazil: Overview and Perspectives

In the past paragraphs, we discussed which factors are responsible for the maintenance of relatively stable forest-grassland boundaries in areas under the influence of fire in grasslands and which plant traits may be important for woody species establish into grasslands. The available data explain why grasslands have persisted in the region until now, under climatic conditions increasingly favorable for development of forest vegetation, and why forest expansion over grassland is slow: forest species are not able to become established in grassland under frequent fires, and grass biomass promotes frequent fires in a positive feedback loop. Under intensive disturbances, forest species can establish successfully only at safe sites where they are protected from fire. On the other hand, stochastic events related to climate, population level processes, and spatial variability of site conditions may cause differences in spatio-temporal patterns of woody species distribution and density, both in patches or scattered in the grasslands. In general terms, the situation in southern Brazil thus corresponds to the increasing shrub encroachment in grasslands throughout the world in consequence to climatic shifts and changes in management. According to recent models, future climate change is expected to lead to a slight increase in velocity of forest expansion, even under fire.

The data that forms the basis of the studies discussed in this chapter comes from forest-grassland ecotones where grasslands are not used for grazing, mostly situated in conservation units that form ideal laboratories for ecological studies and experimentation. However, both grasslands under fire but without grazing, as on Morro Santana, and grasslands without any disturbance (both fire and grazing), as in the Pró-Mata and Aracuri reserves, are the exception rather than the rule in southern Brazil: the greatest part of the grassland areas is under continuous grazing. Grazing management and grazing intensity may differ greatly—and this will have consequences on possible forest-expansion processes: under intensive grazing (high pressure), survival chances of trees germinating in the grassland should be extremely low due to herbivory and trampling. Under a less intensive grazing, on the other hand, chances for survival of tree individuals might be higher, as grassland heterogeneity will increase, possibly allowing for recruitment of shrubs and trees and formation of forest nuclei, leading thus to an advancement of forests. Traits that influence these processes will differ from ecosystems under fire: while resprouting capacity should be important as well, defense structures against herbivory also should be of high importance, allowing also for facilitation processes. These hypotheses, however, have not been tested so far in southern Brazil. Likewise, the effect of grazing animals entering into forests at the forest-grassland boundary on the structure of forest and forest border has not been studied so far.

At any rate, the discussion of effects of fire and grazing on ecosystem dynamics show how many natural processes at the forest-grassland interface in southern Brazil are marked by human action. Human induced disturbance apparently has maintained grassland-forest ecotones over the past millennia, and will continue to do so even under conditions of (anthropogenic) climate change. The study of dynamics and mechanisms of forest-grassland ecotones in southern Brazil thus continues to be an interesting field of research.

References

Adelmann W, Overbeck G (2009) Anthropogene Einflüsse auf die Dynamik artenreicher Wald-Graslandmosaike in Porto Alegre, Südbrasilien. Laufener Spezialbeiträge 2:175–180

Adelmann W, Overbeck G, Zellhuber A, Müller SC, Forneck ED, Drösler M, Pfadenhauer J, Pillar VD, Porto ML (2011) Supporting the ecosystem approach within urban development planning of Porto Alegre, Brazil. In: Paulsch A, Paulsch C (eds) Das Übereinkommen über die Biologische Vielfalt (CBD)—ein Einstieg für Wissenschaftler. ibn - Institut für Biodiversität Netzwerk e.V., Regensburg, pp 113–133

Anderson JM, Anderson HM, Archangelsky S, Bamford M, Chandra S, Dettman M, Hill R, McLoughlin S, Rossler O (1999) Patterns of Gondwana plant colonization and diversification. J Afr Earth Sci 28:145–167

Archer S (1990) Development and stability of grass/woody mosaics in a subtropical savanna parkland, Texas, USA. J Biogeogr 17:453–462

Behling H (2002) South and southeast Brazilian grasslands during Late Quaternary times: a synthesis. Palaeogeogr Palaeoclimatol Palaeoecol 177:19–27

Behling H, Pillar VD, Orlóci L, Bauermann SG (2004) Late Quaternary *Araucaria* forest, grassland (Campos), fire and climate dynamics, studied by high-resolution pollen, charcoal and multivariate analysis of the Cambará do Sul core in southern Brazil. Palaeogeogr Palaeoclimatol Palaeoecol 203:277–297

Behling H, Pillar VD, Orlóci L, Bauermann SG (2005) Late Quaternary grassland (Campos), gallery forest, fire and climate dynamics, studied by pollen, charcoal and multivariate analysis of the São Francisco de Assis core in western Rio Grande do Sul (southern Brazil). Rev Palaeobot Palynol 133:235–248

Behling H, Pillar VD, Müller SC, Overbeck GE (2007) Late-Holocene fire history in a forest-grassland mosaic in southern Brazil: implications for conservation. Appl Veg Sci 10:81–90

Bellingham PJ, Sparrow AD (2000) Resprouting as a life history strategy in woody plant communities. Oikos 89:409–416

Blanco CC (2011) Modelo de simulação da dinâmica de vegetação em paisagens de coexistência campo-floresta no sul do Brasil. Doctor, Programa de Pós-Graduação em Ecologia, Universidade Federal do Rio Grande do Sul, Porto Alegre

Boldrini II (2009) A flora dos campos do Rio Grande do Sul. In: Pillar VD, Müller SC, Castilhos ZMS, Jacques AVA (eds) Campos Sulinos: Conservação e uso sustentável da biodiversidade. Ministério do Meio Ambiente, Brasília, pp 63–77

Boldrini II, Eggers L (1996) Vegetação campestre do sul do Brasil: resposta e dinâmica de espécies à exclusão. Acta Bot Bras 10:37–50

Bond WJ, Midgley GF (2000) A proposed CO_2-controlled mechanism of woody plant invasion in grassland and savannas. Glob Chang Biol 6:865–869

Bond WJ, Midgley GF (2001) Ecology of sprouting in woody plants: the persistence niche. Trends Ecol Evol 16:45–51

Bond WJ, van Wilgen BW (1996) Fire and plants, 14th edn, Population and community biology series. Chapman & Hall, London

Box E (1986) Some climatic relationships of the vegetation of Argentina, in global perspective. Veröff. Geobot. Inst. Stift Rübel Zür 91:181–216

Cabral AC, De Miguel JM, Rescia AJ, Schmitz MF, Pineda FD (2003) Shrub encroachment in Argentinean savannas. J Veg Sci 14:145–152

Cabrera AL, Willink A (1973) Biogeografia de America Latina. Secretaria General de la Organización de los Estados Americanos, Washington, DC

Carlucci MB, Duarte LdS, Pillar VD (2011) Nurse rocks influence forest expansion over native grassland in southern Brazil. J Veg Sci 22:111–119

Clarke PJ (2002a) Habitat insularity and fire response traits: evidence from a sclerophyll archipelago. Oecologia 132:582–591

Clarke PJ (2002b) Habitat islands in fire-prone vegetation: do landscape features influence community composition? J Biogeogr 29:677–684

Cordeiro JLP, Hasenack H (2009) Cobertura vegetal atual do Rio Grande do Sul. In: Pillar VD, Müller SC, Castilhos ZMS, Jacques AVA (eds) Campos Sulinos: Conservação e Uso Sustentável da Biodiversidade. Ministério do Meio Ambiente, Brasília, pp 285–299

Cornelissen JHC, Cerabolini B, Castro-Díey P, Villar-Salvador P, Montserrat-Martí G, Puyravaud JP, Maestro M, Werger MJA, Aerts R (2003) Functional traits of woody plants: correspondence of species rankings between field adults and laboratory-grown seedlings? J Veg Sci 14:311–322

Cruz Ruggiero PG, Batalha MA, Pivello VR, Meirelles ST (2002) Soil-vegetation relationships in cerrado (Brazilian savanna) and semideciduous forest, southeastern Brazil. Plant Ecol 160:1–16

Dadalt LP (2007) Composição e riqueza de espécies em manchas de vegetação lenhosas na Serra do Sudeste, RS, Brasil. Bacharel Ciências Biológicas, Universidade Federal do Rio Grande do Sul, Porto Alegre

Dadalt LP (2010) Padrões de diversidade da vegetação lenhosa da região do Alto Camaquã, Rio Grande do Sul, Brasil. Mestrado, PPG-Ecologia, Universidade Federal do Rio Grande do Sul, Porto Alegre

DeSimone SA, Zedler PH (2001) Do shrub colonizers of southern Californian grassland fit generalities for other woody colonizers? Ecol Appl 11:1101–1111

Díaz S, Cabido M, Zak M, Carretero EM, Araníbar J (1999) Plant functional traits, ecosystem structure and land-use history along a climatic gradient in central-western Argentina. J Veg Sci 10:651–660

Drewa PB, Platt WJ, Moser EB (2002) Fire effects on resprouting of shrubs in headwaters of southeastern longleaf pine savannas. Ecology 83:755–767

Duarte LS, Machado RE, Hartz SM, Pillar VD (2006a) What saplings can tell us about forest expansion over natural grasslands. J Veg Sci 17:799–808

Duarte LS, Santos MMG, Hartz SM, Pillar VD (2006b) The role of nurse plants in *Araucaria* forest expansion over grassland in South Brazil. Austral Ecol 31:520–528

Duemig A, Schad P, Rumpel C, Dignac M-F, Kögel-Knabner I (2008) *Araucaria* forest expansion on grassland in the southern Brazilian highlands as revealed by ^{14}C and $\delta^{13}C$ studies. Geoderma. doi:10.1016/j.geoderma.2007.06.005

Eiten G (1972) The cerrado vegetation of Brazil. Bot Rev 38:201–341

Eriksen W (1978) Ist das Pampaproblem gelöst? Naturwiss Rundsch 31:142–148

Esther A, Groeneveld J, Enright NJ, Miller BP, Lamont BB, Perry GLW, Blank FB, Jeltsch F (2010) Sensitivity of plant functional types to climate change: classification tree analysis of a simulation model. J Veg Sci 21:447–461

Fidelis A (2008) Fire in subtropical grassland in southern Brazil: effects on plant strategies and vegetation dynamics. Dissertation (Dr. rer. nat.), Wissenschaftszentrum Weihenstephan, Technische Universität München, Freising

Fidelis A, Delgado-Cartay MD, Blanco CC, Müller SC, Pillar VD, Pfadenhauer J (2010a) Fire intensity and severity in Brazilian Campos grasslands. Interciencia 35:739–745

Fidelis A, Müller SC, Pillar VD, Pfadenhauer J (2010b) Population biology and regeneration of forbs and shrubs after fire in Brazilian Campos grasslands. Plant Ecol 211:107–116

Furley PA (1999) The nature and diversity of neotropical savanna vegetation with particular reference to the Brazilian cerrados. Glob Ecol Biogeogr Lett 8:223–241

Garcia Martinez P (2005) Caracterización química y física de los suelos del Morro Santana (Porto Alegre, Rio Grande do Sul, Brasil). Chair of Soil Sciences, Technische Universität München, Freising

Gautreau P (2010) Rethinking the dynamics of woody vegetation in Uruguayan Campos, 1800–2000. J Hist Geogr 36:194–204
Gautreau P, Lezama F (2009) Clasificación florística de los bosques y arbustales de las sierras del Uruguay. Ecol Austral 19:81–92
Geiger EL, Gotsch SG, Damasco G, Haridasan M, Franco AC, Hoffmann WA (2011) Distinct roles of savanna and forest tree species in regeneration under fire suppression in a Brazilian savanna. J Veg Sci 22:312–321
Guillet B, Achoundong G, Happi JY, Beyalas VKK, Bonvallot J, Riera B, Mariotti A, Schwartz D (2001) Agreement between floristic and soil organic carbon isotope ($^{13}C/^{12}C$, ^{14}C) indicators of forest invasion of savannas during the last century in Cameroon. J Trop Ecol 17:809–832
Guo Q (2001) Early post-fire succession in California chaparral: changes in diversity, density, cover and biomass. Ecol Res 16:471–485
Haylock MR, Peterson TC, Alves LM, Ambrizzi T, Anunciação YMT, Baez J, Barros VR, Berlato MA, Bidegain M, Coronel G, Corradi V, Garcia VJ, Grimm AM, Karoly D, Marengo JA, Marino MB, Moncunill DF, Nechet D, Quintana J, Rebello E, Rusticucci M, Santos JL, Trebejo I, Vincent LA (2006) Trends in total and extreme South American rainfall in 1960–2000 and links with sea surface temperature. J Climate 19:1490–1512
Hermann J-M (2007) Pioneer woody species in southern Brazilian grasslands: life-history traits and population dynamics. Dissertation (Dr. rer. nat.), Wissenschaftszentrum Weihenstephan, Technische Universität München, Freising
Higgins SI, Bond WJ, Trollope WSW (2000) Fire, resprouting and variability: a recipe for grass–tree coexistence in savanna. Journal of Ecology 88: 213–229.
Hochberg ME, Menaut JC, Gignoux J (1994) The influences of tree biology and fire in the spatial structure of the West African savannah. J Ecol 82:217–226
Hoffmann WA (1996) The effects of fire and cover on seedling establishment in a neotropical savanna. J Ecol 84:383–393
Hoffmann WA (1998) Post-burn reproduction of woody plants in a neotropical savanna: the relative importance of sexual and vegetative reproduction. J Appl Ecol 35:422–433
Hoffmann WA, Moreira AG (2002) The role of fire in population dynamics of woody plants. In: Oliveira PS, Marquis RU (eds) The cerrados of Brazil. Ecology and natural history of a neotropical savanna. Columbia University Press, New York, pp 159–177
Hoffmann WA, Orthen B, Nascimento PKVd (2003) Comparative fire ecology of tropical savanna and forest trees. Funct Ecol 17:720–726
Holl KD (2002) Effect of shrubs on tree seedling establishment in an abandoned tropical pasture. J Ecol 90:179–187
Hueck K (1966) Die Wälder Südamerikas. Fischer, Stuttgart
IBGE (2004) Mapa da vegetação do Brasil e Mapa de biomas do Brasil. IBGE, Río de Janeiro
IPCC (2007) Climate change 2007: the physical science basis. Contribution of Working Group I to the Fourth Assessment Report of the Intergovernmental Panel on Climate Change. Cambridge University Press, Cambridge
Kauffman JB, Cummings DL, Ward DE (1994) Relationships of fire, biomass and nutrient dynamics along a vegetation gradient in the Brazilian cerrado. J Ecol 82:519–531
Kern A (1997) Paleopaisagens e povoamento pré-histórico do Rio Grande do Sul. In: Kern A, Jacobus A, Ribeiro PM, Copé S, Schmitz PI, Naue G, Becker IB (eds) Arqueologia Pré-histórica do RS (2nd ed). Mercado Aberto Porto Alegre, pp 13–61
Klein RM (1975) Southern Brazilian phytogeographic features and the probable influence of Upper Quaternary climate changes in the floristic distribution. Boletim Paranaense de Geociências 33:67–88
Knoop WT, Walker BH (1985) Interactions of woody and herbaceous vegetation in a southern African savanna. J Ecol 73:235–253
Kuintcher A, Buriol GA (2001) Clima do estado do Rio Grande do Sul segunda a classifiação climática de Köppen e Thornthwaite. Discip. Sci., Ser. Cienc. Exatas 2:171–182

Langevelde FV, Van de Vijver CADM, Kumar L, Koppel Jvd, Ridder Nd, Andel JV, Skidmore AK, Hearne JW, Stroosnijder L, Bond WJ, Prins HHT, Rietkerk M (2003) Effects of fire and herbivory on the stability of savanna ecosystems. Ecology 84:337–350

Leite PF, Klein RM (1990) Vegetação. In: IBGE (ed) Geografia do Brasil: Região Sul. Instituto Brasileiro de Geografia e Estatística, Rio de Janeiro, pp 113–150

Li X, Wilson SD (1998) Facilitation among woody plants establishing in an old field. Ecology 79:2694–2705

Lindman CAM (1906) A Vegetação no Rio Grande do Sul. EDUSP/Itatiaia, São Paulo/Belo Horizonte

Lloret F, Verdu M, Flores-Hernandez N (1999) Fire and resprouting in Mediterranean ecosystems: insights from an external biogeographical region, the mexical shrubland. Am J Bot 86: 1655–1661

Marengo JA (2007) Cenários de mudanças climáticas para o Brasil em 2100. Ciênc Ambiente 34:97–114

Marengo JA, Camargo CC (2008) Surface air temperature trends in southern Brazil for 1960–2002. Int J Climatol 28:893–904

Martin L, Fournier M, Mourguiart P, Siefeddine A, Turcq B (1993) Southern oscillation signal in South American palaeoclimatic data of the last 7000 years. Quat Res 39:338–346

Müller SC (2005) Padrões de espécies e tipos funcionais de Plantas lenhosas em bordas de floresta e Campo sob influência do fogo. Tese de Doutorado, Programa de Pós-Graduação em Ecologia, Universidade Federal do Rio Grande do Sul, Porto Alegre

Müller SC, Forneck ED (2004) Forest-grassland mosaics in the hills of Porto Alegre city: a study case of forest expansion patterns in Santana hill, Rio Grande do Sul, Brazil. In: MLP (ed) Workshop—Proteção e manejo da vegetação natural da região de Porto Alegre, com base em pesquisas de padrões e dinâmica da vegetação. Programa de Pós-Graduação em Ecologia/ UFRGS, Porto Alegre, pp 29–37

Müller SC, Overbeck GE, Pillar VD, Pfadenhauer J (2003) Comunidades de espécies lenhosas sob a ação do fogo em vegetação campestre no morro Santana, Porto Alegre, RS. In: Claudino-Sales V, Tonini IM, Correia Dantas EW (eds) VI Congresso de Ecologia do Brasil: Anais de trabalhos completos. Editora da Universidade Federal do Ceará, Fortaleza, pp 590–591

Müller SC, Overbeck GE, Pfadenhauer J, Pillar VD (2007) Plant functional types of woody species related to fire disturbance in forest-grassland ecotones. Plant Ecol 189:1–14

Nimer E (1990) Clima. In: IBGE (ed) Geografia do Brasil: Região Sul. Instituto Brasileiro de Geografia e estatística, Rio de Janeiro, pp 151–187

Oliveira JM (2003) Padrões e processos espaço-temporais em ecótonos de campos e floresta com Araucária em São Francisco de Paula, RS. Programa de Pós-graduação em Ecologia, Universidade Federal do Rio Grande do Sul, Porto Alegre

Oliveira JM, Pillar VD (2004) Vegetation dynamics on mosaics of Campos and *Araucaria* forest between 1974 and 1999 in southern Brazil. Community Ecol 5:197–202

Oliveira-Filho AT, Fontes MAL (2000) Patterns of floristic differentiation among Atlantic forests in southeastern Brazil and the influence of climate. Biotropica 32:793–810

Oliveira-Filho AT, Ratter JA (2002) Vegetation physionomies and woody flora of the Cerrado biome. In: Oliveira PS, Marquis RU (eds) The cerrados of Brazil. Ecology and natural history of a neotropical savanna. Columbia University Press, New York, pp 91–120

Overbeck GE (2005) Effect of fire on vegetation dynamics and plant types in subtropical grassland in southern Brazil. Dissertation (Dr. rer. nat.), Wissenschaftszentrum Weihenstephan, Technische Universität München, Freising

Overbeck GE, Müller SC, Pillar VD, Pfadenhauer J (2005) Fine-scale post-fire dynamics in southern Brazilian subtropical grassland. J Veg Sci 16:655–664

Overbeck GE, Mueller SC, Fidelis A, Pfadenhauer J, Pillar VD, Blanco CC, Boldrini II, Both R, Forneck ED (2007) Brazil's neglected biome: The South Brazilian Campos. Perspect Plant Ecol Evol Syst 9:101–116

Pausas JG (1999) Response of plant functional types to changes in the fire regime in Mediterranean ecosystems: a simulation approach. J Veg Sci 10:717–722

Pausas JG, Bradstock RA, Keith DA, Keeley JE; GCTE Fire Network (2004) Plant functional traits in relation to fire in crown-fire ecosystems. Ecology 85:1085–1100
Pillar VD (2003) Dinâmica de expansão florestal em mosaicos de floresta e campos no sul do Brasil. In: Claudino-Sales V (ed) Ecossistemas brasileiros: manejo e conservação. Expressão Gráfica e Editora, Fortaleza, pp 209–216
Pillar VD, Quadros FLF (1997) Grassland-forest boundaries in southern Brazil. Coenoses 12:119–126
Puyravaud J-P, Dufour C, Aravajy S (2003) Rain forest expansion mediated by successional processes in vegetation thickets in the Western Ghats of India. J Biogeogr 30:1067–1080
Rambo B (1956) A fisionomia do Rio Grande do Sul. Selbach, Porto Alegre
Rambo B (1957) Regenwald und kamp in Rio Grande do Sul. Sellowia 8:257–298
Rambo B (1961) Migration routes of the South Brazilian rain forest. Pesqui Bot 5:1–54
Roques KG, O'Connor TG, Watkinson AR (2001) Dynamics of shrub encroachment in an African savanna: relative influences of fire, herbivory, rainfall and density dependence. J Appl Ecol 38:268–280
Santos MMG, Oliveira JM, Müller SC, Pillar VD (2011) Chuva de sementes de espécies lenhosas florestais em mosaicos de floresta com Araucária e campos no Sul do Brasil. Acta Bot Bras 25:160–167
Scheiter S, Higgins SI (2009) Impacts of climate change on the vegetation of Africa: an adaptive dynamic vegetation modelling approach. Glo Change Biol 15: 2224–2246
Schüle W (1990) Landscapes and climate in prehistory: interactions of wildlife man and fire. In: Goldammer JG (ed) Fire in the Tropical Biota. Springer-Verlag Berlin, pp 273–318
Schüler-da-Silva A (2009) Dinâmica espaço-temporal de *Araucaria angustifolia* (Bertol.) Kuntze em campos no sul do Brasil. Mestrado, Programa de Pós-Graduação em Ecologia, Universidade Federal do Rio Grande do Sul, Porto Alegre
Silva LCR, Anand M (2011) Mechanisms of *Araucaria* (Atlantic) forest expansion in southern Brazil grasslands. Ecosystems 14:1354–1371
Smith FR, Taiton NM (1985) Effects of season of burn on shrub survival, regeneration and structure in the Natal Drakensberg. J Grassl Soc S Afr 2:4–10
Sternberg LDSL (2001) Savanna-forest hysteresis in the tropics. Glob Ecol Biogeogr Lett 10:369–378
Teixeira MB, Coura-Neto AB, Pastore U, Rangel Filho ALR (1986) Vegetação. In: IBGE (ed) Levantamento de recursos naturais: Folha SH.22 Porto Alegre e parte das Folhas SH.21 Uruguaiana e SI.22 Lagoa Mirim. IBGE, Rio de Janeiro, pp 541–632
Tyler CM (1995) Factors contributing ot postfire seedling establishment chaparral: direct and indirect effects of fire. J Ecol 83:1009–1020
Van Auken OW (2000) Shrub invasions of North American semiarid grasslands. Annu Rev Ecol Syst 31:197–215
Vesk PA, Westoby M (2004) Sprouting ability across diverse disturbances and vegetation types worldwide. J Ecol 92:310–320
Waechter JL (2002) Padrões geográficos na flora atual do Rio Grande do Sul. Ciênc Ambiente 24:93–108
Walter H (1967) Das Pampaproblem in vergleichend ökologischer Betrachtung und seine Lösung. Erdkunde 21:181–202
Walter H (1971) Ecology of tropical and subtropical vegetation. Oliver & Boyd, Edinburgh
Weiher E, van der Werf A, Thompson K, Roderick M, Garnier E, Eriksson O (1999) Challenging Theophrastus: a common core list of plant traits for functional ecology. J Veg Sci 10:609–620
Westoby M (1998) A leaf-height-seed (LHS) plant ecology strategy scheme. Plant Soil 199:213–227
Westoby M, Falster DS, Moles AT, Vesk PA, Wright IJ (2002) Plant ecological strategies: some leading dimensions of variation between species. Annu Rev Ecol Syst 33:125–159

Chapter 8
Long-Term Perspectives on Tropical Forest–Savanna Dynamics in Lowland Bolivia from the Last Ice Age Until Present

Francis E. Mayle and Bronwen S. Whitney

8.1 Introduction

Tropical forest–savanna ecotones are of considerable interest to biologists due to their high habitat heterogeneity (*beta* diversity), importance in rainforest speciation (Smith et al. 1997), and, for those climatically controlled ecotones, the manner of their response to future climate change (IPCC 2007). Forest–savanna transitions may be governed by any number of factors, including soil type, fire, hydrology, disturbance (both natural and anthropogenic), as well as climate, in particular the seasonality of precipitation (Furley et al. 1992). Determining the underlying control for a forest–savanna transition in a particular location may therefore not be straightforward and may differ markedly among forest–savanna ecotones, even when in relatively close geographic proximity.

A palaeoecological approach that investigates the long-term history of forest–savanna dynamics spanning centuries and millennia has the potential to reveal the key underlying driver(s) of such vegetation shifts that are not resolvable from ecological studies alone, which span years to decades. For example, palaeoecological data can demonstrate whether a forest–savanna ecotonal shift observed over recent decades merely constitutes a minor short-term oscillation about an essentially stable ecotone, or instead is part of a much longer term (e.g. millennial-scale) unidirectional ecotonal shift forced by climate change.

A consideration of archaeological data, in particular pre-Columbian (before 1492) earthworks in seasonally flooded landscapes of tropical South America, can shed light not only on whether the spatial patterns of forest–savanna mosaics are hydrologically controlled (i.e. depth and duration of flooding) but also on whether these changes in hydrology are natural or instead a function of large-scale landscape

F.E. Mayle (✉) • B.S. Whitney
School of GeoSciences, University of Edinburgh, Drummond Street, Edinburgh EH89XP, UK
e-mail: Francis.Mayle@ed.ac.uk

R.W. Myster (ed.), *Ecotones Between Forest and Grassland*,
DOI 10.1007/978-1-4614-3797-0_8, © Springer Science+Business Media New York 2012

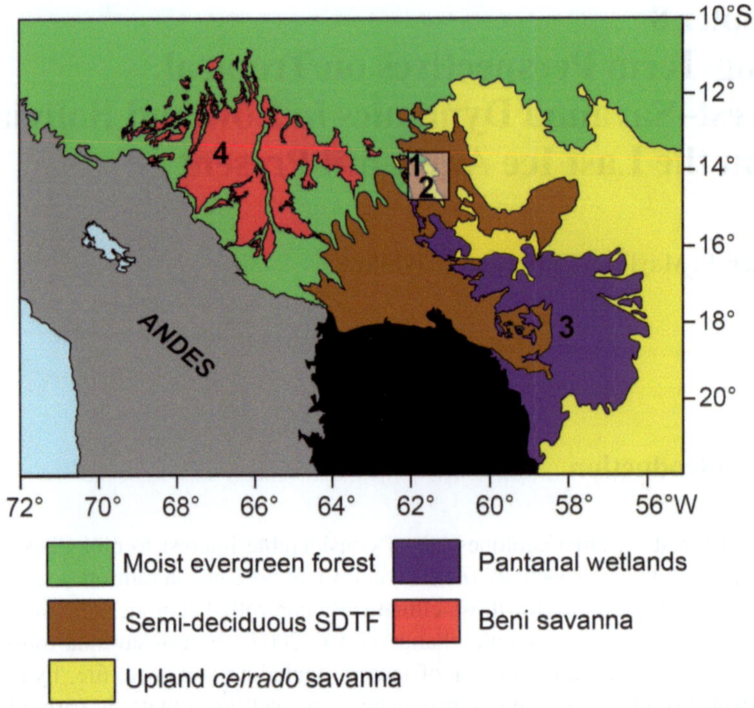

Fig. 8.1 Location of case studies of four types of forest–savanna boundaries discussed in this chapter. Key ecosystem classes of lowland Bolivia and western Brazil are based on data from Olson et al. (2001). Detail of *inset box* is presented in Fig. 8.2

modification by pre-Columbian earthmoving cultures (e.g. via construction of raised fields, canals, and large habitation mounds; e.g. Erickson 2010; Lombardo et al. 2011).

8.2 Aims and Approach

This chapter examines the long-term histories of four contrasting forest–savanna environments in lowland Bolivia (Fig. 8.1), chosen to allow the respective roles of glacial–Holocene climate change, geomorphology, and pre-Columbian (before 1492) land-use to be investigated in terms of shaping the current forest–savanna mosaic in SW Amazonia and the Pantanal savanna wetlands. The four types of Bolivian tropical forest–savanna transitions/ecotones examined are (1) humid rainforest—savanna wetland (type 1 ecotone), (2) humid rainforest—upland (*cerrado*) savanna (type 2 ecotone), (3) seasonally dry tropical forest (SDTF)—savanna wetland (type 3 ecotone), and (4) savanna wetland—forest island mosaic (type 4 ecotone). The long-term pattern and underlying drivers of vegetation dynamics for each of these forest–savanna transitions will be considered by reference to previously published palaeoenvironmental and archaeological data.

8.3 Noel Kempff Mercado National Park

Noel Kempff Mercado National Park (NKMNP) is a 15,230 km^2 biological reserve that straddles forest–savanna ecotones in NE Bolivia on the Precambrian Shield near the SW margin of Amazonia, adjacent to the Brazilian States of Rondônia and Mato Grosso. It was designated a UNESCO World Heritage Site in 2000 due to the diversity of intact tropical forest and savanna ecosystems that it contains and consequent high habitat heterogeneity or *beta* diversity. The lowland, western half of the park, is dominated by *terra firme* (non-flooded), riparian, and seasonally flooded humid evergreen forests, which have sharp boundaries with expanses of seasonally flooded savanna wetlands (type 1 ecotone). Toward the southern boundary of the park lies the ecotone between humid Amazonian rainforest and SDTF (semi-deciduous Chiquitano Dry Forest). The eastern half of the park comprises the Huanchaca plateau, a table mountain between 600 and 900 m above sea level (a.s.l.), composed of Precambrian sandstone and quartzite (Litherland and Power 1989). This plateau is blanketed by a mosaic of humid evergreen rainforest and *cerrado* (upland) savanna (type 2 ecotone). Full details of the ecosystems within this park are given in Killeen and Schulenberg (1998).

The gradual ecotonal transition—spanning tens of kilometres—between the southernmost limit of humid evergreen Amazonian rainforest and semi-deciduous dry forest across the southern border of NKMNP (Fig. 8.2) coincides with a strong rainfall gradient, pointing to climatic control, with mean annual precipitation approximately 1,500 mm per annum (pa) and a 4 months dry season between June and September when mean monthly precipitation is less than 30 mm. Precipitation falls mainly in the austral summer, originating from a combination of deep-cell convective activity over the Amazon Basin and southerly extension of the Intertropical Convergence Zone (ITCZ) due to peak insolation in the austral summer (Bush and Silman 2004). Although the mean annual temperature is 25–26 °C, temperatures frequently decrease to 10 °C for several days at a time during the dry season (June–July–August) when cold dry Patagonian air masses (*surazos*) reach the area.

However, the sharpness of the forest–savanna transitions within NKMNP point to non-climatic controls at these finer spatial scales, where forest–savanna mosaics are strongly influenced by edaphic factors, hydrology, and fire. The importance of edaphic factors is clearly evident on the Huanchaca plateau. Where the substrate is rocky, and soils very thin, open savanna (*campo cerrado*) predominates, but where soils are sufficiently deep, evergreen forest islands are found. In contrast, the adjacent lowland peneplain beneath the plateau to the west (200–250 m a.s.l.) is blanketed by thick Tertiary and Quaternary alluvial sediments, on which are growing humid evergreen rainforests. The patches of seasonally flooded savannas that

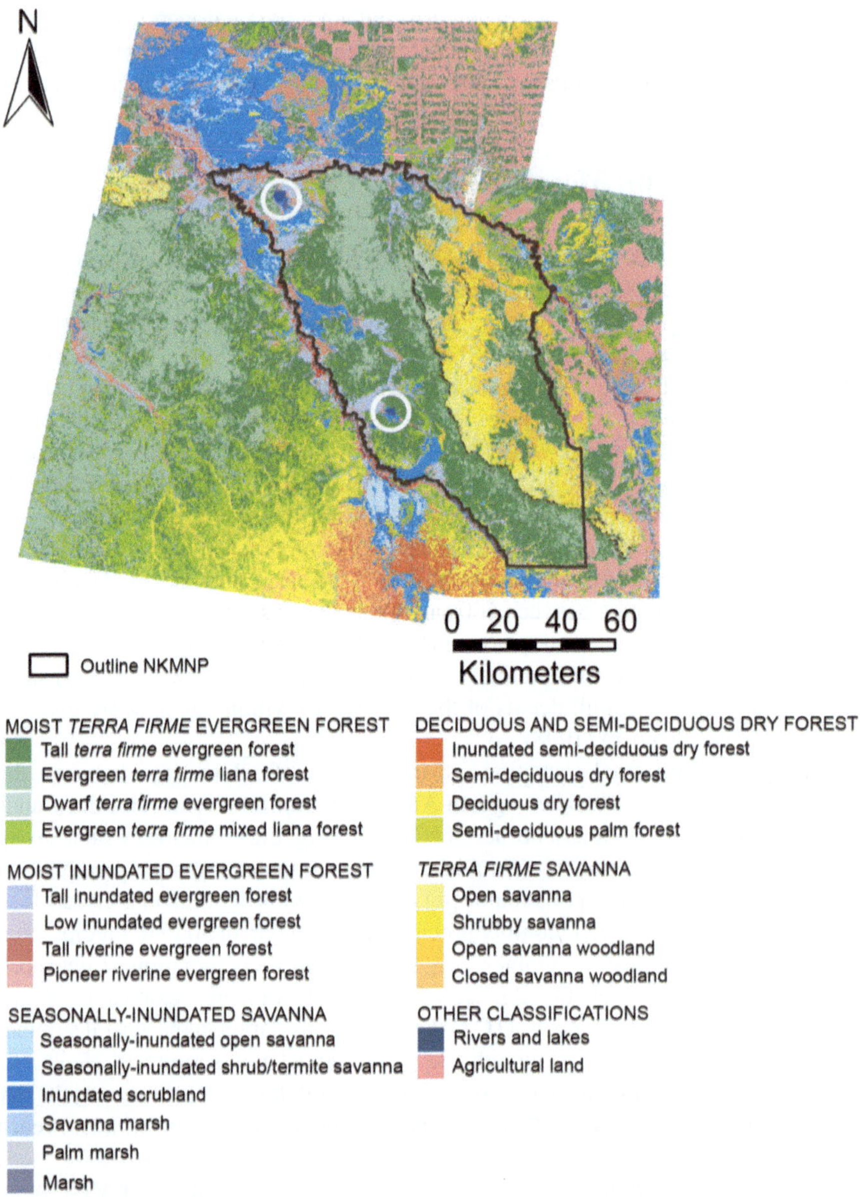

Fig. 8.2 Vegetation classification of Noel Kempff Mercado National Park (NKMNP) (derived from Landsat TM data) (modified from Killeen and Schulenberg 1998). Laguna Chaplin and Laguna Bella Vista are marked with *circles* (reproduced from Mayle et al. 2007)

interrupt this lowland rainforest are found primarily in association with the blackwater rivers—and likely owe their spatial arrangement to local edaphic conditions and micro-topography, as well as hydrological conditions associated with the river flood regime (Killeen and Schulenberg 1998).

In some areas of the park, rainforest is clearly encroaching into savanna, as old relict savanna trees can be found beyond the forest–savanna boundary among closed-canopy rainforest. However, in other areas, savanna appears to be encroaching into forest, with savanna fires creating sharp rainforest–savanna boundaries. Of interest to biologists is the predominant long-term process—savanna expansion into rainforest, or vice versa—and the underlying driver(s). An examination of palaeoecological data can potentially address these issues.

8.3.1 Glacial–Holocene History of Lowland Rainforest, Dry Forest, and Savanna Ecotones

Laguna Chaplin (14 °28′S, 61 °04′W) is a large flat-bottomed lake that lies below the Huanchaca plateau, surrounded by tall humid evergreen rainforest, approximately 30 km north of the ecotone with the semi-deciduous Chiquitano Dry Forest (Fig. 8.2). Full details of the fossil pollen and charcoal data from this site can be found in Burbridge et al. (2004) and Mayle et al. (2000). Unlike the rainforest-dominated catchment of today, over most of the past 50,000 years this lake was surrounded by a mosaic of semi-deciduous dry forest and savanna communities under climatic conditions drier than present. Throughout the glacial period, between approximately 50,000 and 11,400 year BP (calendar years before present), the dominance of Poaceae and Cyperaceae pollen, together with *Mauritia/Mauritiella* palm pollen, indicates that low-lying areas which are today dominated by seasonally flooded rainforest were instead covered by seasonally flooded termite savannas and savanna marsh. Consistent presence of *Paullinia/Roupala*, *Celtis*, *Machaerium*-type, and *Erythroxylum* pollen suggests that semi-deciduous dry forests may have occupied higher ground presently covered by *terra firme* evergreen rainforest. However, absence or negligible abundance of pollen of *Curatella americana* and *Anadenanthera*, which are important elements in present-day termite savannas and dry forests, respectively, shows that these glacial-age ecosystems were floristically distinct from those that occupy the region today. These Pleistocene (glacial-age) plant communities are consistent with reduced precipitation (e.g. longer or more severe dry season) and lowered atmospheric CO_2 levels, both of which would have favoured expansion of C_4 grasses and sedges and drought-adapted savanna and dry forest tree species. Continuous presence of pollen of the predominantly Andean genus *Podocarpus* throughout the glacial sediments corroborates other palaeoclimatic evidence (e.g. Stute et al. 1995) for an Amazon-wide cooling of approximately 5 °C at the Last Glacial Maximum (LGM; approximately 21,000 year BP).

The Holocene history (past 11,500 years) of forest–savanna dynamics of lowland NKMNP comes not only from the Laguna Chaplin record but also from pollen data from Laguna Bella Vista, a lake of similar size located approximately 100 km north of Chaplin and also surrounded by tall humid evergreen rainforest. Through the early-mid Holocene, the catchment of both sites was dominated by seasonally flooded savannas in the immediate vicinity of the lakes and around the floodplains of the neighbouring rivers and streams—habitats which are today occupied by seasonally flooded evergreen forest. Evidence for this comes from pollen assemblages dominated by Cyperaceae, Poaceae, *Mauritia/Mauritiella*, and *C. americana*. That the landscape was open at this time is corroborated by minimal pollen percentages of the wind-pollinated Moraceae family (<10 %), which dominates rainforest communities in the region today (Gosling et al. 2005). Well-drained (*terra firme*) areas around both lakes were covered by a mosaic of upland (*cerrado*) savannas (e.g. *C. americana*) and SDTF (*Anadenanthera*, *Astronium*, *Gallesia*). Presence of *C. americana*, *Anadenanthera*, and *Gallesia* in these Holocene assemblages suggests that, unlike those of the Pleistocene, these Holocene savanna and dry forest communities were floristically similar to those of today.

This mosaic of savanna and SDTF in the catchments of both Chaplin and Bella Vista in the early-mid Holocene (until approximately 6,800–3,200 year BP) is coincident with lower-than-present lake levels at Lake Titicaca (which lies at a similar latitude on the Altiplano of the Andes to the west) (Baker et al. 2001), providing strong evidence that these vegetation communities reflect lower mean annual precipitation (<1,500 mm) and a longer dry season (>5 months with <100 mm) than today. The gradual rise in Moraceae pollen percentages, beginning between 6,800 and 3,200 year BP at Bella Vista, and approximately 2,800 year BP at Chaplin, coinciding with a decline in relative abundance of savanna and SDTF taxa, reflects an expansion of humid evergreen rainforest communities into NKMNP (Burbridge et al. 2004; Burn et al. 2010). The timing of this trend correlates with rising lake levels at Lake Titicaca (Cross et al. 2000), demonstrating that this expansion of humid evergreen rainforest was likely driven by increasing precipitation. The catchments of these lakes were dominated by tall, closed-canopy rainforest by 2,000 year BP at Bella Vista and 650 year BP at Chaplin, when pollen assemblages were similar to present. Recent modern pollen rain studies (Burn et al. 2010) and advances in Moraceae pollen taxonomy (Burn and Mayle 2008) reveal that this late Holocene ecotonal shift involved expansions in geographic extent of both riparian, seasonally flooded, and *terra firme* upland, humid evergreen rainforest. Although the relative contributions of atmospheric CO_2 concentrations, precipitation, temperature, and geomorphology as controls of glacial-age vegetation in NKMNP are difficult to discern, it is clear that a precipitation increase was the primary driver of this late Holocene rainforest expansion not only because of the close correlation with Andean lake-level histories (e.g. Baker et al. 2001) but also because CO_2 concentrations and temperature remained relatively constant through the Holocene. Further support for precipitation as the dominant long-term control for this rainforest expansion comes from the application of a climate–vegetation model (Punyasena 2008), based on the modern abundance distributions of 154 Neotropical rainforest families, to the Chaplin pollen dataset. This model, employing family-level analysis, suggests that the savanna/dry forest pollen assemblages signify early-mid Holocene mean annual precipitation approximately 300–400 mm/year below present (Punyasena et al. 2008).

It is possible, though, that some of this forest expansion in the lowlands of NKMNP was only indirectly driven by climate and that local climate-driven

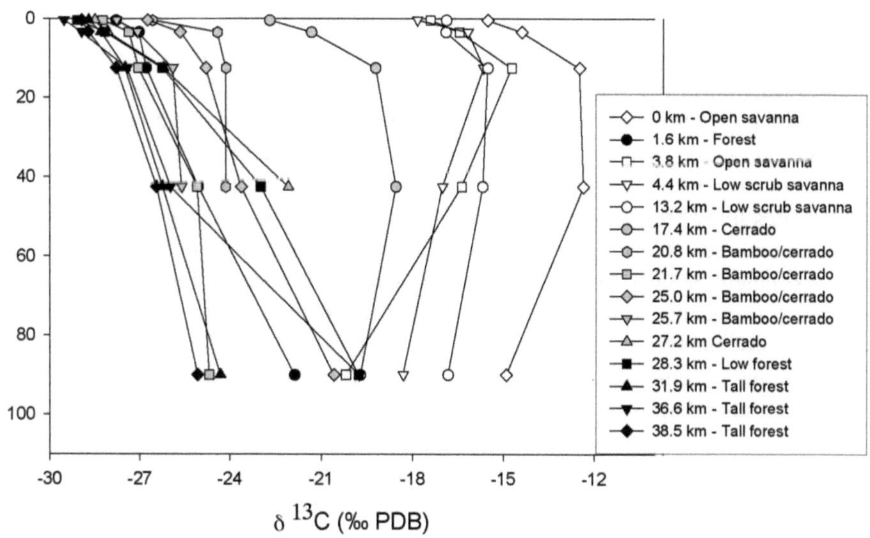

Fig. 8.3 Stable carbon isotope data from soil pits sampled across an ecotonal boundary between open *cerrado* savanna and tall forest on the Huanchaca plateau, NKMNP, northeastern Bolivia (modified from Killeen et al. 2003)

geomorphologic changes played a more direct role. For example, increased precipitation may have led to higher rates of deposition and aggradation in the floodplains, thereby changing the hydrological conditions to facilitate expansion of riparian forest in the floodplains and on the levées (Mayle et al. 2007).

8.3.2 Late Holocene History of the Rainforest–Cerrado Savanna Ecotone on the Huanchaca Plateau

One might expect that the strong edaphic control upon the spatial distribution of forest vs. *cerrado* savanna atop the Huanchaca plateau observed today would mean that forest–savanna boundaries were essentially static, and that the heavily weathered sandstone rocks supporting only a thin veneer of soil can only sustain open savanna (Killeen and Schulenberg 1998). However, stable carbon isotope studies have shown that this forest–savanna mosaic is far from static. Killeen et al. (2003) examined the $\delta^{13}C$ profiles of 15 soil pits (each 1 m deep) dug along a 60 km transect in the centre of the plateau spanning open savanna (*campo cerrado*), savanna woodland (*cerrado* and *cerradão*), and high closed-canopy evergreen forest. Changes in the $\delta^{13}C$ profiles reflect changes in the relative proportions of C_4 savanna grasses vs. C_3 woody vegetation, with progressively lighter isotopic values indicative of increasing abundance of trees (Fig. 8.3). $\delta^{13}C$ values of high forest soil surface samples range between −29.5 and −28.5‰, while those from savanna are heavier but cover a wide range across a gradient of tree density and C_4 grass cover; i.e. −26.5 (dense savanna woodland, *cerradão*) to −15.5‰ (open savanna grassland). The humin fraction of the soil samples at the base of the pits (100 cm depth) dates to approximately 3,000 year BP (S. Panfil, unpublished data). The savanna soil profiles reveal continuous savanna cover over at least the past 3,000 years, although they appear to have been more open in the middle of the profiles. The soil profiles for the forest sites show progressively heavier $\delta^{13}C$ values with increased depth, suggesting that a more open forest (e.g. *cerradão*) or savanna woodland (e.g. *cerrado*) existed approximately 3,000 year BP, which was gradually invaded by forest tree species, eventually resulting in the closed-canopy tall forest that surrounds these soil pits today. Consequently, there appears to have been progressive forest encroachment into savanna in ecotonal areas on the plateau over the last few millennia, consistent with the pollen data from the lakes below the plateau. It therefore appears that the late Holocene precipitation increase forced ecotonal changes not only in the lowlands of NKMNP but also on the plateau as well, at least in areas where edaphic conditions permitted such responses (Mayle et al. 2007).

8.3.3 Implications for Understanding Patterns of Biodiversity and the Rainforest–Savanna Mosaic in NKMNP

The *terra firme* humid evergreen rainforests of NKMNP have lower *alpha* diversity (local species richness) than rainforest communities in more central parts of the Amazon Basin, a pattern which is true not only for tree species (Killeen 1998) but also for birds (Bates et al. 1998) and mammals (Emmons 1998). For example, a Rapid Assessment Program (RAP) report (Killeen and Schulenberg 1998) found that several species usually common in Amazonian rainforests were unexpectedly rare (e.g. frugivorous bats and howler monkeys (*Alouatta*)) or absent (e.g. squirrel monkey (*Saimiri*) and titi monkey (*Callicebus*)). It has been hypothesized (Killeen and Schulenberg 1998) that the lower *alpha* diversity of these rainforests may be due to drought stress in this ecotonal region, lying at the precipitation threshold of many Amazonian species. While this may be true, our fossil pollen data point to an additional or alternative explanation, in that the lower-than-expected *alpha* diversity of these rainforests may be a function of their relatively recent establishment in the park within the last one to three millennia, whereby they may still be undergoing succession and species recruitment and have yet to fill all their available niches (Mayle et al. 2007).

In contrast to the young lowland rainforests, the *cerrado* savannas atop the Huanchaca plateau are highly diverse in comparison with other South American savannas, especially with respect to their avifauna (Bates et al. 1998; Da Silva and Bates 2002). The soil stable carbon isotope data are consistent with the hypothesis that this high biodiversity is due to their long geological history and isolation.

Although there is evidence for both encroachment of lowland rainforest into savanna and vice versa over recent decades (e.g. from satellite imagery; Killeen 1998), the pollen data from Chaplin and Bella Vista indicate that rainforest expansion into savanna is part of a long-term uni-directional trend of climate-driven rainforest expansion over at least 130 km (distance between Bella Vista and the current rainforest ecotone) that began at least 3,000 year BP (Mayle et al. 2000). At the forest–savanna ecotone near Brasília (15 °57′S, 47 °56′W), over 1,000 km eastward, Ratter (1992) has observed giant, relic *cerrado* trees growing 4–5 km within rainforest as evidence of recent forest encroachment into savanna. The Holocene palaeovegetation data from NKMNP, as well as from other sites in ecotonal areas across southern and eastern Amazonia (e.g. Carajas; Absy et al. 1991) suggest that this too could be part of a late Holocene climate-driven trend.

However, while the pollen and stable carbon isotope data indicate substantive climate-driven rainforest encroachment into savanna, both in the lowlands and atop the Huanchaca plateau, geomorphology and edaphic conditions are clearly overriding factors in certain physical settings. Where soils are sufficiently thick, rainforest does grow atop the Huanchaca plateau, but large expanses of the plateau are rocky open ground with virtually no soil (Killeen and Schulenberg 1998)—a habitat which clearly cannot, and likely never has, supported rainforest. The essentially treeless *campo cerrado* that grows in such settings is clearly edaphically controlled and has likely grown atop this plateau for millions of years.

8.4 The Chiquitano Dry Forest and Pantanal Wetlands

A key uncertainty with the Chaplin fossil pollen record from NKMNP is the extent to which reduced rainforest cover during the last glacial period was a function of lower temperatures, lower precipitation, lower carbon dioxide concentrations, or some combination of the three. However, the much higher resolution, and better dated, fossil pollen record from Laguna La Gaiba (LLG) permits identification of the key climatic parameter limiting tropical forest growth. This site is located at a critical biogeographic position straddling the Bolivian/Brazilian border, between the Chiquitano Dry Forest to the west, and seasonally flooded savannas of the Pantanal to the east. This large, flat-bottomed lake is surrounded by closed-canopy semi-deciduous tropical dry forest on upland well-drained hills and slopes which border much of the lake, while the low-lying seasonally flooded portion of the lake basin connects with the Paraguay river and neighbouring seasonally flooded savannas of the Pantanal wetland to the east (Fig. 8.4a). Lake level at LLG is strongly controlled by the flood regime of the Paraguay river and Pantanal savanna wetland, which is in turn a function of seasonal passage of the South American Summer Monsoon (SASM) (Alho 2005).

Whitney et al. (2011) describe in detail the pollen- and diatom-based palaeoclimatic history of this site, spanning the past 45,000 years. Prior to 12,500 year BP, diatom, sedimentological, and macrofossil data indicate that the lake was much

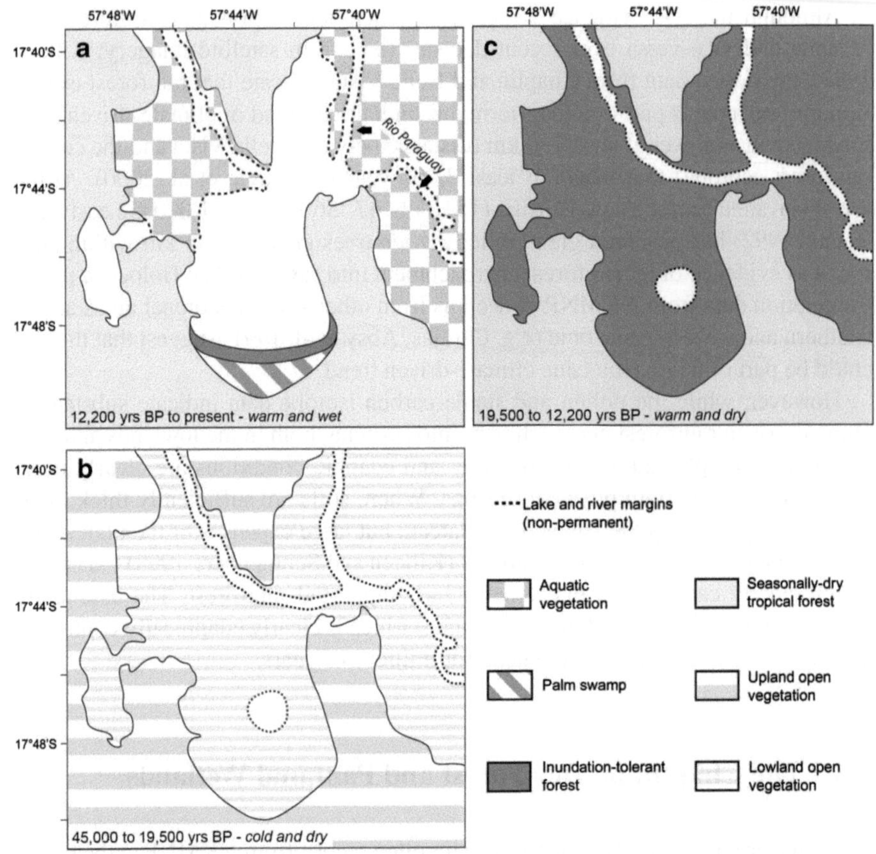

Fig. 8.4 Schematic of environmental change at Laguna La Gaiba (LLG), western Pantanal (seasonally dry tropical forest-wetland savanna ecotone) from 45,000 year BP until present. (**a**) The modern lake environment; seasonally dry tropical forest (SDTF) grows on the steep Amolar hills that surround the lake, but because of the hydrological linkages between the Pantanal and LLG, the pollen signature at this site also reflects the Pantanal savanna wetlands. (**b**) Full glacial; a colder- and drier-than-present climate resulted in reduced flooding in the Pantanal and favoured open, savanna-like vegetation. (**c**) Late-glacial period; increasing temperatures enabled the establishment of tropical forest. Seasonally inundated forest invaded low-lying terrain around the lake, and likely elsewhere in the Pantanal, and SDTF became established on higher, non-flooded terrain. Flooding in the Pantanal remained low due to a drier climate (modified from Whitney et al. 2011)

smaller and shallower than today, indicating that the SASM must have been significantly weaker than present. Between 45,000 and approximately 19,500 year BP the lake catchment was an almost treeless open grassland (Fig. 8.4b), the pollen assemblages dominated by grass, sedge, and *Alternanthera* herbs. However, at approximately 19,500 year BP the upland herbaceous community was replaced by a tropical dry forest community (*Maclura* type [Moraceae], Urticaceae, *Acalypha*, *Celtis*, *Astronium*), while the herbaceous flora of the seasonally flooded exposed lake shallows was simultaneously replaced by a riparian forest community dominated by the inundation-tolerant tree species *Symmeria paniculata* and *Piranhea trifoliata* (Fig. 8.4c).

Several independent lines of evidence show that this forest expansion was driven by deglacial warming. First, precipitation can be discounted as a causal factor because lake levels remained low until 12,500 year BP, when rising precipitation at the onset of the Holocene drove a sharp increase in lake level (replacement of shallow water by deep water diatom communities) that drowned the floodplain forest community dominated by *S. paniculata* and *P. trifoliata* (Whitney et al. 2011). Second, Antarctic ice-core data (Monnin et al. 2001) indicate that no significant increase in carbon dioxide concentrations occurred around 19,500 year BP, so this potential vegetation control can also be ruled out. This leaves deglacial warming as the most parsimonious explanation, for which strong support comes from temperature-driven deglaciation in the Southern Hemisphere tropical Andes (Seltzer et al. 2002). Further corroboratory evidence for a temperature-driven ecotonal shift comes from application of a likelihood model of modern plant climatic distributions (Punyasena 2008; Punyasena et al. 2008) to the LLG fossil pollen dataset, which gives a temperature rise of 4 °C at 19,500 year BP (Whitney et al. 2011). These findings strongly imply that the climate was too cold to support tropical forest taxa around LLG during the LGM (prior to 19,500 year BP). They also strongly suggest that the absence of rainforest in NKMNP during the LGM is also primarily due to cold glacial temperatures rather than insufficient precipitation or decreased water-use efficiency due to low carbon dioxide concentrations.

8.4.1 Implications for Understanding Patterns of Biodiversity in the Chiquitano Dry Forest and Pantanal Wetlands

8.4.1.1 The Chiquitano Dry Forest

The fossil pollen data from LLG have a bearing on understanding some interesting features of the current biodiversity of the Chiquitano Dry Forest of eastern Bolivia and its biogeographic history. The Chiquitano Dry Forest is arguably the largest intact swathe of SDTF remaining in the Neotropics (Parker et al. 1993; Killeen et al. 2006), and despite exhibiting some of the highest *alpha* (local) diversity of any Neotropical SDTF (Killeen et al. 1998), levels of endemicity are significantly lower here than in other SDTF communities, such as those in western Mexico, and in particular, the inter-Andean valleys of Peru where endemism is recorded to be as high as 38 % of all recorded plant species (Gentry 1995; Pennington et al. 2006, 2010). A likely explanation for the discrepancy in number of endemics among these two SDTF communities is their difference in antiquity. Whereas high endemicity in the inter-Andean dry forests reflects several million years of geographical isolation (Pennington et al. 2006, 2010), the absence of SDTF taxa in the catchment of LLG prior to 19,500 year BP shows that the Chiquitano SDTF is very young by comparison and was absent from eastern Bolivia during full glacial times due to the significantly cooler temperatures that then prevailed. Much greater antiquity would be required for endemic tree species to evolve (Hughes and Eastwood 2006).

Furthermore, the LLG pollen record sheds light on the biogeographic history of SDTF in tropical South America, the current disjunct distribution of which has been hypothesized by some authors to represent refugia or remnants of a much greater, contiguous geographic distribution that existed during drier intervals of the Pleistocene (the Pleistocene Dry Forest Arc (PDFA); Prado and Gibbs 1993; Pennington et al. 2000). The PDFA hypothesis is supported by the distribution of several key SDTF taxa, such as *Anadenanthera colubrina* (Fabaceae) and *Astronium urundeuva* (Anacardiaceae), common to all these isolated dry forest communities, implying past linkages (Prado and Gibbs 1993). Although recent floristic comparisons have highlighted greater dissimilarity between the ancient dry forests of the inter-Andean valleys and the adjacent lowland SDTF of Bolivia (Pennington et al. 2006), implying a separate biogeographic history, recent genetic analysis of *A. urundeuva* populations suggests there may have been Pleistocene linkages among disjunct blocks of SDTF in the South American lowlands, including the Chiquitano Dry Forest (Caetano et al. 2008). However, the pollen record of LLG demonstrates that, rather than extending far back into the last glacial period, the history of *A. urundeuva* at this site dates back only to 19,500 year BP, which marks the beginning of deglacial warming in tropical South America (Seltzer et al. 2002, Bush et al. 2004), and *A. colubrina*, a key pollen indicator of SDTF (Gosling et al. 2009), migrated to the region as recently as 10,000 year BP. Thus the pollen data point to a biogeographic history of recent (Holocene) *expansion* in the geographic ranges of key dry forest taxa such as *A. colubrina*, as argued by Mayle (2004), rather than *contraction* from a formerly larger Pleistocene distribution. Furthermore, Whitney et al. (2011) show that low glacial temperatures prior to 19,500 year BP restricted the distribution of SDTF in eastern lowland Bolivia, and likely also elsewhere in South America at similar latitudes. Therefore, during cool glacial periods, lowland SDTF probably persisted as small isolated patches under favourable microclimates.

8.4.1.2 The Pantanal Savanna Wetlands

The seasonally flooded savanna wetlands that dominate the Pantanal comprise the largest tropical wetland on Earth (135,000 km^2). The close hydrological linkage between LLG and the adjacent Pantanal wetlands means that significant reductions in Pleistocene water levels recorded at LLG (relative to present) are likely to have corresponded with a significant reduction in flooding of the Pantanal. Insights into the impact of such a decline in flooding upon the Pantanal wetland savannas can be gleaned from the impact of precipitation fluctuations over recent decades. From 1960 to 1973, reduced precipitation in the Pantanal resulted in an average 3 m lowering of the maximum stage of the Paraguay River and a consequent 50 % reduction in the total area of inundation of the Pantanal, compared with preceding decades

(Hamilton et al. 1996; Hamilton 2002). This major reduction in flooding enabled forest encroachment upon the savanna floodplains (Hamilton 2002), but a return to higher precipitation levels in the mid-1970s caused renewed widespread flooding which drowned the forests, the dead stumps of which are still visible in water-logged areas today (Hamilton 2002).

This observed link between precipitation and forest–savanna dynamics over recent decades in the Pantanal strongly implies that, during the late-glacial period (approximately 19,500–12,500 year BP), when lowered lake levels enabled flood-plain trees *S. paniculata* and *P. trifoliata* to invade the seasonally exposed shorelines of LLG, similar floodplain forest communities must also have expanded their cover throughout a drier Pantanal basin, with a consequent major decline in the geographic extent of inundated savanna. It therefore appears that the forest–savanna mosaic in the Pantanal basin is highly dynamic over all timescales, from inter-decadal to glacial–interglacial. We hypothesize that this dynamism in flood regime and continual shifting in forest–savanna extent is an important contributory factor to the low levels of endemism (Prance and Schaller 1982; Alho 2005) in the Pantanal flora. Major changes in hydrology and temperature, operating over millennial timescales, means that environmental selective pressures do not operate in the same direction long enough to enable speciation through natural selection (Bennett 1990; McGlone 1996; Willis and Niklas 2004). Given that the Pantanal biota is not geographically isolated, its flora comprising taxa common in adjacent biogeographical regions (humid rainforest, Chiquitano SDTF, Bolivian and Brazilian savannas) (Prance and Schaller 1982), spread of plant taxa to and from these adjacent biomes allowed for rapid assembly, disassembly, and re-assortment of constituent species of vegetation communities in the Pantanal throughout glacial–interglacial cycles.

8.5 Forest–Savanna Mosaic of the Seasonally Flooded Beni Basin

The Beni basin of northern lowland Bolivia is characterized by a vast swathe of seasonally flooded savanna (the Llanos de Moxos), approximately 160,000 km^2 in area, interspersed with a complex mosaic of forest islands and riverine gallery forests. Numerous white-water rivers and hundreds of shallow, rectilinear lakes cover the landscape. The spatial distribution of forest and savanna is closely related to micro-topography. The impermeable clay sediments of the extremely flat basin support only treeless savanna and are unsuitable for tree growth because they are flooded throughout the rainy season and are baked hard during the dry season. Trees grow only as forest islands (*islas*) upon topographic highs (mounds) which are raised sufficiently to escape flooding.

The extent to which this forest–savanna mosaic is natural, or instead a human artefact, has been debated since the 1960s (e.g. Denevan 1966; Beck 1983; Erickson 1995, 2000). Many forested mounds are clearly natural, such as sinuous river levées supporting ribbons of gallery forest, or small regularly spaced termite mounds supporting clumps of trees <10 m diameter. However, large areas of forest are demonstrably artificial in the sense that they only exist because pre-Columbian (before 1492) indigenous peoples built the underlying *terra firme* mounds that support them. The most obvious examples include the colonization by trees of abandoned pre-Columbian raised fields, appearing as linear parallel forest strips within open savanna, which can be clearly seen in Google Earth imagery north of Santa Ana de Yacuma, west of the Mamoré river.

However, raised fields have also been found beneath extensive areas of tropical forest long assumed by many tropical ecologists to be essentially pristine "old growth", such as the forested region between Trinidad and San Borja (J. Iriarte, 2010, personal communication). Such forests must therefore be secondary forest regrowth that arose after abandonment of the raised fields around the time of European Contact in the region. Elsewhere, such as in the Casarabe region east of Trinidad, forest islands associated with river levées also appear superficially to be natural, but have in fact been constructed, or at least built up extensively, in pre-Columbian times by earth-moving cultures into large habitation mounds, as evident by the dense accumulations of ceramics throughout excavated mound sections (Erickson and Balée 2006; Lombardo and Prumers 2010). These forests must therefore also be post-contact (after 1492) secondary forest regrowth.

Further evidence for a strong anthropogenic influence upon the forest–savanna mosaic comes from the Baures region in the NE of the Beni, where many forest islands among seasonally flooded savanna are covered by large geometric ring-ditches or "geoglyphs" (100–300 m in diameter, with ditches up to 4.5 m deep and 10 m wide) hidden beneath the forest (Erickson 2010). Although it is unclear whether such mounds were forested prior to geoglyph construction, the geoglyphs clearly demonstrate that the overlying forest in this region is also secondary forest growth, most likely post-dating European Contact.

Clearly, across much of the Beni basin, the process of tropical forest expansion is not a long-term and natural climate-driven process akin to forest expansion during the late Holocene in NKMNP (increasing precipitation) and during deglaciation at LLG (increasing temperature), but is instead a function of landscape modification over a vast geographic scale by pre-Columbian earthmound cultures. In some regions of the Beni, forest expansion was only possible because humans created new *terra firme* topography for this to happen—e.g. forest growth upon abandoned raised fields among open seasonally inundated savanna. Elsewhere, forest expansion may simply have been due to an ongoing process of secondary forest succession over recent centuries following abandonment of previously forested landscape that was cleared for construction of raised fields (e.g. between Trinidad and San Borja) (J. Iriarte, 2010, personal communication) or ring-ditches/geoglyphs (Baures; Erickson 2010).

Although the fine-scale spatial configuration of the forest–savanna mosaic across much of the Beni has clearly been heavily influenced by pre-Columbian cultures, either through deforestation and subsequent abandonment and forest regrowth (as demonstrated elsewhere in the Neotropics; e.g. Myster 2008) or through creation of new *terra firme* habitat (e.g. raised fields), evidence for vertebrate endemicity demonstrates that forest–savanna mosaics (albeit of changing configuration) have characterized the Beni for many thousands and possibly millions of years. Examples of vertebrates endemic to the *Llanos de Moxos* (forest–savanna mosaic) of the Beni are as follows: the Beni anaconda (*Eunectes beniensis*), the blue-throated macaw (*Ara glaucogularis*), and the Beni and Olalla's titi monkeys (*Callicebus modestus* and *Callicebus olallae*, respectively). The Beni anaconda lives only in the seasonally flooded savannas, while the macaw and titi monkeys are restricted to forest islands. The latter are most significant because they are not found in larger contiguous forests or extensive gallery forests (Mayle et al. 2007).

8.5.1 Implications for Understanding Patterns of Biodiversity and the Rainforest–Savanna Mosaic in the Beni

The extent to which pre-Columbian indigenous peoples altered the floristic composition of the rainforests of NKMNP is unknown, but the widespread evidence for wholesale landscape transformation in the Beni basin (through construction of raised fields, habitation mounds, geoglyphs, canals and causeways) is clear and indisputable (Erickson and Balée 2006; Erickson 2010). Rather than having a negligible impact upon rainforests, as commonly assumed (Meggers 1971), it is therefore highly likely that the high population density required to engineer the physical landscape through massive programs of earthwork construction had a similarly profound impact upon the species composition (*alpha* diversity) of the forests via preferential selection of economically important species. If so, one would predict today's forests to have a strong legacy of pre-Columbian management/disturbance. This hypothesis is supported by botanical inventories of forest communities atop the large pre-Columbian habitation mound complex Ibibate around the village Casarabe by Erickson and Balée (2006). These authors found that as much as 45 and 50 % of the tree species of the forested mound complex, and in the adjacent *pampas*, respectively, were fruit-bearing, economically important species for the Siriono indigenous tribes living among the forests today, and were likely to have been similarly important to the ancient earth-moving cultures too. The edible palm species, *Astrocaryum murumuru* var. *murumuru*, comprises as much as 25 % of all trees in the mound forest inventories. Similarly, *Sorocea guilleminiana* (Moraceae) occurs only in the mound inventories where the ancient mound-builders created habitat required for this culturally and economically prized taxon. These peoples left a strong anthropogenic legacy not only on the floristic composition of the mound forests but also on the neighbouring local *pampas* (seasonally flooded savanna), which were found to have higher *alpha* diversity than savanna communities further away from the forest mounds.

When attempting to understand the patterns and causes of floristic diversity among rainforests in the Beni basin, tropical ecologists need to appreciate that such diversity likely contains a strong legacy of pre-Columbian human disturbance when indigenous populations were much higher than those of today.

8.6 Conclusions

The different case studies considered in this chapter demonstrate the dynamic nature of different kinds of tropical forest–savanna ecotones in lowland Bolivia across a range of timescales, extending from the last glacial period to the late Holocene. Fossil pollen data from NKMNP show that the rainforest–savanna ecotone of southern Amazonia is currently at its most southerly extent over at least the last 50,000 years and that rainforest has been expanding southward and encroaching into a mosaic of savannas and tropical dry forests through the late Holocene (the last approximately 3,000 years). This long-term ecotonal shift was driven by increasing austral summer precipitation associated with a progressively stronger SASM, in turn driven by increasing insolation in tune with the precession orbital cycle.

Late Quaternary fossil pollen data from the margin between the Bolivian Chiquitano Dry Forest and Pantanal wetland show that temperatures at the height of the last glacial period (prior to 19,500 year BP) were too cold to support tropical trees, and that easternmost lowland Bolivia, which is today covered by SDTF, was instead covered by largely treeless grassland. Millennial-scale changes in the flooding regime of the Pantanal likely influenced the abundance of wetland savanna vs. inundation-tolerant floodplain forest, particularly at the onset of the Holocene, when rising precipitation caused major flooding.

Examination of the long-term history of the ecosystems in the Beni basin reveals that climate change has not been the only driver of long-term forest–savanna dynamics. Pre-Columbian (before 1492) indigenous cultures played a key role in forming the current forest–savanna mosaic of the Beni basin by creating micro-topography (earthmounds) available for tree colonization following post-Contact (after 1492) earthmound abandonment. Rather than comprising "old growth" primary forest, many forest expanses across the Beni overlie a variety of earth-mounds, including raised fields, ring-ditches, and large habitation mounds, and are therefore secondary forest regrowth.

An appreciation of both natural (climate) and human-driven disturbance over past centuries and millennia is therefore needed to understand the underlying reasons for the current forest–savanna mosaic across lowland Bolivia. Forest–savanna ecotones have clearly responded dynamically to past climate change and are likely to do so again in response to global warming over the twenty-first century (Mayle et al. 2007). A policy of landscape corridors spanning forest–savanna ecotones would be needed to allow any future climate-driven ecotonal shifts. Rather than assuming that any form of human land-use is inherently detrimental to the conservation of forest biodiversity in some imagined pristine state (the Yellowstone National Park model), a key lesson from the forest–savanna mosaic of the Beni is that the biodiversity that conservationists seek to conserve may hold a strong legacy of past human land-use, spanning centuries and millennia. The Beni basin is home to several indigenous tribes—the Moxo, Movima, Canichana, Baure, and Itonama. Conservationists need to appreciate the importance of land-use practices of such tribes in shaping current patterns of biodiversity among these forest–savanna mosaics.

References

Absy ML, Cleef A, Fournier M, Martin L, Servant M, Sifeddine A, Silva MF, Soubiès F, Suguio K, Turcq B, Van der Hammen T (1991) Mise en évidence de quatre phases d'ouverture de la forêt dense dans le sud-est de l'Amazonie au cours des 60,000 dernières années. Première comparaison avec d'autres régions tropicales. C R Acad Sci Paris 312:673–678

Alho CJR (2005) The Pantanal. In: Fraser LH, Keddy PA (eds) The world's largest wetlands: ecology and conservation. Cambridge University Press, Cambridge

Baker PA, Seltzer GO, Fritz SC, Dunbar RB, Grove MJ, Tapia PM, Cross SL, Rowe HD, Broda JP (2001) The history of South American tropical precipitation for the past 25,000 years. Science 291:640–643

Bates JM, Stotz DF, Schulenberg TS (1998) Avifauna of Parque Nacional Noel Kempff Mercado. In: Killeen TJ, Schulenberg TS (eds) A biological assessment of Parque Nacional Noel Kempff Mercado, Bolivia, RAP working papers 10. Conservation International, Washington, DC

Beck SG (1983) Vegetationsökologische Grundlagen der Viehwirtschaft in den Überschwemmungs-Savannen des Río Yacuma (Departamento Beni, Bolivien). Diss Bot 80:1–186

Bennett RD (1990) Milankovitch cycles and their effects on species in ecological and evolutionary time. Paleobiology 16:11–21

Burbridge RE, Mayle FE, Killeen TJ (2004) Fifty-thousand-year vegetation and climate history of Noel Kempff Mercado National Park, Bolivian Amazon. Quat Res 61:215–230

Burn MJ, Mayle FE (2008) Palynological differentiation between genera of the Moraceae family and implications for Amazonian palaeoecology. Rev Palaeobot Palynol 149:187–201

Burn MJ, Mayle FE, Killeen TJ (2010) Pollen-based differentiation of Amazonian rainforest communities and implications for lowland palaeoecology in tropical South America. Palaeogeogr Palaeoclimat Palaeoecol 295:1–18

Bush MB, Silman MR (2004) Observations on Late Pleistocene cooling and precipitation in the lowland Neotropics. J Quat Sci 19:677–684

Bush MB, Silman MR, Urrego DH (2004) 48,000 years of climate and forest change in a biodiversity hot spot. Science 303:827–829

Caetano S, Prado D, Pennington RT, Beck S, Oliveira-Filho A, Spichiger R, Naciri Y (2008) The history of seasonally dry tropical forests in eastern South America: inferences from the genetic structure of *Astronium urundeuva* (Anacardiaceae). Mol Ecol 17:3147–3159

Cross SL, Baker PA, Seltzer GO, Fritz SC, Dunbar RB (2000) A new estimate of the Holocene lowstand level of Lake Titicaca, central Andes, and implications for tropical palaeohydrology. The Holocene 10:21–32

Da Silva JMC, Bates JM (2002) Biogeographic patterns and conservation in the South American Cerrado: a tropical savanna hotspot. Bioscience 52:225–233

Denevan WM (1966) The aboriginal cultural geography of the Llanos de Mojos of Bolivia. Ibero Americana 48:1–60

Emmons LH (1998) Mammal fauna of Parque Nacional Noel Kempff Mercado. In: Killeen TJ, Schulenberg TS (eds) A biological assessment of Parque Nacional Noel Kempff Mercado, Bolivia, RAP working papers 10. Conservation International, Washington, DC

Erickson CL (1995) Archaeological methods for the study of ancient landscapes of the Llanos de Mojos in the Bolivian Amazon. In: Stahl P (ed) Archaeology in the American Tropics: current analytical methods and applications. Cambridge University Press, Cambridge

Erickson CL (2000) An artificial landscape-scale fishery in the Bolivian Amazon. Nature 408:190–193

Erickson CL (2010) The transformation of environment in landscape: the historical ecology of monumental earthwork construction in the Bolivian Amazon. Diversity 2:618–652

Erickson CL, Balée W (2006) The historical ecology of a complex landscape in Bolivia. In: Balée W, Erickson CL (eds) Time and complexity in historical ecology. Columbia University Press, New York, pp 187–233

Furley PA, Proctor J, Ratter JA (1992) The nature and dynamics of forest-savanna boundaries. Chapman and Hall, London

Gentry AH (1995) Diversity and floristic composition of Neotropical dry forests. In: Bullock SH, Mooney SA, Medina E (eds) Seasonally dry tropical forests. Cambridge University Press, Cambridge

Gosling WD, Mayle FE, Tate NJ, Killeen TJ (2005) Modern pollen-rain characteristics of tall terra firme moist evergreen forest, southern Amazonia. Quat Res 64:284–297

Gosling WD, Mayle FE, Tate NJ, Killeen TJ (2009) Differentiation between Neotropical rainforest, dry forest, and savannah ecosystems by their modern pollen spectra and implications for the fossil pollen record. Rev Palaeobot Palynol 153:70–85

Hamilton SK (2002) Hydrological controls of ecological structure and function in the Pantanal wetland (Brazil). In: McClain ME (ed) The ecohydrology of South American rivers and wetlands, IAHS special publication no. 6. IAHS Press, Centre for Ecology and Hydrology, Oxfordshire

Hamilton SK, Sippel SJ, Melack JM (1996) Inundation patterns in the Pantanal wetland of South America determined from passive microwave remote sensing. Arch Hydrobiol 137:1–23

Hughes C, Eastwood R (2006) Island radiation on a continental scale: exceptional rates of plant diversification after uplift of the Andes. Proc Natl Acad Sci USA 103:10334–10339

IPCC (2007) Climate change 2007: the physical science basis. Contribution of Working Group I to the Fourth Assessment Report of the Intergovernmental Panel on Climate Change. Cambridge University Press, Cambridge

Killeen TJ (1998) Vegetation and flora of Noel Kempff Mercado National Park. In: Killeen TJ, Schulenberg TS (eds) A biological assessment of Parque Nacional Noel Kempff Mercado, Bolivia, RAP working papers 10. Conservation International, Washington, DC

Killeen TJ, Schulenberg TS (1998) A biological assessment of Parque Nacional Noel Kempff Mercado, Bolivia, RAP working papers 10. Conservation International, Washington, DC

Killeen TJ, Jardim A, Mamani F, Rojas N (1998) Diversity, composition and structure of a tropical semideciduous forest in the Chiquitania region of Santa Cruz, Bolivia. J Trop Ecol 14:803–827

Killeen TJ, Siles TM, Grimwood T, Tieszen LL, Steininger MK, Tucker CJ, Panfil S (2003) Habitat heterogeneity on a forest-savanna ecotone in Noel Kempff Mercado National Park (Santa Cruz, Bolivia): implications for the long-term conservation of biodiversity in a changing climate. In: Bradshaw GA, Marquet PA (eds) How landscapes change: human disturbance and ecosystem fragmentation in the Americas, Ecological studies 162. Springer, Berlin

Killeen TJ, Chavez E, Pena-Claros M, Toledo M, Arroyo L, Caballero J, Correa L, Guillen R, Quevedo R, Saldias M, Soria L, Uslar Y, Vargas I, Steininger M (2006) The Chiquitano dry forest, the transition between humid and dry forest in Eastern Lowland Bolivia. In: Pennington RT, Lewis GP, Ratter JA (eds) Neotropical savannas and seasonally dry forests: plant diversity, biogeography, and conservation, vol 69. CRC Press, Boca Raton, pp 213–233

Litherland M, Power G (1989) The geological and geomorphic evolution of Serrania Huanchaca (Eastern Bolivia): the lost world. J South Am Earth Sci 2:1–17

Lombardo U, Canal-Beeby E, Fehr S, Veit H (2011) Raised fields in the Bolivian Amazon: a prehistoric green revolution or a flood risk mitigation strategy? J Archaeol Sci 38:502–512

Lombardo U, Prümers H (2010) Pre-Columbian human occupation patterns in the eastern plains of the Llanos de Moxos, Bolivian Amazonia. J Archaeol Sci 37:1875–1885

Mayle FE (2004) Assessment of the Neotropical dry forest refugia hypothesis in the light of palaeoecological data and vegetation model simulations. J Quat Sci 19:713–720

Mayle FE, Burbridge R, Killeen TJ (2000) Millennial-scale dynamics of southern Amazonian rain forests. Science 290:2291–2294

Mayle FE, Langstroth RP, Fisher RA, Meir P (2007) Long-term forest-savannah dynamics in the Bolivian Amazon: implications for conservation. Philos Trans R Soc B 362:291–307

McGlone MS (1996) When history matters: scale, time, climate and tree diversity. Glob Ecol Biogeogr 5:309–314

Meggers BJ (1971) Amazonia: man and culture in a counterfeit paradise. Aldine, Chicago
Monnin E, Indermuhle A, Dallenbach A, Fluckiger J, Stauffer B, Stocker TF, Raynaud D, Barnola JM (2001) Atmospheric CO_2 concentrations over the last glacial termination. Science 291:112–114
Myster RW (2008) Post-agricultural succession in the Neotropics. Springer, New York
Olson DM et al (2001) Terrestrial ecoregions of the world: a new map of life on Earth. Bioscience 51:933–938
Parker TA, Gentry AH, Foster RB, Emmons LH, Remsen JV (1993) The lowland dry forests of Santa Cruz, Bolivia: a global conservation priority, RAP working papers 4. Conservation International, Washington, DC
Pennington RT, Prado DE, Pendry CA (2000) Neotropical seasonally dry forests and Quaternary vegetation changes. J Biogeogr 27:261–273
Pennington RT, Lewis GP, Ratter JA (2006) An overview of the plant diversity, biogeography and conservation of Neotropical savannas and seasonally dry forests. In: Pennington RT, Lewis GP, Ratter JA (eds) Neotropical savannas and seasonally dry forests: plant diversity, biogeography, and conservation, vol 69. CRC Press, Boca Raton, pp 1–29
Pennington RT, Lavin M, Särkinen T, Lewis GP, Klitgaard BB, Hughes CE (2010) Contrasting plant diversification histories within the Andean biodiversity hotspot. Proc Natl Acad Sci USA 31:13783–13787
Prado DE, Gibbs PE (1993) Patterns of species distributions in the dry seasonal forests of South America. Ann Mo Bot Gard 80:902–927
Prance GT, Schaller GB (1982) Preliminary study of some vegetation types of the Pantanal, Mato Grosso, Brazil. Brittonia 34:228–251
Punyasena SW (2008) Estimating Neotropical palaeotemperature and palaeoprecipitation using plant family climatic optima. Palaeogeogr Palaeoclimat Palaeoecol 265:226–237
Punyasena SW, Mayle FE, McElwain JC (2008) Quantitative estimates of glacial and Holocene temperature and precipitation change in lowland Amazonian Bolivia. Geology 36:667–670
Ratter JA (1992) Transitions between cerrado and forest vegetation in Brazil. In: Furley PA, Proctor J, Ratter JA (eds) Nature and dynamics of forest-savanna boundaries. Chapman and Hall, London
Seltzer GO, Rodbell DT, Baker PA, Fritz SC, Tapia PM, Rowe HD, Dunbar RB (2002) Early warming of tropical South America at the Last Glacial-interglacial transition. Science 296:1685–1686
Smith TB, Wayne RK, Girman DJ, Bruford MW (1997) A role for ecotones in generating rainforest biodiversity. Science 276:1855–1857
Stute M, Forster M, Frischkorn H, Serejo A, Clark JF, Schlosser P, Broecker WS, Bonani G (1995) Cooling of tropical Brazil (5°C) during the Last Glacial Maximum. Science 269:379–383
Whitney BS, Mayle FE, Punyasena SW, Fitzpatrick KA, Burn MJ, Guillen R, Chavez E, Mann D, Pennington RT, Metcalfe SE (2011) A 45 kyr palaeoclimate record from the lowland interior of tropical South America. Palaeogeogr Palaeoclimat Palaeoecol 307:177–192
Willis KJ, Niklas KJ (2004) The role of Quaternary environmental change in plant macroevolution: the exception or the rule? Philos Trans R Soc B 359:159–172

Chapter 9
Using Paleoecological Data to Define Main Vegetation Dynamics Along the Savanna–Forest Ecotone in Colombia: Implications for Accurate Assessment of Human Impacts

Juan-Carlos Berrio, Hanne Wouters, Henry Hooghiemstra, Andrew S. Carr, and Arnoud Boom

9.1 Introduction

In recent decades there has been increasing interest, from scientists of many disciplines, in the origins and dynamics of tropical savanna–forest boundaries. These boundaries are rarely present as a smooth gradient from tropical forests to scattered trees and open grassland (Bond and Parr 2010); rather, they are often patchy and irregular, occurring where at first sight no driver for an ecosystem shift is apparent (Sarmiento 1984). In general terms, savanna ecosystems cover approximately 40% of the tropics or 23 million km^2 (Cole 1986; Gardner 2006) and host around one-fifth of the world's human population (Young and Solbrig 1993). These people are imparting a growing impact on savanna systems, as agriculture and other subsistence activities occupy increasingly larger land areas (Gardner 2006).

A fundamental research question, yet to be fully addressed, concerns the main environmental drivers creating savannas within what appear to be otherwise continuous and extended tropical forest landscapes. Seasonal variability in climate, spatial variation in soil properties in relation to regional geology and geomorphology have all been identified as potential explanations (Cole 1982). The origin of the tropical savannas is however still debated, a debate which increasingly revolves around the extent to which these biomes are a natural or anthropogenic in origin and/or whether they represent a transition process or a stable ecosystem state (Cavelier et al. 1998; Marchant 2010).

J.-C. Berrio (✉) • H. Wouters • A.S. Carr • A. Boom
Department of Geography, University of Leicester, University Road, Leicester LE1 7RH, UK
e-mail: jcb34@leicester.ac.uk

H. Hooghiemstra
Department of Palynology and Paleo/Actuo-Ecology, Institute for Biodiversity and Ecosystem Dynamics, University of Amsterdam, Amsterdam, The Netherlands
e-mail: H.Hooghiemstra@uva.nl

R.W. Myster (ed.), *Ecotones Between Forest and Grassland*,
DOI 10.1007/978-1-4614-3797-0_9,

There are several reasons why forest–savanna boundaries or ecotones are of special interest. First, these boundaries are characterized by high levels of biodiversity, as they encompass a mosaic of different ecosystems (Mayle et al. 2007; Ndidi and Suleiman 2009). It has been proposed that these ecotones play a crucial role in tropical forest speciation (Bush 1994; Smith et al. 1997). Second, it has been observed that these boundaries are not stable and are likely to shift in response to climate change over a range of timescales, as well as under anthropogenic influence (Mayle et al. 2007). Worldwide, forest–savanna boundaries are thought to have varied greatly over geological timescales as a consequence of climate fluctuations associated with the glacial–interglacial cycles of the Quaternary Period (Mariotti and Peterschmitt 1994; Cavelier et al. 1998; Van der Hammen and Hooghiemstra 2000). Paleoecological evidence indicates periodic replacement of forests by savannas when climate conditions became more arid (Pennington et al. 2004a, b; Pérez-Garcia and Meave 2006; Rull 2008). The mechanisms behind these shifts, however, are still poorly understood and the available data thus far shed little light on how these boundaries will respond to future climate change (Mayle et al. 2007).

Recent work has suggested that small differences in soil properties, topography, and climate may determine the presence of either savanna or forest in a specific area (Prance 2006). Furthermore, paleoecological data have demonstrated that many tropical savannas represent natural ecosystems and are not the result of anthropogenic influence (e.g., Wijmstra and van der Hammen 1966; Van der Hammen 1974; Berrio et al. 2002b; Montoya et al. 2011). Nonetheless, climate, natural and anthropogenic fires, soil properties, and drainage are thought to be the key factors in the maintenance of contemporary savannas (Sarmiento 1984; Furley 2004; Cavelier et al. 1998). Both natural and anthropogenic fires play an important role in the distribution of the forest–savanna boundary (Bush et al. 2008), widening the savanna–forest boundary as a result of a loss in the tree density and habitat discontinuity (Hoffman and Franco 2003). Such impacts have extensive effects on species composition and change across the ecotones within a savanna ecosystem (Williams et al. 1998). When the fire approaches the savanna–forest ecotone it kills forest seedlings close to savanna boundary before it is extinguished where dry grass is replaced by the moist leaf litter of the forest (Ratnam et al. 2011). It is thought that suppression of fires will result in the expansion of forest into the savanna (Biddulph and Kellman 1998), and it has been further suggested that under present environmental conditions and in the absence of fire, tropical forests of South America would be able to invade savannas by as much as 200 km at certain savanna–forest boundaries (Hirota et al. 2010). This situation is somewhat more complicated though, as the suppression of fires also leads to a short-term increase in fuel loading in the savannas, favoring the occurrence of high-intensity fires at a later stage (Bilbao and Medina 1996; Bond et al. 2004).

On the other hand, savannas can invade forests when the latter are felled and burned for agricultural needs. It has been shown that, at least for the past 3,000 years, this has encouraged the invasion of forest by savannas. Evidence for this is observed in paleoecological data from the Colombian savannas (Wijmstra and van der Hammen 1966; Berrio et al. 2000; Behling and Hooghiemstra 2000). In addition, Cavelier et al. (1998) argue that the savannas of the Sierra Nevada de Santa Marta in Colombia originate from the time when indigenous tribes frequently used fire to the clear land. In this area, tree patches are scattered among the savannas, indicating the existence of a much larger area of moist forest in the past.

In South America, many other grasslands can be found that resemble natural savannas, which are, in fact, the result of agricultural transformations in areas of tropical forest (Boada 2006). Examples include the Gran Pajonal in Peru (Scott 1977) and the Caribbean Cordillera in Venezuela (Tamayo 1943; Vareschi 1968; Cavelier et al. 1998). In these areas, the progress of pre-Columbian (pre-Columbian refers to before 1492) and post-Columbian agriculture undoubtedly has brought about the deforestation of large areas, creating secondary savannas which might be easily confused with the natural savannas (Camacho 1994; Berrio et al. 2002a).

As previously mentioned, paleoecological data such as pollen records may be used to address the question of whether or not savannas are naturally occurring ecosystems. Such data can also be applied to investigate when and how savanna–forest boundaries shifted during recent millennia and whether these shifts are the signature of a large-scale climate fluctuations or rather are a function of smaller, local-scale biome replacement process (Mayle et al. 2007; Mayle and Power 2008). Based on the distinct pollen classes which are representative of the savanna and forest species, changes in the ecotone position can be reconstructed over time. Differences in the stable carbon isotope (^{13}C) ratios of savanna (primarily grasses using C_4 photosynthesis) and forest (C_3 photosynthesis) vegetation can also be used in a similar manner, as these ratios, derived from the plants are generally preserved in sedimentary organic matter (Pessenda et al. 1998a, b; Silva et al. 2008). Furthermore, such data can be coupled to charcoal records to unravel how resilient or responsive these boundaries are to variations in fire regime (Bush and Silman 2007; Bush et al. 2007).

In the neotropics, where the focus of this chapter lies, major savanna systems are found in the Cerrados and the Roraima of Brazil, the Llanos de Moxos in Bolivia and the Llanos of Colombia and Venezuela (Sarmiento 1984; Cavelier et al. 1998, Fig. 9.1). The Llanos Orientales of Colombia and the Orinoco Llanos of Venezuela represent the main tropical savanna of northern South America (Berrio 2002, Fig. 9.2).

For this chapter, we focus on the Colombian Llanos Orientales and the savanna–forest ecotones found in the area. In the northwest these savannas are bordered by the Andes, while to the southwest an ecotone of mosaic savannas and tropical forests can be found before the dense vegetation of the Amazon rainforest appears. More locally within the Llanos, smaller-scale ecotones are found between savannas and patches of gallery forest scattered amongst the landscape.

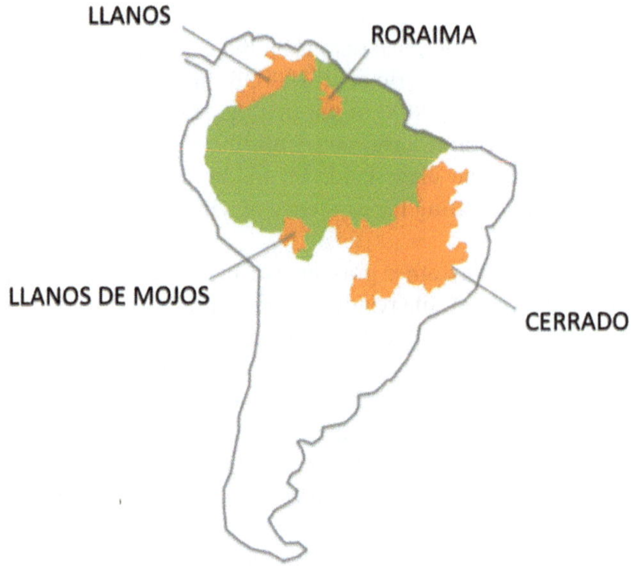

Fig. 9.1 Main savanna systems of South America

Fig. 9.2 The Llanos Orientales of Colombia are part of the Llanos savanna system covering part of Colombia and Venezuela (after Berrio et al. 2002b)

9.2 General Characteristics of the Colombian Llanos Orientales

9.2.1 Location

The Llanos Orientales are located in the northeast of Colombia, between 2.5 to 7°N and 74 to 67°W (Botero 1999). Together with the Orinoco Llanos of Venezuela, the Llanos Orientales constitute the main neotropical savanna system north of the Equator (Sarmiento 1984), stretching over around 1,243,200 km^2 (Behling and Hooghiemstra 1998). The area lies at an altitude ranging from 200 to 400 m above sea level (Behling and Hooghiemstra 1998; Romero-Ruiz et al. 2010), and is topographically diverse, with different natural drainage systems and a marked NE-SW environmental gradient (Malanson 1993; San-José et al. 2010).

9.2.2 Physiography and Geomorphology

Goosen (1971) attributes the presence of the current surface of the Llanos to the transportation of alluvial sediments from the Eastern Cordillera during the Pleistocene. The area is located on a vast geosyncline spreading from the Andes to the highlands of the Guiana Shield in the northeast of the continent. The Colombian Llanos Orientales extends over a large portion of the Precambrian basement (Guiana Shield) which dates to the Proterozoic Eon ca. 1,700 Ma. The main physiographic units of this rather flat plain are piedmont and alluvial overflow plains. The latter were deposited during the Cretaceous Period, and include some marine sediment at the western part of the Colombian Llanos. The main alluvial terraces were formed during the Miocene (25-to-12 Ma), while Quaternary sediments derived from the Eastern Cordillera have been deposited as alluvial plains and fans, some of which has been reworked by aeolian processes, including a veneer of loessic material in the southwestern limits of the Llanos (Goosen 1971; Iriondo 1997; Botero 1999; Rippstein et al. 2001; Naiman et al. 2005). Aeolian land forms, primarily source bordering parabolic dunes associated with NW/SE draining river systems, become increasingly prominent landscape features towards the northeast and into Venezuela (Goosen 1971; Roa 1979; Khobzi 1981). Vegetation associated with these dunes is often visually distinct, reflecting the increased drainage capacity of this sandy substrate. The soils of the Llanos Orientales are generally acid and poor in nutrients (Blydestein 1967). However, the gallery forests, which are found as patches of trees scattered among the savannas, are typically marked by greater nutrient and water levels (Silva et al. 2008).

Several rivers cross the Llanos as they run from the Andes towards the Rio Orinoco drainage system that ultimately flows into the Atlantic Ocean. The largest of these comprise the Rios Meta, Vichada, Guaviare, and Inírida (Flantua et al. 2007). The Rio Arauca borders the Llanos to the north while the Rio Ariari and Rio

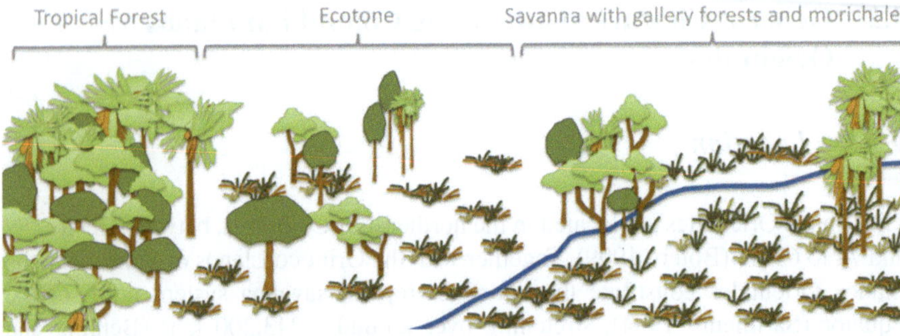

Fig. 9.3 An illustrated overview of a savanna–forest ecotone. A dense tropical forest zone (*left*) which gradually moves towards the transition zone (ecotone) with some stands of trees scattered amongst the landscape (*middle*) which become more open savanna vegetation followed by gallery forests and morichales (*Mauritia flexuosa* palms) (*right*)

Guaviare intercept the ecosystem in the south (Behling and Hooghiemstra 1998). To the northeast, the border between Colombia and Venezuela marks the political transition from the Colombian Llanos Orientales to the Venezuelan Orinoco Llanos (Fig. 9.3).

9.2.3 Climate

The Inter-Tropical Convergence Zone (ITCZ) and rain from the Atlantic Ocean dominate the climate of the Llanos, driving the distinct rainfall seasonality, and wet and dry seasons that the region experiences. The area is marked by a warm and rainy season from April to November as the ITCZ moves southwestwards, while the dry season ranges from November to February (Rangel et al. 1995). In the southern and southwestern part of the Llanos (Provinces Meta and Vichada), however, this dry season is shorter than in the north of the region (province of Arauca) (Behling and Hooghiemstra 2000). Annual precipitation ranges from 1,200 to 2,800 mm, of which 90% falls during the wet season. Higher precipitation rates are observed towards the Colombian Amazon in the south and southwest of the region (Berrio et al. 2000; Haug et al. 2001). The mean annual temperature lies around 26–27 °C, but daily oscillations may vary up to 15 °C with maximum daily mean temperatures of 33.6 °C and a minimum daily mean temperature of 19 °C (Rangel et al. 1995).

9.2.4 Vegetation

The Llanos Orientales are not represented by a homogeneous area of grassy vegetation (Fig. 9.3), but encompass a variety of savanna ecosystems with a range of plant communities and ecotones (San-José et al. 2010). Within the Llanos of Colombia, different types of savanna can be found, including floodplain savannas, seasonally flooded savannas, and relict savanna forests (Berrio et al. 2002b). Several authors have provided general information on the flora and ecology of the area (e.g., Goosen 1971; Cuatrecasas 1989; Rangel et al. 1995). Blydestein (1967) documented ten different savanna communities in the Llanos. The differing composition and distribution of the distinct savanna plant communities reflects a range of environmental factors, including physical, e.g., soil type and landscape units (Veneklaas et al. 2005), and climatic factors (Behling and Hooghiemstra 1998). The total annual precipitation and the length of the dry seasons are particularly important climate variables, influencing the vegetation distribution of the area. During prolonged dry seasons the forest ecotone retreats and is replaced by grassy savanna (Wille et al. 2003). Scattered among the grassy vegetation of the savannas, small patches or strips of gallery forest can be found, typically along rivers or on the river plains. These forests have a species composition similar to that of the Amazon forest, rendering their biodiversity much higher than their surroundings. Dense stands of the *Mauritia flexuosa* palm, often referred to as "morichales" by the local population, typically follow water courses or border lakes and swampy areas (Sarmiento 1984).

The open savanna areas are predominately covered by grassy species belonging to the Poaceae, Cyperaceae, and Xyridaceae families. The more important genera of the grassy savanna are *Andropogon*, *Aristida*, *Axonopus*, *Eragrostis*, *Paspalum*, and *Ctenium*, with *Trachypogon vestitus* (Poaceae) and *Axonopus purpusii* being present in the seasonally inundated savanna plains, while *Paspalum pectinatum* and *T. vestitus* are more dominant in the well-drained savannas. Typical trees and shrub genera of the gallery forest and ecotone transitions are *Clitoria*, *Miconia*, *Pavonia*, *Sida*, *Tibouchina*, and *Waltheria*. In general the most important tree families of the forest and scrub formations are Apocynaceae, Burseraceae, Chrysobalanaceae, Clusiaceae, Combretaceae, and Melastomataceae, while characteristic low trees are *Byrsonima crassifolia*, *Curatella americana*, and *Palicourea rigida*. Aquatic vegetation present in the flooded savannas, springs, lakes, and the so-locally-called "esteros" include *Eichhornia azurea* (Pontederiaceae) and *Pistia stratiotes* (Araceae). Other important aquatic taxa are *Eriocaulon* sp., *Limnocharis flava*, *Neptunia natans*, *Polygala* sp., *Rynchospora* sp., *Syngonanthus* sp., *Typha angustifolia*, and *Utricularia* sp. For more specific details of the plant species composition of the savannas of Colombia readers should refer to Cuatrecasas (1989), Pinto-Escobar (1993), Rangel et al. (1995), and San-José et al. (2010).

9.3 Approaches to Reconstructing Savanna–Forest Ecotone Dynamics

Beyond the time ranges of historical or documentary records, sedimentary archives and associated proxy evidence for paleoecological change must analyzed in order to reconstruct savanna–forest dynamics. Such information provides a longer-term perspective on vegetation dynamics, particularly for periods prior to significant

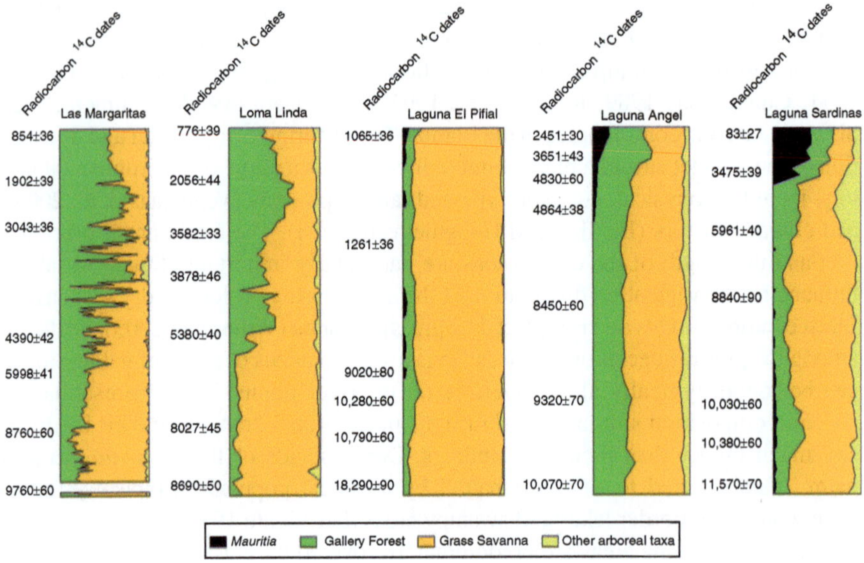

Fig. 9.4 Selection of principal published pollen records from the Colombian Llanos Orientales. From *left* (west) to *right* (east) are shown: Laguna Las Margaritas and Laguna Loma Linda, Laguna El Piñal, Laguna Angel and Laguna Sardinas (after Berrio et al. 2002b; Behling and Hooghiemstra 2000; Wille et al. 2003)

human impacts (e.g., Willis and Birks 2006). The study of fossil pollen in particular has provided useful insights into long-term vegetation and climatic shifts. Typically, sedimentary archives such as lakes and wetlands may be cored, with chronological control provided by the application of radiocarbon dating. Depending on the length, depositional rate, and type of sedimentary record in question, these data have enabled the consideration of past ecological changes over a range of timescales comprising decades (Groot et al. 2011) to thousands of years (e.g., Berrio et al. 2002b).

Soil profiles may also preserve information on past vegetation change, as the organic matter within them may be preserved from periods before the present (see below).

9.4 Fossil Pollen Analysis

Several sedimentary records from the Colombian Llanos Orientales have provided detailed paleoenvironmental reconstructions for the savanna biome of northern South America (Hooghiemstra and Berrio 2007). Figure 9.4 shows a selection of stacked pollen diagrams from a total of ten studied records, representing five different lakes located along a 500 km transect from west-to-east in the Colombian savannas.

The pollen records from Lagunas Las Margaritas (3°23′N, 73°26′W, 290 masl; Wille et al. 2003) and Loma Linda (3°18′N, 73°23′W, 310 masl; Behling and Hooghiemstra 2000) lie near to (ca. 100 km) the Colombian eastern cordillera and about 30 km away from one another. Lagunas Sardinas (4°58′N, 69°28′W, 80 masl) and El Angel (4°28′N, 70°34′W, 200 masl; Behling and Hooghiemstra 1998) lie at the eastern margins of the Colombian Llanos, while those from Laguna El Piñal (4°04′N, 70°14′W, 180 masl; Behling and Hooghiemstra 1999) are located near to the center of the Colombian Llanos Orientales.

These pollen records show evidence of distinctive climatic and vegetation change since the Last Glacial Maximum (LGM) ca. 21,000 years before present and throughout the Holocene (the last 11,500 years). The oldest pollen record (Laguna El Piñal) dates back to ca. 21,800 cal years BP (Before Present) and is indicative of drier climate conditions during the LGM, consistent with other evidence from South America (Peterson et al. 2000; Mayle 2004; Punyasena et al. 2008). The Younger Dryas cold reversal was clearly evident at lakes Sardinas (11,600 cal years BP), El Angel (12,300 cal years BP) and El Piñal (12,000 cal years BP) and is manifested in an increase of gallery forest (Fig. 9.4); similar environmental conditions and ecological responses have been recorded at Lake Chonita in the Venezuelan Gran Sabana (Montoya et al. 2011). The beginning of wetter conditions were recorded at 7,800 cal years BP across the Llanos Orientales (Hooghiemstra and Berrio 2007) promoting the development of gallery forest elements such as *Alchornea*, *Acalypha*, *Cecropia*, *Croton*, and species from the Moraceae/Urticaceae and Melastomataceae/Combretaceae families (Eden 1974; Behling and Hooghiemstra 1999; Berrio et al. 2002b; Marchant et al. 2006). The significant presence of open savanna (mainly Poaceae) and few savanna shrub elements (*Byrsonima* and *Curatella*) is indicative of a prolonged dry period that lasted ca. 3,000 years between early and middle Holocene period (Wirrman and De Oliveira-Almeida 1987).

From mid-to-late Holocene (the last 4,000 years) the Colombian Llanos Orientales experienced a more significant increase in moisture conditions (Behling and Hooghiemstra 1999; Berrio et al. 2002b; Wille et al. 2003; Hooghiemstra and Berrio 2007) as a result of the intensive rainfall due to a more rapid change in the migratory position of the Intertropical Convergence zone (Haug et al. 2001; Bird et al. 2011). Such wetter conditions are not reflected at El Piñal pollen record indicating a low abundance of gallery forest in the surrounding areas. Nonetheless, most studied lakes along the 500 km transect clearly indicate the expansion of gallery forest and the development of "moriche" palm *M. flexuosa* as a result of the increased precipitation in the region (Berrio et al. 2002b; Velez et al. 2005).

Pollen records from lakes Loma Linda (Behling and Hooghiemstra 1999) and Las Margaritas (Wille et al. 2003) clearly show a smooth transition of the savanna–forest ecotone during the mid-late Holocene. The strategic location of both lakes with respect to the Amazon rainforest ecosystem (80 and 100 km southwards respectively), and the sedimentary settings of the cores allow for a detailed reconstruction of the savanna–forest boundary during the last ca. 5,000 years. Nearby, lies Laguna Agua Sucia (3°359 S, 73°319 W, 300 masl) from which the pollen record shows a notable forest–savanna transition since middle Holocene that was initially interpreted as the "origin" of the neotropical savannas (Wijmstra and Van der Hammen 1966).

Overall, it is seen that pollen records from sites situated in close proximity to the forest–savanna ecotones display vegetation changes in response of climate change over timescales of 10^3–10^5 years. Interestingly, lakes Loma Linda and Las Margaritas (Fig. 9.4) both provide a relatively high-resolution record of the forest–savanna ecotone during the Holocene. Important pollen types indicative of this transition are Poaceae and Asteraceae (grass savanna) and *Cecropia*, *Didymopanax*, Moraceae/Urticaceae, Leguminosae, and Sapindaceae (Wille et al. 2003). The gallery forests of the Llanos Orientales tend to show a very positive response (i.e., forest expansion) to wetter conditions, as driven by the annual position of the ITCZ (Hooghiemstra and Berrio 2007).

9.5 Stable Carbon Isotope Analysis

Aside from the previously discussed application of palynology, stable carbon isotope data derived from soils and sediments may also be used to reconstruct savanna–forest ecotones. The natural difference in stable carbon isotope composition of plant tissue derived from plants using the C_3 (Calvin–Benson) and C_4 (Hatch–Slack) photosynthetic pathways offers a method to reconstruct past vegetation dynamics, as well as climate conditions over a range of timescales (Lloyd et al. 2008). Typical $\delta^{13}C$ values for C_4 grasses lie within the range −15 to −10‰, while $\delta^{13}C$ values of C_3 trees and shrubs are typically −30 to −24‰ (Smith and Epstein 1971). These isotopic differences are preserved in soil organic matter (SOM) as plant debris are buried and preserved within the soil profile over time (Boutton et al. 1998; Silva et al. 2008). SOM comprises a complex mixture of organic components; therefore, bulk isotopes data are not generally used to directly calculate percentage C_3/C_4; however, trends in isotope data can be seen as indicative of changing vegetation types. In many savanna ecosystems around the world this has proved an extremely useful approach, and has been used to understand SOM dynamics associated with vegetation change (e.g., Jackson et al. 2002) and the long-term development of Savanna ecosystems (Cerling et al. 1993). Ratnam et al. (2011) emphasize the utility of this technique, stating that for most savanna systems it is the domination of C_4 grasses in the understory that differentiates savannas from forest, regardless of the level of tree cover in the savanna biome. Such approaches have also proved useful in African savanna environments (e.g., Gillson et al. 2004). In the neotropics, a rare exception to this rule is the Brazilian cerrado, where C_3 vegetation dominates the savannas (Lloyd et al. 2008; Edwards et al. 2010; Ratnam et al. 2011).

This stable isotope approach has been applied successfully at several forest–savanna boundaries (Fig. 9.5). For example, Boutton et al. (1998) demonstrated that C_3 woodlands have invaded C_4 savannas over the past century in the Rio Grande

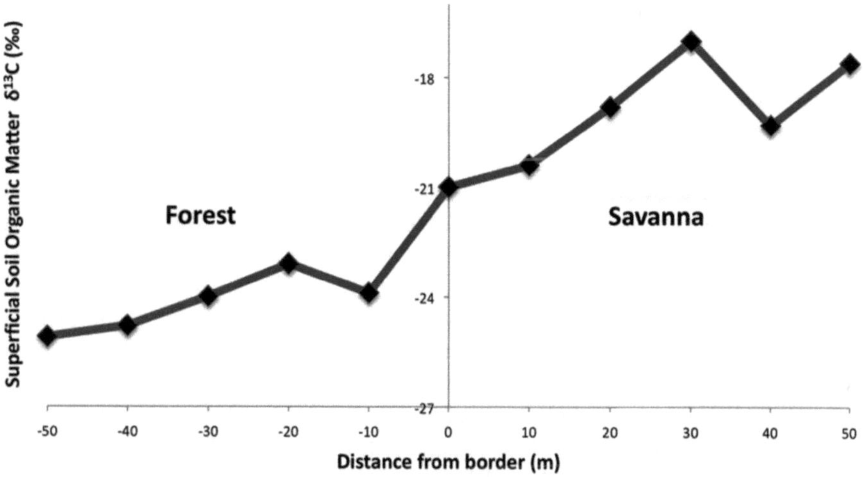

Fig. 9.5 Changes in $\delta^{13}C$ of superficial soil organic matter at a forest–savanna ecotone of the Brazilian cerrados (after Silva et al. 2010)

Plains of Texas. Mariotti and Peterschmitt (1994) also employed this method to demonstrate the recent invasion of the Indian savannas by tropical forest. In Madagascar, Grubb (2003) applied the technique to reveal how savannas in hyper-seasonal locations invade evergreen forest after a fire episode. This method has not thus far been applied to the Colombian Llanos, although it has been proved valuable in other parts of the country and continent. In the Colombian Andes for example, Boom et al. (2001, 2002) demonstrated that decreased atmospheric pCO_2 during glacial times favored the expansion of C_4 grasses, while C_3 plants were able to spread when atmospheric pCO_2 increased and climate conditions became more humid. In Bolivia, Mayle et al. (2007) recorded an ecotone shift in the Noel Kempff Mercado National Park near the Llanos de Moxos, where rainforest invaded the savannas of the park as a result of more humid conditions. In the Amazon, the past presence of woody savannas in areas now occupied by forests has been demonstrated by Sanaiotti et al. (2001). Silva et al. (2008) sampled the gallery forest–savanna boundaries of the Brazilian cerrado for stable isotope analysis, identifying an expansion of gallery forests from at least 4,000 ^{14}C years BP due to a regional increase in rainfall. More recently the same authors (Silva et al. 2010) explored savanna ecotone dynamics in this region in more detail; this time in areas bordering deciduous and xerophytic forests. While the forest–savanna boundary had generally remained stable, the authors found that the deciduous forest had retreated at least 50 m since around 870 cal year BP. Based on these results, they concluded that deciduous and xerophytic forests may not have respond to climate changes in a similar manner to tropical humid forests, as their flora are adapted to survive strong seasonality in water regimes (Silva et al. 2010).

9.6 Evidence of Human Impacts During the Past

Archaeological evidence in the area suggests that pre-Columbians settled close to the rivers in order to cultivate crops from the mid-Holocene (Marwitt 1973; Reichel-Dolmatoff and Dussan 1974; Mora and Cavelier 1983; Alarcón and Segura 1998).

Archaeological information suggests that Guayavero I (8,080 cal years BP) is one of the earliest sites of the Llanos Orientales (Correal et al. 1990), while at the Maporita (Casanare Province, Colombia) the site dates back to 3,950 cal years BP (Barse 1995). On the Venezuelan side of the Orinoco River, Barse (1990) discovered what seems to be the oldest (ca. 10,400 cal years BP) evidence of human occupation of the Orinoco savannas. Mora and Cavelier (1983) unearthed palaoebotanical remains from the Meta piedmont (site A2 of age 418 cal years BP) belonging to *Zea mays*, *Arachis hypogaea* (Fabaceae), *Guilielma gasipaes* (Arecaceae), and *Anadenanthera peregrina* (Leguminosae). Evidence of organic rich black soils (*Terra Preta*) has also been recorded in the Colombian Llanos Orientales near to Aguazul (Casanare Province) which date back to 1,339 cal years BP (Alarcón and Segura 1998).

Indigenous communities are still present in the Llanos Orientales of today. The main groups are the Guahibos, Chiricoas, Yaruro and the Guamos who were still typically hunter-gatherers groups previous to the contact with Spanish colonists (Hernández de Alba 1948). Some of these groups might be the same people who were present 380 cal years BP at Catanga in Casanare (Mora and Cavelier 1989).

Estimations of the human impacts on the savanna ecosystem during the Holocene period have not yet been determined in any detail. We rely mainly on palynological evidence to infer human impact on the environment. Unfortunately, charcoal analysis has been neglected in most of the Llanos Orientales sedimentary records (Behling and Hooghiemstra 1999; Berrio et al. 2000), although it has great potential as a proxy for human impact (either natural and/or induced) and environmental change in the neotropical savannas (Dull 2004; Armenteras et al. 2006; De Toledo and Busk 2008). Today significant parts of the Llanos in the Meta province are used for extensive agriculture and have experienced a massive deforestation (72.6% human-related) between 1939 and 1997 (Madriñan et al. 2007).

Many of the Llanos Orientales pollen records show evidence of "morichales" expansion (*M. flexuosa*-dominated) since middle Holocene (last 4,000 years BP), which is indicative of the cultivation of moriche palm mainly for local consumption and handcrafting (Castaño-Arboleda et al. 2007). At present this represents some of the best evidence of human alteration the Llanos landscape and ecology (Berrio 2002).

9.7 Summary

In this chapter we have explored some of the environmental drivers that create savannas in the neotropics using case studies from the Colombian Llanos Orientales. These savannas are typically controlled by the annual distribution of the ITCZ, which creates a complex suit of vegetation types such as gallery forests, "morichales," and open grass savanna.

Botanical and paleoecological studies have shown that forest–savanna ecotones are characterized by high biodiversity and that they are sensitive to climate change over a range of timescales. During the LGM savanna vegetation expanded, seemingly as a result of drier environmental conditions. Towards the Holocene and modern times climate became wetter allowing expansion of gallery forest.

Finally the complex relation between climate and savanna ecotones cannot be seen as separate from human interference. Throughout this region it seems that humans have affected the natural ecotone dynamics since at least the early Holocene, primarily through burning and clearance of forest.

References

Alarcón J, Segura LA (1998) Rescate Arqueológico en el Municipio de Aguazul, Casanare. Fundación de Investigaciones Arqueológicas Nacionales del Banco de la República, Bogotá, Colombia

Armenteras D, Romero M, Galindo G (2006) Vegetation and fire in the savannas of the Llanos Orientales of Colombia. World Resour Rev 17(4):531–543

Barse W (1990) Preceramic occupations in the Orinoco River valley. Science 250:1388–1390

Barse W (1995) El período arcaico en el Orinoco y su contexto en el norte de Sudamerica. In: Cavelier I, Mora S (eds) Ambito y Ocupaciones Tempranas de la América Tropical. Instituto Colombiano de Antropología y Fundación Erigaie, Bogotá

Behling H, Hooghiemstra H (1998) Late Quaternary palaeoecology and palaeoclimatology from pollen records of the savannas of the Llanos Orientales in Colombia. Palaeogeogr Palaeoclimatol Palaeoecol 139:251–267

Behling H, Hooghiemstra H (1999) Environmental history of the Colombian savannas of the Llanos Orientales since the Last Glacial Maximum from lake records El Piñal and Carimagua. J Paleolimnol 21:461–476

Behling H, Hooghiemstra H (2000) Holocene Amazon rainforest–savanna dynamics and climatic implications: high-resolution pollen record from Laguna Loma Linda in eastern Colombia. J Quat Sci 15(7):687–695

Berrio JC (2002) Lateglacial and Holocene vegetation and climatic change in lowland Colombia. PhD thesis, University of Amsterdam

Berrio JC, Hooghiemstra H, Behling H, van der Borg K (2000) Late Holocene history of savanna gallery forest from Carimagua area, Colombia. Rev Palaeobot Palynol 111:295–308

Berrio JC, Boom A, Botero P, Herrera LF, Hooghiemstra H, Romero F, Sarmiento G (2002a) Multi-interdisciplinary evidence of the Holocene history of a cultivated floodplain area in the wetlands of northern Colombia. Veg Hist Archaeobot 10:161–174

Berrio JC, Hooghiemstra H, Behling H, Botero P, Van der Borg K (2002b) Late-Quaternary savanna history of the Colombian Llanos Orientales from Lagunas Chenevo and Mozambique: a transect synthesis. Holocene 12(1):35–48

Biddulph J, Kellman M (1998) Fuels and fire at savanna-gallery forest boundaries in southeastern Venezuela. J Trop Ecol 14:445–461

Bilbao B, Medina E (1996) Types of grassland fires and nitrogen volatilization in tropical savannas of Venezuela. Biomass Burning and Global Change, vol 2. Biomass Burning in South America, Southeast Asia and Temperate and Boreal Ecosystems, and the Oil Fires of Kuwait, 2. pp 569–574

Bird BW, Abbott MB, Vuille M, Rodbell DT, Stansell ND, Rosenmeier MF (2011) A 2,300-year-long annually resolved record of the South American summer monsoon from the Peruvian Andes. Proc Natl Acad Sci USA. doi:10.1073/pnas.100371910

Blydestein J (1967) Tropical savanna vegetation of the Llanos of Colombia. Ecology 48:1–15

Boada AM (2006) Patrones de asentamiento regional y sistemas de agricultura intensive en Cota y Suba, Sabana de Bogotá (Colombia). Fundacion de investigaciones Arqueológicas Nacionales. Banco de la República, Bogota, pp 181

Bond WJ, Parr CL (2010) Beyond the forest edge: ecology, diversity and conservation of the grassy biomes. Biol Conserv 143:2395–2404

Bond WJ, Woodward FI, Midgley GF (2004) The global distribution of ecosystems in a world without fire. New Phytol 165:525–538

Boom A, Mora G, Cleef AM, Hooghiemstra H (2001) High altitude C_4 grasslands in the northern Andes: relicts from glacial conditions? Rev Palaeobot Palynol 115:147–160

Boom A, Marchant R, Hooghiemstra H, Sinninghe Damsté JS (2002) CO_2 and temperature-controlled altitudinal shifts of C_4 and C_3 dominated grasslands allow reconstruction of palaeoatmospheric pCO_2. Palaeogeogr Palaeoclimatol Palaeoecol 177(1–2):151–168

Botero P (1999) Paisajes fisiográficos de la Orinoquia-Amazonia (ORAM) Colombia. Análisis Geográficos (IGAC, Colombia) 27–28. pp 1–361 (and maps)

Boutton TW, Archer SR, Midwood AJ, Zitzer SF, Bol R (1998) $\delta^{13}C$ values of soil organic carbon and their use in documenting vegetation change in a subtropical savanna ecosystem. Geoderma 82:5–41

Bush MB (1994) Amazonian speciation: a necessarily complex model. J Biogeogr 21:5–17

Bush MB, Silman MR (2007) Amazonian exploitation revisited: ecological asymmetry and the policy pendulum. Front Ecol Environ 5(9):457–465

Bush MB, Silman MR, Toledo MB, Listopad C, Gosling WD, Williams C, de Oliveira PE, Krisel C (2007) Holocene fire and occupation in Amazonia: records from two lake districts. Philos Trans R Soc B 362:209–218

Bush MB, Silman MR, McMichael C, Saatchi S (2008) Fire, climate change and biodiversity in the Amazonia: a Late-Holocene perspective. Philos Trans R Soc B 363:1795–1802

Camacho JH (1994) Sabanas Naturales de Colombia. Banco de Occidente Credencial, Bogotá

Castaño-Arboleda N, Cárdenas-Lopez D, Rodríguez-Otavo E (2007) Ecología, aprovechamiento y manejo sostenible de nueve especies de plantas del departamento del Amazonas, generadoras de productos maderables y no maderables. Instituto Sinchi & Corpoamazonía, Bogotá

Cavelier J, Aide TM, Santos C, Eusse AM, Dupuy JM (1998) The savannization of moist forests in the Sierra Nevada de Santa Marta, Colombia. J Biogeogr 25:901–912

Cerling T, Wang Y, Quade J (1993) Expansion of C_4 ecosystems as indicator of global ecological change in the Late Miocene. Nature 361:344–345

Cole MM (1982) The influence of soils, geomorphology and geology on the distribution of plant communities in savanna ecosystems. In: Huntley BJ, Walker BH (eds) Ecology of tropical savannas. Springer, Berlin

Cole MM (1986) The savannas: biogeography and geobotany. Academic, London

Correal G, Piñeros F, Van der Hammen T (1990) Guayabero I: un sitio precerámico de la localidad Angosturas II, San José del Guaviare. Caldasia 16(77):245–254

Cuatrecasas J (1989) Aspectos de la vegetación natural en Colombia. Perez Arbelaezia 2(8):155–283

De Toledo MB, Busk MB (2008) A Holocene pollen record of savanna establishment in coastal Amapá. An Acad Bras Cienc 80(2):341–351

Dull RA (2004) A Holocene record of Neotropical savanna dynamics from El Salvador. J Paleolimnol 32:219–231

Eden MJ (1974) Palaeoclimatic influences and the development of savanna in Southern Venezuela. J Biogeogr 1(2):95–109

Edwards EJ, Osborne CP, Strömberg CAE, Smith SA, C4 Grasses Consortium (2010) The origins of C_4 grasslands: integrating evolutionary and ecosystem science. Science 328:587–591

Flantua SGA, van Boxel JH, Hooghiemstra H, van Smaalen J (2007) Application of GIS and logistic regression to fossil pollen data in modelling present and past spatial distribution of the Colombian savanna. Clim Dynam 29:697–712

Furley P (2004) Tropical savannas. Prog Phys Geogr 28:581–598

Gardner TA (2006) Tree–grass coexistence in the Brazilian cerrado: demographic consequences of environmental instability. J Biogeogr 33:448–463

Gillson L, Waldron S, Willis KJ (2004) Interpretation of soil $\delta^{13}C$ as an indicator of vegetation change in African savannas. J Veg Sci 15:339–350

Goosen D (1971) Physiography and soils of the Llanos Orientales, Colombia. Publicaties van het Fysisch-Geografisch en Bodemkunding Laboratorium van de Universiteit van Amsterdam 20. pp 1–200

Groot MHM, Bogotá RG, Lourens LJ, Hooghiemstra H, Vriend M, Berrio JC et al (2011) Ultra-high resolution pollen record from the northern Andes reveals rapid shifts in montane climates within the last two glacial cycles. Clim Past 7:299–316

Grubb PJ (2003) Interpreting some of the outstanding features of the flora and vegetation of Madagascar. Perspect Plant Ecol Evol Syst 6:125–146

Haug G, Hughen KA, Sigman DM, Peterson LC, Röhl U (2001) Southward migration of the Intertropical Convergence Zone through the Holocene. Science 293:1304–1308

Hooghiemstra H, Berrio JC (2007) Pollen records, Late Pleistocene, South America. In: S.A. Elias (ed) Encyclopaedia of quaternary science, vol 2. Elsevier, San Diego, pp 2648–2660

Hirota M, Nobre C, Oyama MD, Bustamante MMC (2010) The climatic sensitivity of the forest, savanna and forest–savanna transition in tropical South America. New Phytol 187:707–719

Hoffman WA, Franco AC (2003) Comparative growth analysis of tropical forest and savanna woody plants using phylogenetically independent contrasts. J Ecol 91:475–484

Hernández de Alba G (1948) The Achaguas and their Neighbors. In: J.H. Steward (ed.) Handbook of South American Indians, vol 5. Smithsonian Institute Bureau of American Ethnology, Bulletin 143, 399–412

Iriondo M (1997) Models of deposition of loess and loessoids in the upper Quaternary of South America. J S Am Earth Sci 10:71–79

Jackson RB, Banner JL, Jobbágy EG, Pockman WT, Wall DH (2002) Ecosystem carbon loss with woody plant invasion of grasslands. Nature 418(6898):623–626

Khobzi J (1981) Los campos de dunas del norte de Colombia y de los llanos de la Orinoquía (Colombia y Venezuela). Revista CIAF 6:257–292

Lloyd J, Bird MI, Vellen L, Miranda AC, Veenendaal EM, Djagbletey G, Miranda HS, Cook G, Farquhar GD (2008) Contributions of woody and herbaceous vegetation to tropical savanna ecosystem productivity: a quasi-global estimate. Tree Physiol 28:451–468

Madriñan LF, Etter A, Boxall GD, Ortega-Rubio A (2007) Tropical alluvial forest fragmentation in the eastern lowlands of Colombia (1939–1997). Land Degrad Dev 18:199–208

Malanson G (1993) Riparian landscapes. Cambridge University Press, Cambridge

Marchant R (2010) Understanding complexity in savannas: climate, biodiversity and people. Curr Opin Environ Sustain 2:101–108

Marchant R, Berrio JC, Behling H, Boom A, Hooghiemstra H (2006) Colombian dry moist forest transitions in the Llanos Orientales—a comparison of model and pollen-based biome reconstructions. Palaeogeogr Palaeoclimatol Palaeoecol 234:28–44

Mariotti A, Peterschmitt E (1994) Forest savanna ecotone dynamics in India as revealed by carbon isotope ratios of soil organic matter. Oecologia 97:475–480

Marwitt J (1973) Reconnaissance of the upper Ariari river region. Departament of Meta, Eastern Colombia. 38th Annual meeting of the Society for American Archaeology in San Francisco, San Francisco

Mayle FE (2004) Assessment of the neotropical dry forest refugia hypothesis in the light of palaeo-ecological data and vegetation model simulations. J Quat Sci 19(7):713–720

Mayle FE, Power MJ (2008) Impact of a drier Early–Mid-Holocene climate upon Amazonian forests. Philos Trans R Soc B 363:1829–1838

Mayle FE, Langstroth RP, Fisher RA, Meir P (2007) Long-term forest-savannah dynamics in the Bolivian Amazon: implications for conservation. Philos Trans R Soc B 362:291–307
Montoya E, Rull V, Stansell ND, Bird BW, Nogue S, Vegas-Vilarru T, Abbott MB, Wilmer AD (2011) Vegetation changes in the Neotropical Gran Sabana (Venezuela) around the Younger Dryas chron. J Quat Sci 26(2):207–218
Mora S, Cavelier I (1983) Contrapunteo llanero. Tesis de Grado, Universidad de los Andes, Bogotá, Colombia (inédito)
Mora S, Cavelier I (1989) Resultados preliminares de una prospección en el pie de monte Llanero, Departamento del Meta. Maguaré 5:73–84
Naiman RJ, Décamps H, McClain ME (2005) Riparia: ecology, conservation and management of streamside communities. Elsevier, San Diego
Ndidi EC, Suleiman GA (2009) Species diversity patterns along the forest savanna boundary in Nigeria. Manage Environ Qual Int J 20(1):64–72
Pennington RT, Lavin M, Prado DE, Pendry CA, Pell S, Butterworth CA (2004a) Historical climate change and speciation: neotropical seasonally dry forest plants show patterns of both Tertiary and Quaternary diversification. Philos Trans R Soc B Biol Sci 359: 515–537
Pennington RT, Cronk QCB, Richardson RA (2004b) Introduction and synthesis: plant phylogeny and the origin of major biomes. Philos Trans R Soc B 359:1455–1464
Pérez-Garcia EA, Meave JA (2006) Coexistence and divergence of tropical dry forests and savannas in southern Mexico. J Biogeogr 33:438–447
Pessenda LCR, Gouveia SEM, Aravena R, Gomes BM, Boulet R, Ribeiro AS (1998a) ^{14}C dating and stable carbon isotopes of soil organic matterin forest-savanna boundary areas in Southern Brazilian Amazon Region. Radiocarbon 40:1013–1022
Pessenda LCR, Gomes BM, Aravena R, Ribeiro AS, Boulet R, Gouveia SEM (1998b) The carbon isotope record in soils along a forest-cerrado ecosystem transect: implications for vegetation changes in Rondonia State, southwestern Brazilian Amazon region. Holocene 8:631–635
Peterson LC, Haug GH, Hughen KA, Röhl U (2000) Rapid changes in the hydrologic cycle of the tropical Atlantic during the Last Glacial. Science 290:1947–1951
Pinto-Escobar P (1993) Vegetación y flora de Colombia. Editorial Guadalupe, Bogotá
Prance GT (2006) Tropical savannas and seasonally dry forests: an introduction. J Biogeogr 33:385–386
Punyasena SW, Mayle FE, McEwain JC (2008) Quantitative estimates of glacial and Holocene temperature and precipitation change in lowland Amazonian Bolivia. Geology 36(8):667–670
Rangel JO, Sánchez H, Lowy P, Aguilar M, Castillo A (1995) Región de la Orinoquía. In: Rangel JO (ed) Colombia Diversidad Biótica 1. Universidad Nacional de Colombia, Santafé de Bogotá, pp 233-238
Ratnam J, Bond WJ, Fensham RJ et al (2011) When is a 'forest' a savanna, and why does it matter? Glob Ecol Biogeogr. doi:10.1111/j.1466-8238.2010.00634.x
Reichel-Dolmatoff G, Dussan A (1974) Un sistema de agricultura Prehistórica de los Llanos Orientales. Revista Colombiana de Antropología 17:198–200
Rippstein G, Escobar G, Motta F (2001) Agroecología y biodiversidad de la sabana en los llanos Orientales de Colombia. Centro de Agricultura Tropical—CIAT 322
Roa P (1979) Estudio de los Medanos de los Llanos centrales de Venezuela: Evidencias de un clima desertico. Acta Biologica Venezuelica 10:19–49
Romero-Ruiz M, Etter A, Sarmiento A, Tansey K (2010) Spatial and temporal variability of fires in relation to ecosystems, land tenure and rainfall in savannas of northern South America. Glob Chang Biol 16:2013–2023
Rull V (2008) Speciation timing and Neotropical biodiversity: the Tertiary-Quaternary debate in the light of the molecular phylogenetic evidence. Mol Ecol 17(11):2722–2729
Sanaiotti TM, Martinelli LA, Victoria RL, Trumbore SE, Camargo PB (2001) Past vegetation changes in Amazon savannas determined using carbon isotopes of soil organic matter. Biotropica 34(1):2–16
San-José J, Montes R, Mazorra MA, Matube N (2010) Heterogeneity of the inland water–land palm ecotones (morichals) in the Orinoco lowlands, South America. Plant Ecol 208(2):259–269
Sarmiento G (1984) The ecology of neotropical savannas. Harvard University Press, Cambridge

Scott GAJ (1977) The role of fire in the creation and maintenance of savanna in the Montana of Peru. J Biogeogr 4:143–167

Silva LCR, Sternberg L, Haridasans M, Hoffmann WA, Miralles-Whilhelm F, France AC (2008) Expansion of gallery forests into central Brazilian Savannas. Glob Chang Biol 14:2108–2118

Silva LCR, Haridasan M, Sternberg LSL, Franco AC, Hoffmann WA (2010) Not all forests are expanding over central Brazilian savannas. Plant Soil 333:431–442

Smith BN, Epstein S (1971) Two categories of $^{13}C/^{12}C$ ratios of higher plants. Plant Physiol 47:380–384

Smith TB, Wayne RK, Girman DJ, Bruford MW (1997) A role for ecotones in generating rain-forest biodiversity. Science 276:1855–1857

Tamayo F (1943) Notas de ecología venezolana. Proceso de despoblación y reposición vegetal en las colinas de Caracas. Anales del Instituto Pedagógico. No. 1. Caracas

Van der Hammen T (1974) The Pleistocene changes of vegetation and climate in tropical South America. J Biogeogr 1:3–26

Van der Hammen T, Hooghiemstra H (2000) Neogene and Quaternary history of vegetation, climate, and plant diversity in Amazonia. Quat Sci Rev 19:725–742

Vareschi V (1968) Las sabanas del Valle de Caracas. Acta Botánica Venezuelica 4:425–522

Velez MI, Wille M, Hooghiemstra H, Metcalfe S (2005) Integrated diatom-pollen based Holocene environmental reconstruction of lake Las Margaritas, eastern savannas of Colombia. Holocene 15(8):1184–1198

Veneklaas EJ, Fajardo A, Obregon S, Lozano J (2005) Gallery forest types and their environmental correlates in a Colombian savanna landscape. Ecogr Pattern Divers Ecol 28(2):236–252

Warner H, Beer J, Bütikofer J, Crowley TJ et al (2008) Mid- to Late Holocene climate change: an overview. Quat Sci Rev 27:1791–1828

Wijmstra TA, Van der Hammen T (1966) Palynological data on the history of tropical savannas in northern South America. Leidse Geologische Mededelingen 38:71–90

Wille M, Hooghiemstra H, Van Geel B, Behling H, de Jong A, Van der Borg K (2003) Submillennium-scale migrations of the rainforest-savanna boundary in Colombia: ^{14}C wiggle-matching and pollen analysis of core Las Margaritas. Palaeogeogr Palaeoclimatol Palaeoecol 193:201–223

Williams RJ, Cook GD, Gill AM, Moore PHR (1998) Fire regime, fire intensity and tree survival in a tropical savanna in northern Australia. Aust J Ecol 24(1):50–59

Willis KJ, Birks HJB (2006) What is natural? The need for a long-term perspective in biodiversity conservation. Science 314:1261–1265

Wirrman D, De Oliveira-Almeida LF (1987) Low Holocene level (7700 to 3650 years ago) of lake Titicaca (Bolivia). Palaeogeogr Palaeoclimatol Palaeoecol 59:315–323

Young MD, Solbrig OT (1993) Providing an environmentally sustainable, economically profitable, and socially equitable future for the world's savannas. The world's savannas: economic driving forces, ecological constraints, and policy options for sustainable land use. UNESCO, Paris and The Panthenon Publishing Group, New York

Part III
Boral and Cloud Forest-Grassland Ecotones: Páramo and Alpine Grasslands

Chapter 10
The Dynamic History of the Upper Forest Line Ecotone in the Northern Andes

Henry Hooghiemstra, Juan Carlos Berrio, Mirella H.M. Groot, Raul G. Bogotá-A, Marcela Moscol Olivera, and Zaire González-Carranza

10.1 Introduction

In the Andean cordilleras very conspicuous ecotones can be found. The transition from continuous upper montane forest to treeless herbaceous vegetation, regionally known as "páramo" (Cleef 1981; Luteyn 1999) is known as the "upper forest line" (UFL) or "timber line" (Holtmeier 2009). Above the UFL trees may occur forming small patches with diameters of ten to several hundreds of metres. The elevation where individual trees find their altitudinal limits is at significantly higher altitudes and this limit reflects the "upper tree line". In the Colombian Andes the upper tree line, most formed by dwarf trees of *Polylepis*, may be up to 800 m above the UFL. Therefore, it is relevant to differentiate between both ecotones. Across the Andes the altitudinal position of the UFL varies much depending latitude (Schmithüsen 1976) but in the area under consideration located between 0 and 11°N the altitudinal position of the UFL mostly varies between 3,000 and 3,800 m. In this paper we focus on the tropical Andes of northern Ecuador and Colombia. The spatial and temporal dynamics of the UFL ecotone, which has scientific and economic relevance is explored and discussed.

Knowledge of altitudinal distributions of individual plant taxa in tropical mountains is still insufficient and is subject to vigorous discussions (Cárdenas et al. 2011;

H. Hooghiemstra (✉)
Department of Paleoecology and Landscape Ecology, Institute for Biodiversity and Ecosystem Dynamics, University of Amsterdam, Amsterdam, The Netherlands
e-mail: H.Hooghiemstra@uva.nl

J.C. Berrio
Department of Geography, University of Leicester, University Road, Leicester LE1 7RH, UK
e-mail: jcb34@leicester.ac.uk

M.H.M. Groot • R.G. Bogotá-A • M.M. Olivera • Z. González-Carranza
Department Paleoecology and Landscape Ecology, Institute for Biodiversity and Ecosystem Dynamics, University of Amsterdam, Science Park 904, 1098 XH Amsterdam, The Netherlands

R.W. Myster (ed.), *Ecotones Between Forest and Grassland*, 229
DOI 10.1007/978-1-4614-3797-0_10, © Springer Science+Business Media New York 2012

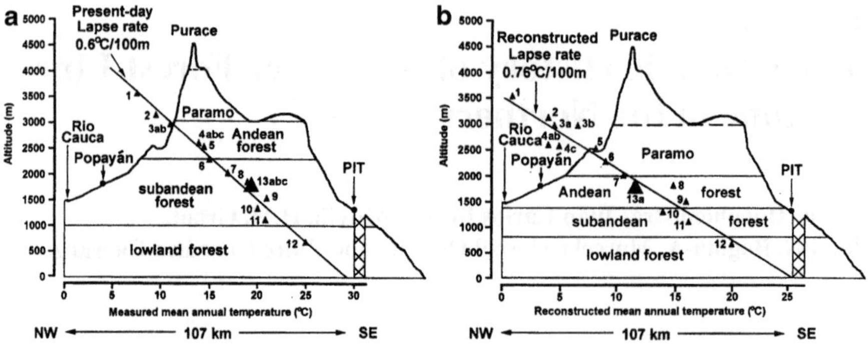

Fig. 10.1 Altitudinal distribution of main vegetation belts along a cross section through the Puracé Volcano in the southern Colombian Andes. (**a**) Preseny-day distribution from modern vegetation survey studies; (**b**) last glacial maximum distribution inferred from pollen records. The altitudinal temperature gradient (lapse rate) for both time-slices is shown (for further explanation see Wille et al. 2001) (after Wille et al. (2001) with permission)

Punyasena et al. 2011). Model studies are helpful to explore the sensitivity of ecosystems and ecotones to various climatic variables. For tropical ecosystems the development from the bioclimatic model (Smith et al. 1992) to the coupled climate–vegetation models (e.g. Marchant et al. 2002a, b) significantly improved understanding of the contribution of various drivers to environmental change. From the paleoecological point of view the improved spatial resolution of studied sites (Marchant et al. 2009) and the high temporal resolution of an increasing number of studies documented a new level of understanding of ecological change in the tropics (Flenley 1992).

In the northern Andes the altitudinal vegetation distribution has been studied in great detail and we refer to the main phytosociological studies further on in this paper. Initial studies of the altitudinal distribution of main plant species were provided by Von Humboldt and Bonpland (1807) and Cuatrecasas (1934, 1958). Van der Hammen (1974) placed the altutudinal vegetation distribution in a paleo-ecological perspective. Van't Veer and Hooghiemstra (2000) identified along the altitudinal range the most responsive taxa to climate change. Wille et al. (2001) defined more precisely for southern Colombia the altitudinal migration of main vegetation belts since the last glacial maximum (LGM) and the different levels of compression (Fig. 10.1). The transition from tropical lowland vegetation (forest or savanna) to lower montane forest is located between 800 and 1,000 m altitude and mainly constrained by humidity and temperature. The transition from lower montane to upper montane forest is around 2,300 m altitude and mainly constrained by a rapidly increasing frequency of night frost. In the northern Andes the uppermost limit of continuous forest (UFL) is located between 3,200 and 3,800 m and constrained by a mean annual temperature (MAT) of ~9.5 °C (e.g. Van der Hammen and González 1963; Hooghiemstra 1984; Van't Veer and Hooghiemstra 2000;

Groot et al. 2011). There is increasing evidence that changes in atmospheric pCO_2 are also an important factor for the altitudinal position of the UFL (Boom et al. 2002; Marchant et al. 2002a, b; Groot et al. 2011). The lowermost zone of páramo vegetation is dominated by shrub and is known as subpáramo. The transition from shrubby subpáramo to herbaceous grasspáramo may occur at some 50–300 m above the UFL and is mainly constrained by humidity and intensity of night frost (Cleef 1981). The transition from grasspáramo to superpáramo reflects the highest altitudinal ecotone and is mainly constrained by intensity of night frost, mechanical damage to the vegetation by falling rocks, UV radiation and lack of substrate.

Species' climatic and environmental tolerances reflect a conservative character. Under changing climate conditions plant species move to track their preferred climatic conditions, known as their ecological envelope. Plants respond to climate change in species specific ways (Chen et al. 2011), hence specific taxa may only occur at particular altitudinal intervals. Therefore, species that currently form robust associations may move at different rates and in different directions to form new associations in the future. For the same reason plant associations in the past may differ from present ones making phytosociological studies ephemeral (Gleason 1926; see also Dickinson and Murphy 1998). However, many plant species move in concert to climate change and therefore higher-ranked plant associations, such as upper montane forest and páramo vegetation respond as a quasi-unit. Individual species may deviate from current altitudinal ranges. Therefore plant associations constituting the present-day UFL are subject to change and efforts to recognise past UFL positions in pollen records should allow that various plant species may dominate at the forest–páramo transition. In the Colombian Andes *Polylepis* formed extensive woods at the UFL between 38,000 and 32,000 calibrated years before present (cal yr BP) (Bogotá-A. et al. 2011b) and this setting did not return up to present-day.

Many vegetation studies in the northern Andes have substantiated the characteristics of major ecotones at various altitudes. However, this information is mainly descriptive and a better understanding of the major physiological constraints of ecotones, expressed in climatological variables, is needed to improve quantitative reconstructions of past UFL changes and to feed combined climate–vegetation models (Marchant et al. 2002a, b). Understanding the impact of climate change and greenhouse gases on the biology of plants varies from insufficient to absent. This explains that coupled climate–vegetation models are not fed with information at the species level rather with information at the level of plant functional types or traits (Young and León 2007).

In the following paragraphs aspects of ecotone forest such as present-day floral composition, changes in composition during Holocene and late Pleistocene time, and estimations of rates of change are presented and discussed. Here we focus on three well studied areas (Fig. 10.1): Guandera Reserve in northern Ecuador (00°36′N, 77°42′E, 3,400 m alt.), Lake La Cocha in southern Colombia (01°08′N, 77°09′W; 2,780 m alt.) and Lake Fúquene in central Colombia (05°27′N, 73°46′W, 2,540 m alt.).

10.2 The Upper Forest Line in the Northern Ecuadorian Andes

At many places in northern Ecuador the natural altitudinal position of the UFL is unknown due to severe deforestation. In the literature an UFL position around 4,000 m alt. (Laegaard 1992; Lauer et al. 2001) and an UFL position around 3,600–3,700 m have been claimed (Bakker et al. 2008). The consequences are significant. If the UFL is around 3,650 m páramo vegetation at higher elevations reflects a natural ecosystem, and forest plantations above 3,700 m damage the páramo vegetation. If the UFL is around 4,000 m páramo vegetation has expanded downslope due to deforestation and should be considered anthropogenic. In that case forest plantations between 3,700 and 4,000 m can be considered as a type of reforestation. The most recent reconstructions of the natural position of the UFL was attained using vegetation analysis, modern pollen rain analysis, and pollen and biomarker records along an altitudinal gradient.

In Guandera Reserve near the Ecuadorian–Colombian border, vegetation plots were studied along altitudinal gradients between 3,000 and 4,000 m. Distinct altitudinal ranges of trees and páramo herbs were documented and the specific elevation with maximal cover identified (Bakker et al. 2008) (Fig. 10.2). Most trees have a maximum range up to ~3,700 m, while páramo herbs have their altitudinal ranges from 3,400 to >4,000 m. This altitudinal distribution documents continuous forest cover reach up to ~3,700 m maximally where vegetation cover is changing into herbaceous páramo vegetation with scattered patches of trees. Analysis of successional stages in ecotone forest showed that the present-day UFL had maximally shifted downslope over 50 m due to incidental occurrence of fire, grazing and clear cutting (Moscol Olivera and Cleef 2009a) (Fig. 10.3).

Modern pollen rain was studied along the vegetational transect between 3,000 and 4,000 m. At 3,650 m a sharp transition from herb-dominated spectra to tree-dominated spectra was registered reflecting the elevation of the UFL (Moscol Olivera et al. 2009).

Sediment sequences with adequate chronological control showed changes of the altitudinal distribution of individual plant taxa and ecotones. A pollen record from 3,400 m elevation (Bakker et al. 2008) showed during the last 6,000 years the UFL shifted between 3,200 and 3,650 m, maximally up to 3,700 m. This result supports the low UFL hypothesis. A sediment record from 3,810 m elevation showed continuous presence of páramo during the last 3,000 years supporting the view that the natural undisturbed UFL was significantly below 3,810 m (Moscol Olivera and Hooghiemstra 2010). In conclusion, pollen analysis calibrated by vegetation analysis and modern pollen rain shows that páramo vegetation between 3,650 and 4,000 m elevation reflects a natural ecosystem. The assumption that páramo between 3,650 and 4,000 m is the result of downslope expansion of páramo vegetation after clear cutting (a process locally known as "paramisación") is rejected.

Pollen grains of plants making up upper montane forest and páramo vegetation are partly wind transported (Moscol Olivera et al. 2009). Therefore pollen spectra from páramo sites in particular include the pollen signal of local páramo vegetation

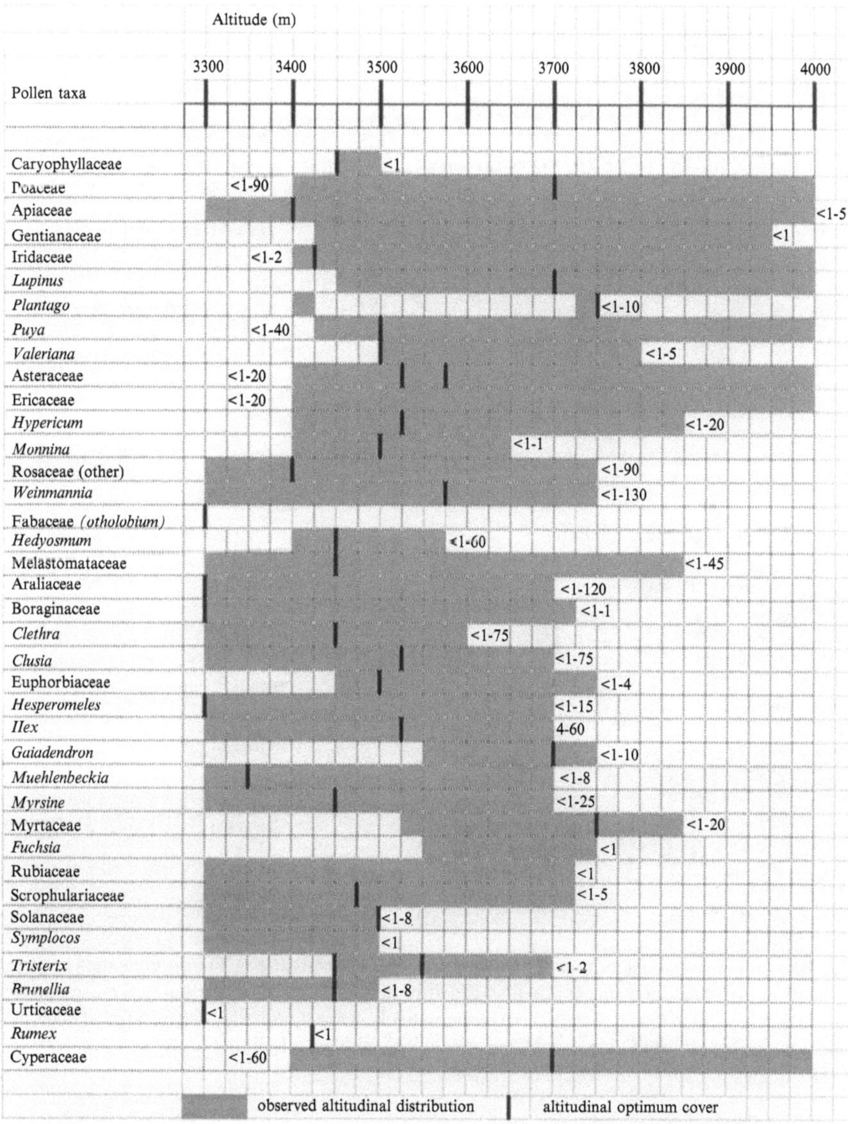

Fig. 10.2 Estimated altitudinal range and altitudinal position of the optimum cover in the present-day vegetation of Guandera reserve (northern Ecuador) of the most important taxa represented in the pollen record from 3,400 m altitude reflecting the last 7,000 years (after Bakker et al. (2008) with permission)

as well as signals of vegetation of the wider area. This hinders precise conclusions about the altitudinal position of the UFL. Sediments of bogs and lakes often include plant macro-remains allowing one to infer whether a site was surrounded by páramo or forest. However, in peaty sediment infills of small depressions, but also in lakes

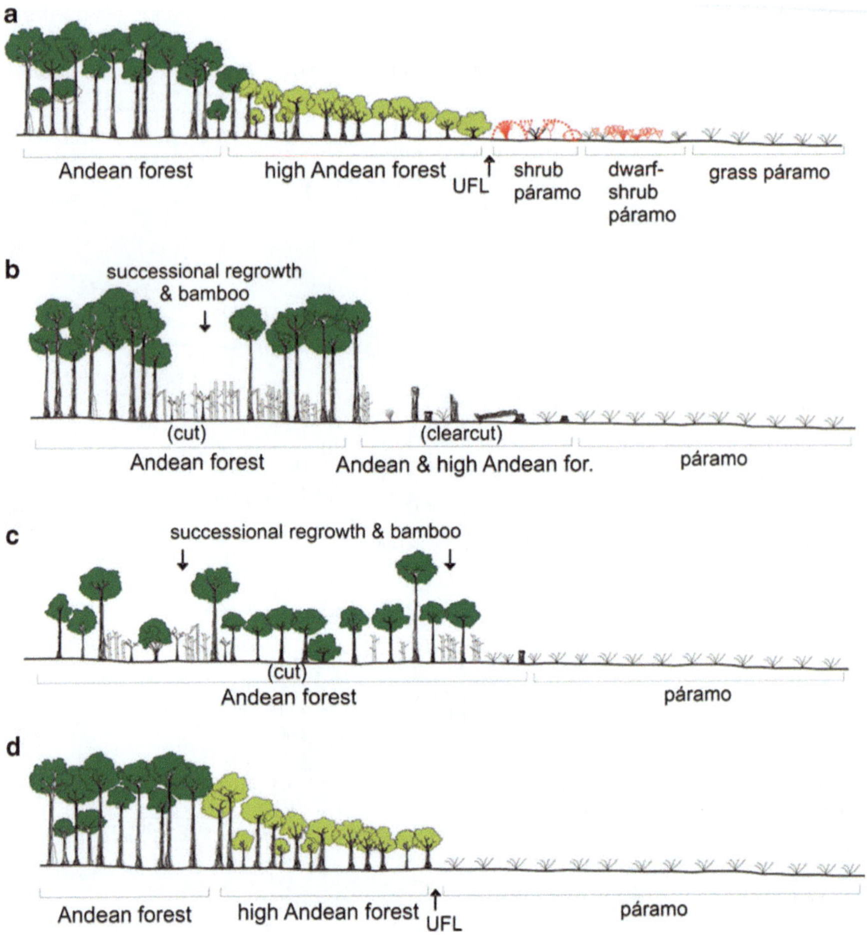

Fig. 10.3 Idealised vegetation profile across the upper forest line (UFL) in Guandera reserve, northern Ecuador (**a**), and successional stages after disturbance (**b**, **c**), leading to the present-day reset UFL at ca. 50 m lower elevation (**d**) with relatively sharp boundaries between forest and open vegetation (after Moscol Olivera and Cleef (2009a, b) with permission)

and in soil profiles in particular, plant macro-remains may be poorly or not conserved preventing precise conclusions about the distance of the studied site to the UFL. A new approach based on molecular biomarker analysis was developed to circumvent this drawback.

Biogeochemical markers are produced by all major plants and are relatively well preserved in soils (Otto and Simpson 2005). Straight-chain lipids of 20–36 carbon atoms originate exclusively from the epicuticular wax layers on leaves and roots of terrestrial higher plants (see Jansen et al. 2008 and references therein). Biomarker ratios of the concentration of a limited number of *n*-alkanes and *n*-alcohols that differ between relevant plant groups are indicative of the local vegetation cover

(Jansen et al. 2008). Biomarker-based vegetation reconstructions were even improved with a model enabling to unravel mixed biomarker patterns (Jansen et al. 2010). We compared biomarker-based and pollen-based records of vegetation change in a peat sequence at 3,400 m elevation, at present-day located at 200 m below the UFL. The pollen-based record includes the ecologically indecisive group of Asteraceae as representatives do occur in forest as well as in páramo. Moreover, Asteraceae are wind pollinated plants and therefore, it is not possible to precisely indicate the moment the UFL migrated across the study site. In contrast, the biomarker record showed precisely the moment biomarkers from trees were replaced by biomarkers from páramo herbs (Jansen et al. 2010), reflecting the moment the UFL passed along the site. Potentially plants may be more precisely characterised by specific biomarkers rather than by their pollen morphology. This is in particular the case for plants belonging to the families of the Poaceae, Asteraceae and Ericaeae. The present calibration sets for biomarker analysis are in need of improvement but first results show the combined pollen-biomarker method has good potential to be developed into a powerful method in reconstructions of past environments.

10.3 The Upper Forest Line in the Southern Colombian Andes

In the south Colombian Andes uppermost montane forest is relatively well preserved. The natural position of the UFL is estimated at 3,550 m (González et al. 2012). The area of Lake La Cocha (2,780 m altitude, 01°08′N, 77°09′W) has mountains up to 3,600 m elevation. This region reflects the southernmost distribution of oak forest (*Quercus humboldtii*) (Hooghiemstra 2006) and therefore the floral composition of ecotone forest is changing substantially across this latitude.

The 12 m deep core Cocha-1 was collected with a hand-operated Russian Corer. Chronological control of the sediments is based on 18 accelerator mass spectroscopy (AMS) ^{14}C ages. The pollen record Cocha-1 shows changes of the altitudinal vegetation distribution, as well as changes in the floristic composition of the ecotone forest during the last 14,000 cal yr BP. This pollen record shows a long-term gradual increase of the UFL from 2,400 m elevation in the period from 11,200 to 10,350 cal yr bp up to 3,600 m in the period from 1,900 to 1,400 cal yr BP (Fig. 10.4). Although fluctuating, the UFL needed some 8,000 years to reach the highest elevations (González et al. 2012). Inferred changes in climatic moisture at Lake La Cocha parallels the history of increasing climatic moisture inferred from the pollen records obtained in the adjacent savannas of the Llanos Orientales (Behling and Hooghiemstra 1999; Berrio et al. 2002). Both areas receive precipitation from trans-Amazonian atmospheric moisture transport. Thus, it is suggested that increasing atmospheric moisture flows, and related increasing MAT and decreasing night frost frequency are in this area the main constraints of the altitudinal position of the UFL.

During the Holocene *Hedyosmum*, *Myrsine* and Urticaceae–Moraceae are continuous representatives of the montane forest near the ecotone. Other taxa start

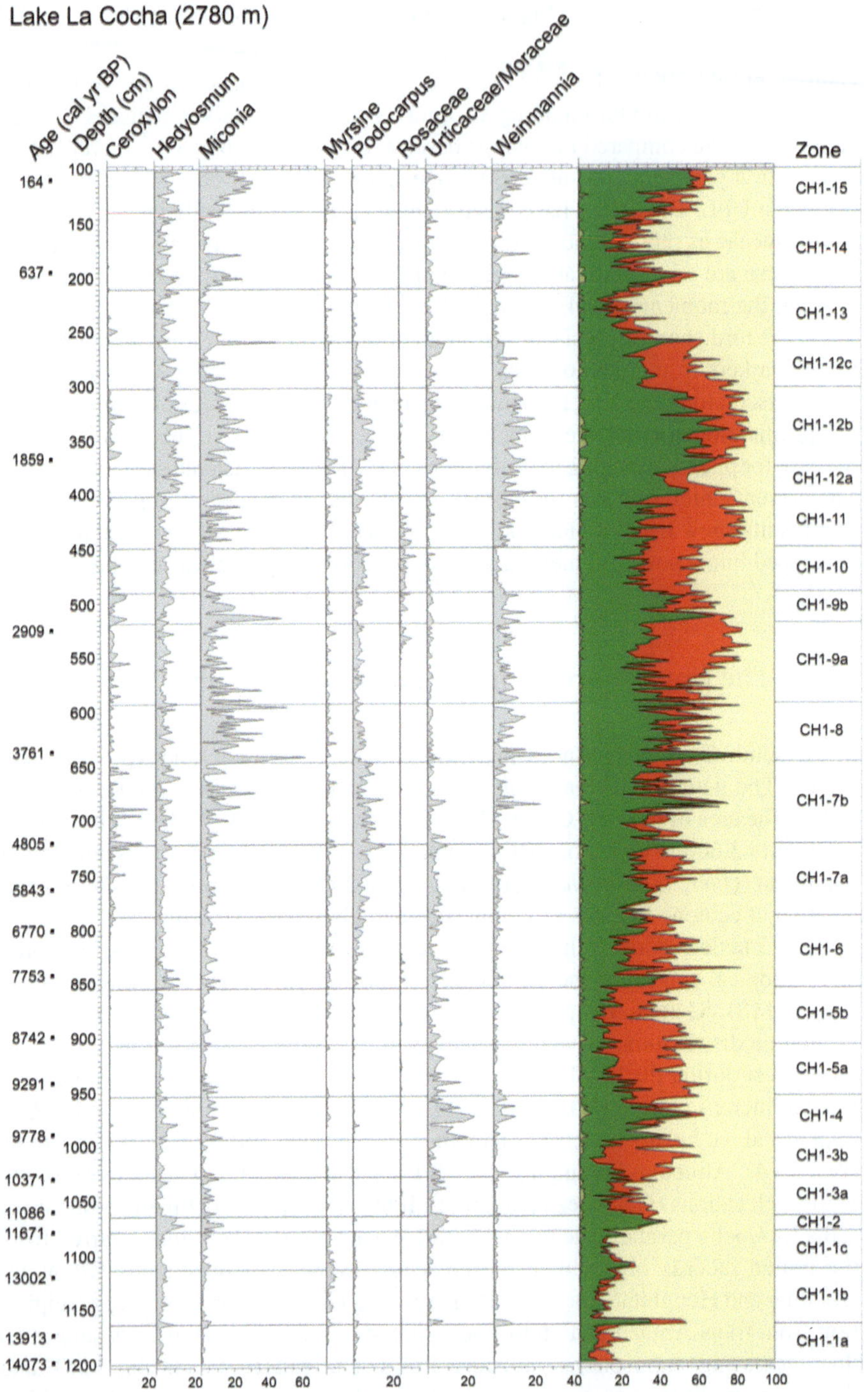

Fig. 10.4 Simplified pollen record of lake La Cocha (2,780 m alt.) showing the main diagram and a selected number of arboreal taxa. The main diagram shows temporal changes in the altitudinal vegetation distribution (subandean forest, Andean forest, the upper forest line ecotone, subpáramo and grasspáramo), and selected arboreal taxa show changes in floral composition of the montane forest during the last 14,000 years (after González-Carranza et al. (2012) with permission)

their share in the ecotone forest at a later moment: the palm *Ceroxylon* and the conifer *Podocarpus* at ~6,450 cal yr BP, *Miconia* and *Weinmannia* at ~4,200 cal yr BP, and representatives of the Rosaceae increased at ~3,000 cal yr BP. The increasing share of *Ceroxylon* and *Hedyosmum* in particular suggest climatic humidity increased in the second half of the Holocene. Asteraceous shrub is abundant in the páramo, but may also have representatives in the ecotone forest. Therefore a part of the Asteraceae may be indicative of forest and, in that case, the UFL reached highest altitudes already at ~3,400 cal yr BP instead of ~1,900 cal yr BP. This interpretation of the age of the first maximum altitude of the UFL after the last glacial coincides with the upslope migration of the UFL in northern Ecuador (Bakker et al. 2008). *Quercus*, an important taxon of the montane forest in the Colombian Andes is neither registered in the La Cocha pollen record nor in the forests surrounding the lake. This confirms the identified southernmost limit of *Quercus* near the Colombian–Ecuadorian border (Hooghiemstra 2006). Similarly, the absence of *Polylepis* in the La Cocha record, as well as in the Guandera record from northern Ecuador, confirms that the distribution of *Polylepis* from ~30°S to ~10°N is interrupted between 2 and 4°N latitude (Baumann 1988). The pollen record shows plant associations forming ecotone forest have varied over time, reflecting that example plant species responded individually to climate change. However, most plant taxa show a concerted response allowing the recognition of main vegetation formations (vegetation belts) restricted to altitudinal intervals.

The pollen record Cocha-1 shows at ~1,400 cal yr BP a sudden start of deforestation and this is strong evidence of the start of the period of human occupation. The pollen record shows that the palm *Ceroxylon* and *Podocarpus* were most prone to clear cutting. Trees belonging to the families of the Urticaceae and Moraceae, and *Weinmannia* have been harvested after ~1,100 cal yr BP in particular. All these taxa may be frequent in the ecotone forest near the UFL and the selection of these taxa for clear cutting by the pre-Columbian people shows that ecotone forest contains a high proportion of economically attractive timber. This observation makes it plausible that a great deal of the current UFL suffered during the last millennia from clear cutting and burning and explains that caution should be taken in accepting the present-day UFL as a natural ecotone. The study in northern Ecuador showed that the relatively undisturbed UFL was still lowered by ~50 m (Fig. 10.3; Moscol Olivera and Cleef 2009a).

10.4 The Upper Forst Line in the Central Colombian Andes

The area of the basins of Bogotá and Fúquene is most intensively studied during the past decades. Van der Hammen and González (1963) characterised the transition from upper montane forest to páramo in terms of changing vegetation, modern pollen rain and MAT. The UFL lies here at 3,200 m, is characterised in the modern pollen rain by ~40% arboreal pollen, and coincides with ~9.5 °C MAT (Van der Hammen and González 1963; Hooghiemstra 1984). Based on these relationships

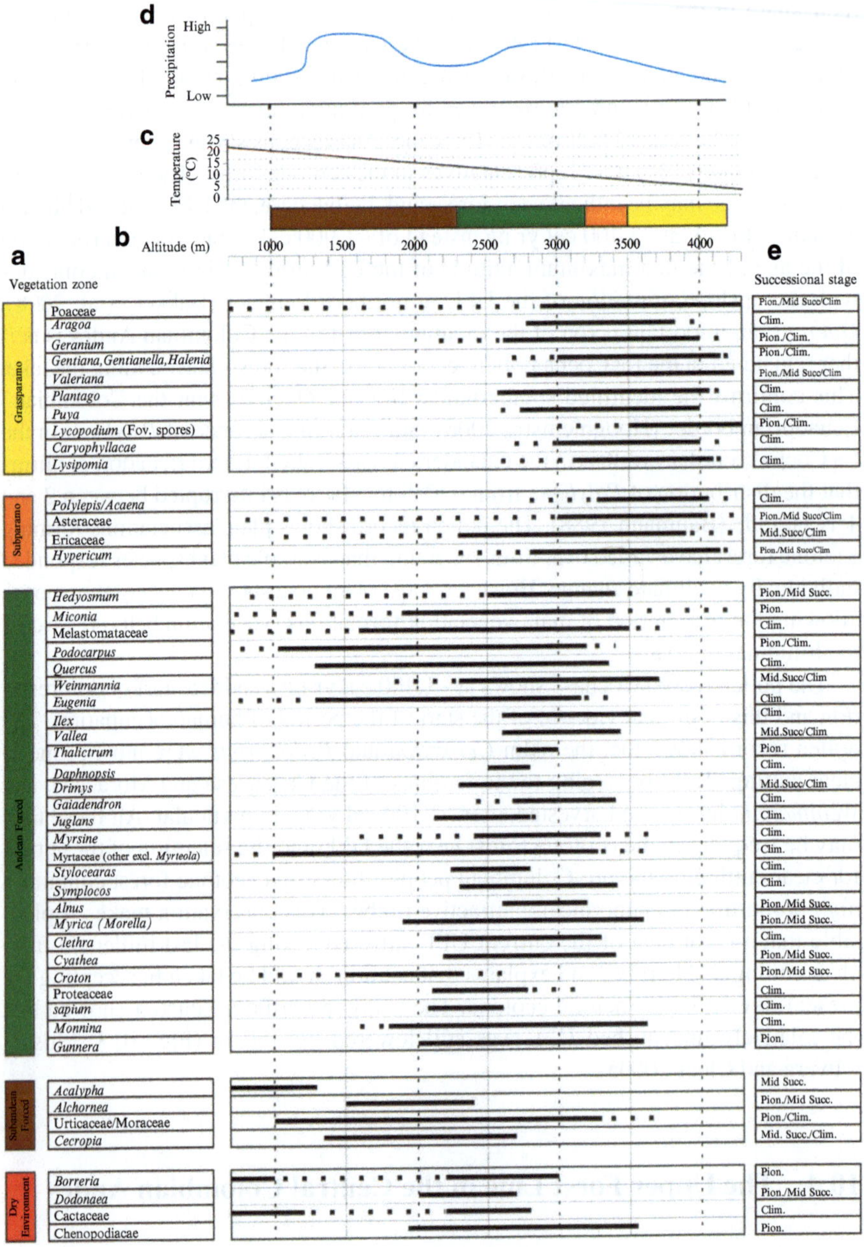

Fig. 10.5 Altitudinal ranges of selected taxa in the area of Lake Fúquene, Eastern Cordillera of Colombia. Taxa are arranged after ecological preference. (**a**) Main ecological groups; (**b**) modern altitudinal range (*dotted line* = full range; *solid line* = inteval of optimum cover), (**c**) mean annual temperature (MAT) along altitudinal gradient, (**d**) estimated anual precipitation (after Bogotá-A. et al. (2011a, b) with permission)

Cleef and Hooghiemstra (1984) provided a concise description of the main forest and páramo associations on the slopes of the Eastern Cordillera. We recognise that present-day vegetation associations are ephemeral, and may have repeatedly changed of floral composition during intervals of the past. Moreover, the taxonomic level of pollen identification is mostly at the genus level, sometimes at the higher ranked family level (e.g. Poaceae, Asteraceae, Ericaceae, Urticaceae–Moraceae), or at the lower ranked species level in case only one, or few species with characteristic pollen morphology do occur. Therefore, palynologists should work with a general framework of the altitudinal vegetation distributions allowing for deviations at site-specific conditions such as the high precipitation levels in the Tatamá area, and anticipate the presence of no-analogue vegetation types in the past.

The altitudinal distribution of pollen taxa selected as most informative to characterise environmental conditions in the area of Lake Fúquene was revisited (Bogotá-A. et al., 2011a) (Fig. 10.5). Most species turnover is located between 3,000 and 3,500 m elevation showing that vegetation at 3,200 m elevation, where we find today the UFL, has much opportunity to vary in composition in space and time.

Lake Fúquene sediments were cored up to 60 m depth in the deepest part of the lake. Pollen, and a suite of other proxies (grain size distributions and geochemical element analysis), were analysed at 1-cm distance along the full length of the composite record (Fq-9C) and show 284,000 years long records. The age model of this record is based on ^{14}C dating and a few biostratigraphical events in the top, and cyclostratigraphy in combination with orbital tuning in the remaining older part of the record (Groot et al. 2011). The temporal resolution is ~60 years between 284 and 27 ka (4,646 samples) and ~400 years for the last 27 ka (102 samples). This new record was compared with earlier palynological records from Lake Fúquene (Bogotá-A. et al. 2011b). It was demonstrated that all records show a consistent and robust sequence of environmental and climatic changes and differences between records could be related to the proximal or distal position of the core with respect to the lake border.

The pollen record shows changes in altitudinal vegetation distributions (Fig. 10.6). The position of the UFL shifted between ~2,000 and ~3,400 m elevation and was used to infer past changes in MAT. The main drivers of long-term climate change were the 41-kyr cycle of the obliquity band and the atmospheric concentration of greenhouse gases (Groot et al. 2011). Century up to millennial scale variation of 2–3.5 °C are registered all over the record and extreme temperature changes up to ~10 °C were registered at the last three glacial Terminations (Groot et al. 2011). Temperature variability in the northern Andes during the last two glacial cycles mimics temperature variation known from Greenland and Antarctic ice cores. The millennium-scale climate variability of the last glacial cycle, known as Dansgaard–Oeschger cycles (DO-cycles), also occurred during the previous glacial cycle and apparently is a stable characteristic of north Andean ecosystem variability. Based on distinct short-lived upslope excursions of the UFL we identified 28 DO-cycles during the last interglacial–glacial cycle (reflecting marine isotope stages 5-1), 20 DO-style cycles during the previous interglacial–glacial cycle (reflecting marine isotope stages 7 and 6), and 6 DO-style cycles during the last part of the

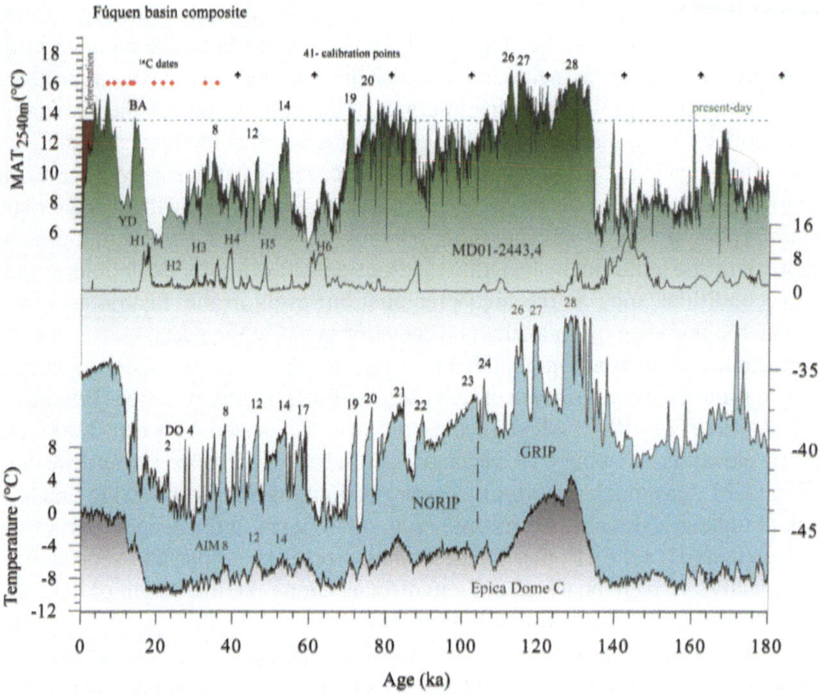

Fig. 10.6 *Top*: reconstruction of the MAT for the northern Andes based on the arboreal pollen percentage (AP%) record reflecting altitudinal migrations of the UFL ecotone. *Bottom*: reconstructed MAT for Greenland (NGRIP and GRIP records) and Antarctica (Epica Dome C record). Numbers 1–28 indicate Dansgaard–Oeschger (DO) cycles reflecting millennium-scale climate change. For additional explanation see Groot et al. 2011 (after Groot et al. (2011) with permission)

period reflecting marine isotope stage 8. The arboreal pollen percentage (AP%) record shows cycles of 119, 35, and 18 kyr, and at sub-orbital time-scale cycles of 10, 8, 7, 5, 4 and 1.49 kyr (Bogotá-A. et al. 2011a) which resemble the periodicities of the DO-cycles (Rahmstorf 2003). Cycles of 10 kyr have been related to half a precession cycle. Cycles of 7 and 5 kyr are interpreted as to reflect the third and fourth harmonics of the precession cycle (Berger et al. 2006). These sub-Milankovitch cycles are reported from low latitude records and explained as the reflection of variations in annual precipitation driven by variation on the strength of the monsoon systems (Tuenter et al. 2007). Changes in montane forest composition are led by *Alnus*, *Myrica*, *Quercus* and *Weinmannia* and these taxa apparently respond most rapidly to change. Later *Podocarpus*, *Miconia* and *Hedyosmum* show a response.

The high resolution pollen record of Lake Fúquene shows a high level of forest dynamics that was unknown so far. Ecotone forest apparently is able to respond at century-scale to rapid climate change. A new framework of past tropical montane forest dynamics is established with significant implications to assess current effects of Global Change. Observed migrations of ecotones and other altitudinally constrained distribution patterns of elements of flora and fauna, are expected to include a natural and an anthropogenic component. Records of UFL changes with decadal resolution spanning the last millennium are required to discriminate between natural trends and possible superimposed anthropogenic effects.

10.5 Discussion

In the frame of the Ecoandes project the altitudinal vegetation distribution was studied along altitudinal transects in several Cordilleras of Colombia. Here the most relevant results with respect to the ecotone reflecting the UFL are summarised and integrated with the previous case studies. For more specific information the reader is referred to the book series.

In the Parque Los Nevados (Central Cordillera, ~5°N) a clear altitudinal structural and floristic gradient can be observed. Upper montane rain forest (UMRF) (sensu Grubb 1974, 1977) is transitional to subalpine rain forest (SARF) which forms the UFL. This subdivision applies for the equatorial Andes. SARF is characterised by trees from 3 to 8 m height dominated by representatives of the Asteraceae, and a conspicuous shrub layer with more than 50% cover (Cleef et al. 1983, 2003). In downslope direction SARF is transitional to UMRF with trees of 20–35 m height with frequent Rosaceae, Brunelliaeae and Cunoniaceae and below this canopy horizons consisting of smaller trees and shrubs. Where condensation leads to maximum humidity briophytes are abundant. In general on the wet slopes of the mountains hepatics are more frequent and at thc dry side of the mountains mosses. A synthesis of observed plant associations in the Parque Los Nevados transect (Central Cordillera) shows an UFL at ~3,600 m elevation (Van der Hammen 2003). In the Tatamá transect at ~5°N (Van der Hammen 2005) reflecting the westernmost extension of the Parque Los Nevados transect montane forest is replaced by shrubby wet bamboo páramo at 3,800 m elevation. This remarkably high position of the UFL is explained by the high precipitation levels which reduce the frequency of night frost significantly. In general, in pollen records from the northern Andes the contribution of asteraceous pollen is shown as reflecting subpáramo vegetation (e.g. Van't Veer and Hooghiemstra 2000). However, at sites where at present-day asteraceous ecotone vegetation is abundant, percentages of Asteraceae are better shown as a separate ecological group potentially indicative of forest or páramo (e.g. Bakker et al. 2008).

For the Cordillera de Santa Marta (~11°N) the most common trees and shrubs of the high Andean forest were listed and include the following taxa: *Abatia*, *Alnus*, *Befaria*, *Berberis*, *Bocconia*, *Brunellia*, *Buddleia*, *Ceroxylon*, *Cinchona*, *Clethra*, *Clusia*, *Cordia*, *Daphnopsis*, *Durantha*, *Drimys*, *Eugenia*, *Gaiadendron*, *Geissanthus*, *Hesperomeles*, *Ilex*, *Ladenbergia*, *Meriania*, *Monochaetum*, *Nectandra*, *Ocotea*, *Oreopanax*, *Palicourea*, *Persea*, *Piper*, *Podocarpus*, *Prunus*, *Psychotria*, *Rapanea* (=*Myrsine*), *Rhamnus*, *Schefflera*, *Symplocos*, *Ternstroemia*,

Tibouchina, *Tournefortia*, *Vallea*, *Veronica*, *Weinmannia* and *Xylosma* (Cleef et al. 1984). Outside the Santa Marta massif and as far south as the Colombian–Ecuadorian border area also *Quercus* is an important element (Hooghiemstra 2006). These taxa have a general distribution in the northern Andes and are relevant for all studies of modern and past vegetation. Over 15 genera are palynologically not recognised due to lacking pollenmorphological keys. Therefore, there is a potential to increase the accuracy of pollen-based reconstructions of the UFL. According to different studies the UFL lies between 3,200 and 3,500 m (Cleef et al. 1984). Of the palms the genera *Ceroxylon* and *Geonoma* may reach up to 3,100 m, thus close to the UFL. Therefore, in palynological studies the record for palms need a specification in order to infer conclusions with respect to paleo-altitude and past temperature. Van der Hammen (1984) summarised the zonal ecosystems of the Santa Marta Massif and concluded that 3,300 m altitude is the best estimate for the position of the UFL.

Summarising UFL positions show an altitudinal range of some 500 m: 3,650 m in Guandera (northern Ecuador), 3,550 m in the area of Lake La Cocha (southern Colombia), 3,200 m in area of Lake Fúquene, 3,600 m in the parque Los Nevados, (Central Cordillera, Colombia), 3,800 m in Tatamá (Western Cordillera) and 3,200 to 3,400–3,500 m in the Cordillera de Santa Marta (northern Colombia). MAT, anual precipitation, fog persistence, edaphic factors, levels of UV radiation, and concentration of atmospheric greenhouse gases have been put forward as forcing factors of the altitudinal position of the UFL. Using an altitudinal temperature gradient of 0.6 °C/100 m UFL displacement the observed maximum 600 m variation in UFL positions potentially reflect a maximum temperature range of ~3.5 °C. However, inter-site variation of UFL positions in the northern Andes seems mainly related to stable topographical differences, or changes in the global setting of climate. This implies that during the period reflected by a pollen records driving factors are relatively stable. Therefore, establishing the natural altitudinal position of the UFL at the study site is sufficient to calibrate UFL changes into a record of past temperature change.

Past altitudinal positions of the UFL are estimated on the basis of the cumulative signals of all trees, i.e. on the basis of AP%. Today, databases on the distribution of neotropical plant taxa have developed into increasingly reliable sources of information (Punyasena et al. 2011). It should be explored if multiple genus-specific estimations of the altitudinal position of the UFL could lead to more accurate reconstructions.

10.6 Conclusions

The UFL is the most distinct ecotone along the slopes of tropical mountains. Notwithstanding regional variation in the northern Andes of the altitudinal position of the undisturbed UFL from 3,200 to 3,800 m, and various factors driving altitudinal change, the AP% record is the best parameter reflecting altitudinal UFL changes

(Groot et al. 2011). Main altitudinally restricted plant formations such as "upper montane forest" and "páramo" are characterised by suites of plant taxa which respond in concert to climate change. However, pollen records also show that species composition of ecotone forest has varied over time, reflecting examples of plant species that responded individually to climate change. The implication is that current plant associations are to some degree ephemeral. The search for categories of plants to reflect climatic parameters should allow for this temporal variation.

Acknowledgements Results presented in this paper were developed in several projects financially supported by NWO-ALW (J.C. Berrio: ALW 854.00.007; M. Vriend †: ALW 854.00.007), WOTRO (J.C. Berrio: WB 84-552; M. Vriend †: WB 84-552; Z. González-Carranza; WB 84-636; M.C. Moscol-Olivera: WAN 84-572), AlBan (R.G. Bogota-A.: E04D0388907CO) and IBED (M. Groot). HH wrote the manuscript with additional comments from JCB and RGBA. A.M. Cleef supported Fig. 10.3 and ZGC prepared Fig. 10.4. Data from the Ecoandes project were used and we thank the main coordinators of this project T. van der Hammen †, A.M. Cleef and O. Rangel.

References

Bakker J, Moscol Olivera M, Hooghiemstra H (2008) Holocene environmental change at the upper forest line in northern Ecuador. Holocene 18:877–893

Baumann F (1988) Geographische Verbreitung und Okologie südamerikanischer Hochgebirgspflanzen. Physische Geographie 28, Geographisches Institut der Universität, Zürich, p 206

Behling H, Hooghiemstra H (1999) Environmental history of the Colombian savannas of the Llanos Orientales since the Last Glacial Maximum from lake records El Piñal and Carimagua. J Paleolimnol 21:461–476

Berger A, Loutre MF, Mélice JL (2006) Equatorial insolation: from precession harmonics to eccentricity frequencies. Clim Past 2:519–533

Berrio JC, Hooghiemstra H, Behling H, Botero P, Van der Borg K (2002) Late-Quaternary savanna history of the Colombian Llanos Orientales from Lagunas Chenevo and Mozambique: a transect synthesis. Holocene 12:35–48

Bogotá-A. RG, Hooghiemstra H, Berrio JC (2011a) An ultra-high resolution multi-proxy record from lake Fúquene (Colombia): orbital to submillennial-scale dynamics of montane vegetation, climate, lake-level changes and sedimentary environments. I: period 284-130 kyr before present. In: Bogotá-A. RG (ed) Pleistocene centennial-scale vegetational, environmental, and climatic change in the Colombian Andes. PhD thesis, University of Amsterdam, p 144

Bogotá-A. RG, Groot MHM, Hooghiemstra H, Lourens LJ, Van der Linden M, Berrio JC (2011b) Rapid climate change from north Andean Lake Fúquene pollen records driven by obliquity: implications for a basin-wide biostratigraphic zonation for the last 284 ka. Quat Sci Rev 30:3321–3337

Boom A, Marchant R, Hooghiemstra H, Sinninghe Damsté JS (2002) CO_2- and temperature controlled altitudinal shifts of C_4- and C_3-dominated grasslands allow reconstruction of paleoatmospheric pCO_2. Palaeogeogr Paloaeoclimatol Palaeoecol 177:151–168

Cárdenas ML, Gosling WD, Sherlock SC, Poole I, Pennington RT, Mothes P (2011) The response of vegetation on the Andean Flank in Western Amazonia to Pleistocene climate change. Science 331:1055–1058

Chen I-C, Hill JK, Ohlemüller R, Roy DB, Thomas CD (2011) Rapid range shifts of species associated with high levels of climate warming. Science 333:1024–1026

Cleef AM (1981) The vegetation of the páramos of the Colombian Cordillera Oriental. Diss Bot 61:p 321

Cleef AM, Hooghiemstra H (1984) Present vegetation of the area of the high plain of Bogotá. In: Hooghiemstra H (ed) Vegetational and climatic history of the high plain of Bogotá, Colombia. Dissertaciones Botanicae 79:42–66, Cramer, Vaduz

Cleef AM, Rangel-Ch. O, Salamanca-V. S (1983) Reconocimiento de la vegetación de la parte alta del transecto Parque los Nevados. In: Studies on Tropical Andean Ecosystems, vol 1, La Cordillera Central Colombiana Transecto Parque Los Nevados. J Cramer Vaduz, pp 150–173

Cleef AM, Rangel-Ch. O, Van der Hammen T, Jaramillo-M. R (1984) La vegetación de las selvas del Transecto Buritaca/The forest vegetation of the Buritaca transect. In: In: Studies on Tropical Andean Ecosystems, vol 2, La Sierra Nevada de Santa Marta (Colombia) Transecto Buritaca-la Cumbre. J Cramer, Berlin-Stuttgart, pp 267–406

Cleef AM, Rangel-Ch. O, Salamanca S (2003) The Andean rain forests of the Parque Los Nevados transect, Cordillera Central, Colombia. In: Studies on Tropical Andean Ecosystems, vol 5, La Cordillera Central Colombiana Transecto Parque los Nevados. J Cramer/Bornträger, Berlin-Stuttgart, pp 79–141

Cuatrecasas J (1934) Observaciones geobotánicas en Colombia. Trabajos Museo Nacional de Ciencias Naturales Serie Botánica 27:144

Cuatrecasas J (1958) Aspectos de la vegetación natural de Colombia. Revista Academía Colombiana de Ciencias Exactas Fisicas y Naturales 10(40):221–264

Dickinson G, Murphy K (1998) Ecosystems; a functional approach. Routledge, London

Flenley JR (1992) Palynological evidence relating to disturbance and other ecological phenomena of rain forests. In: Goldammer JG (ed) Tropical forests in transition; ecology of natural and anthropogenic disturbance processes. Birkhauser, Basel, pp 17–24

Gleason HA (1926) The individualistic concept of the plant association. Torrey Bot Club Bull 53:7–26

González-Carranza Z, Hooghiemstra H, Vélez MI (2012) Major altitudinal shifts in Andean vegetation on the Amazonian flank show temporarily loss of biota in the Holocene. The Holocene, in press

Groot MHM, Bogotá RG, Lourens LJ, Hooghiemstra H, Vriend M, Berrio JC, Tuenter E, Van der Plicht JH, Van Geel B, Ziegler M, Weber SL, and 14 other Fúquene Project Members (2011) Ultra-high resolution pollen record from the northern Andes reveals shifts in montane climates within the last two glacial cycles. Clim Past 7:299–316

Grubb PJ (1974) Factors controlling the distribution of forest-types on tropical mountains: new facts and a new perspective. In: Flenley JR (ed.), Altitudinal zonation in Malesia. Transactions on the Third Aberdeen-Hull Symposium on Malesian Ecology, Hull, 1973, pp 13–45. Miscellaneous Series vol. 13 Dept. Geography, University of Hull, UK

Grubb PJ (1977) Control of forest growth and distribution on wet tropical mountains: with special reference to mineral nutrition. Review of Ecology and Systematics 8:83–107

Holtmeier FK (2009) Mountain timberlines: ecology, patchiness and dynamics. Advances in Global Change Research 36. Springer, Germany

Hooghiemstra H (1984) Vegetational and climatic history of the high plain of Bogotá, Colombia. Diss Bot 79:368

Hooghiemstra H (2006) Immigration of oak into northern South America: a paleo-ecological perspective. In: Kappelle M (ed) Ecology and conservation of neotropical montane oak forests. Ecological Studies, vol 185. Springer, Berlin, pp 17–28

Jansen B, Haussmann NS, Tonneijck FH, Verstraten JM, De Voogt P (2008) Characteristic straight-chain lipid ratios as a quick method to assess past forest-páramo transitions in the Ecuadorian Andes. Palaeogeogr Palaeoclimatol Palaeoecol 262:129–130

Jansen B, Van Loon EE, Hooghiemstra H, Verstraten JM (2010) Improved reconstruction of palaeo-environments through un ravelling of preserved vegetation biomarker patterns. Palaeogeogr Palaeoclimatol Palaeoecol 285:119–130

Laegaard S (1992) Influence of fire in the grass páramo vegetation of Ecuador. In: Balslev H, Luteyn JL (eds) Páramo, an Andean ecosystem under human influence. Academic, London, pp 151–170

Lauer W, Rafiqpoor MD, Theisen I (2001) Physiogeographie, Vegetation und Syntaxonomie der Flora des Páramos de Papellacta (Ostkordillere, Ecuador). Erdwissenschaftliche Forschung 39. Steiner, Stuttgart, Germany, p 144

Luteyn JL (1999) Páramos; a checklist of plant diversity, geographical distribution, and botanical literature. Memoirs of the New York Botanical Garden, Bronx, New York, vol 84. p 278

Marchant R, Behling H, Berrio JC, Cleef A, Duivenvoorden J, Hooghiemstra H, Kuhry P, Melief B, Schreve-Brinkman E, Van Geel B, Van der Hammen T, Van Reenen G, Wille M (2002a) Pollen-based biome reconstructions for Colombia at 3000, 6000, 9000, 12000, 15000 and 18000 ^{14}C yr ago: Late Quaternary tropical vegetation dynamics. J Quat Sci 17:113–129

Marchant R, Boom A, Hooghiemstra H (2002b) Pollen-based biome reconstructions for the past 450,000 yr from the Funza-2 core, Colombia: comparisons with model-based vegetation reconstructions. Palaeogeogr Paloaeoclimatol Palaeoecol 177:29–45

Marchant R, Harrison SP, Hooghiemstra H, Markgraf V, Van Boxel JH, Ager T, Almeida L, Anderson R, Baied C, Behling H, Berrio JC, Burdridge R, Björck S, Byrne R, Bush MB, Cleef AM, Duivenvoorden JF, Flenley JR, De Oliveira P, Van Geel B, Graf KJ, Gosling WD, Haberle S, Van der Hammen T, Hansen BCS, Horn SP, Islebe GA, Kuhry P, Ledru M-P, Mayle FE, Leyden BW, Lozano-García S, Melief ABM, Moreno P, Moar NT, Prieto A, Van Reenen GB, Salgado-Labouriau ML, Schäbitz F, Schreve-Brinkman EJ, Wille M (2009) Pollen-based biome reconstructions for Latin America at 0, 6000 and 18,000 radiocarbon years. Clim Past 5:369–461

Moscol Olivera M, Cleef AM (2009a) Vegetation composition and altitudinal distribution of Andean rain forests in El Angel and Guandera reserves, northern Ecuador. Phytocoenologia 39:175–204

Moscol Olivera M, Cleef AM (2009b) A phytosociological study of the páramo along two altitudinal transects in El Carchi Province, northern Ecuador. Phytocoenologia 39:79–107

Moscol Olivera M, Hooghiemstra H (2010) Three millennia upper forest line changes in northern Ecuador: pollen records and altitudinal vegetation distributions. Rev Palaeobot Palynol 163:113–126

Moscol Olivera M, Duivenvoorden JF, Hooghiemstra H (2009) Pollen rain and pollen representation across a forest-páramo ecotone in northern Ecuador. Rev Palaeobot Palynol 157:285–300

Otto A, Simpson MJ (2005) Degradation and preservation of vascular plant-derived biomarkers in grassland and forst soils from Western Canada. Biogeochemistry 74:377–409

Punyasena SW, Dalling JW, Jaramillo C, Turner BL (2011) Comment on "The response of vegetation on the Andean flank in western Amazonia to Pleistocene climate change" Science 333:1825-b

Rahmstorf S (2003) Timing of abrupt climate change: a precise clock. Geophys Res Lett 30:1510. dio:10.1029/2003GL017115

Schmithüsen J (1976) Atlas zur Biogeographie. Bibliographisches Institut, Mannheim-Wien-Zürich, Meyers Grosser Physischer Weltatlas, vol 3

Smith TM, Smith JB Jr, Shugart HH (1992) Modeling the response of terrestrial vegetation to climate change in the tropics. In: Goldammer JG (ed) Tropical forests in transition; ecology of natural and anthropogenic disturbance processes. Birkhauser, Basel, pp 253–268

Tuenter E, Weber SL, Hilgen FJ, Lourens LJ (2007) Simulating sub-Milankovitch climate variations associated with vegetation dynamics. Clim Past 3:169–180

Van der Hammen T (1974) The Pleistocene changes of vegetation and climate in tropical South America. J Biogeogr 1:3–26

Van der Hammen T (1984) Ecosistemas zonales en el flanco norte de la Sierra Nevada de Santa Marta (Trensecto Buritaca-La Cumbre)/Zonal ecosystems of the northern flank of the Sierra Nevada de Santa Marta (Buritaca-La Cumbre transect). In: Studies on Tropical Andean Ecosystems, vol 2, La Sierra Nevada de Santa Marta (Colombia) Transecto Buritaca-la Cumbre. J Cramer, Berlin-Stuttgartt, pp 589–603

Van der Hammen T (2003) Ecosistemas zonales en los flancos oeste y este de la Cordillera Central Colombiana (transecto Parque Los Nevados)/Zonal ecosystems of the west and east flanks of the Colombian Central Cordillera (Parque Los nevados transect). In: Studies on Tropical Andean Ecosystems, vol 5, La Cordillera Central Colombiana Transecto Parque los Nevados. J Cramer/Bornträger, Berlin-Stuttgart, pp 503–545

Van der Hammen T (2005) Zonal ecosystems of the west and east flanks of the Colombian Western Cordillera (Tatamá Transect). In: Studies on Tropical Andean Ecosystems, vol 6, La Cordillera Occidental Colombiana Transecto Tatamá. J Cramer/Bornträger, Berlin-Stuttgart, pp 935–972

Van der Hammen T, González E (1963) Historia de clima y vegetación del Pleistoceno Superior y del Holoceno de la Sabana de Bogotá. Boletin Geológico 11:189–266

Van't Veer R, Hooghiemstra H (2000) Montane forest evolution during the last 650,000 years in Colombia: a multivariate approach based on pollen record Funza-I. J Quat Sci 15:329–346

Von Humboldt A, Bonpland A (1807) Ideen zu einer Geographie der Pflanzen nebst einem Naturgemälde der Tropenländer. Cotta, Tübingen

Wille M, Hooghiemstra H, Behling H, Van der Borg K, Negret AJ (2001) Environmental change in the Colombian subandean forest belt from 8 pollen records: the last 50 kyr. Veg Hist Archaeobot 10:61–77

Young K, León B (2007) Tree-line changes along the Andes: implications of spatial patterns and dynamics. Philos Trans R Soc Lond B 362:263–272

Chapter 11
Climatic Drivers of Tree Growth and Recent Recruitment at the Pyrenean Alpine Tree Line Ecotone

Enric Batllori, J. Julio Camarero, and Emilia Gutiérrez

11.1 Introduction

Global climate is currently warming at an unprecedented rate with potentially profound and widespread effects on the distributions of plant species and ecological communities (IPCC 2007; Lenoir et al. 2008). Mountain ecosystems and their unique biota are particularly sensitive to such changes (Beniston 2003). In high elevation forests, climate has been considered to be the main limiting factor for tree growth, reproduction and establishment (e.g. Tranquillini 1979; Körner 1998; Ettinger et al. 2011). The upper elevational limit of forest and tree growth on mountain slopes, the alpine tree line ecotone (ATL), represents an abrupt transition in life form dominance and is one of the most prominent vegetation boundaries between ecosystems (Holtmeier 2009). On a global scale, heat deficiency remains the most likely explanation of ATL elevation irrespective of the latitude and the tree line forming species (Körner and Paulsen 2004). Hence, based on the traditional tree line paradigm, warm temperature is favourable to both tree radial growth and reproductive success, and thus the ATL may be exceptional for the potential it offers for the assessment of the impacts of anthropogenic warming on mountain forests.

E. Batllori (✉)
Departament d'Ecologia, Universitat de Barcelona, Avinguda Diagonal 645, 08028 Barcelona, Spain
mail: enric.batllori@gmail.com

Department of Environmental Science, Policy & Management, University of California, Berkeley 94720, USA

J.J. Camarero
ARAID, Instituto Pirenaico de Ecología, CSIC, Avenida Montañana, 1005, 50080 Zaragoza, Spain

E. Gutiérrez
Departament d'Ecologia, Universitat de Barcelona, Avinguda Diagonal 645, 08028 Barcelona, Spain

R.W. Myster (ed.), *Ecotones Between Forest and Grassland*,
DOI 10.1007/978-1-4614-3797-0_11, © Springer Science+Business Media New York 2012

Ongoing climate warming is expected to cause significant changes in the structure, dynamics and position (e.g. upward shifts) of the ATL (Grace et al. 2002; Harsh et al. 2009), which is of concern due to potential loss of biodiversity and ecosystem function as alpine grasslands are replaced by woody species. Indeed, many studies around the globe report important population changes at the ATL during the twentieth century in response to climate change (e.g. Camarero and Gutiérrez 2004; Gamache and Payette 2005; Kullman 2005; Danby and Hik 2007; Batllori and Gutiérrez 2008; Fang et al. 2009; Kharuk et al. 2010; Liang et al. 2011). However, despite broad-scale constraints on tree lines imposed by global-scale thermal trends, inverse relationships between growth and climate warming have also been described (e.g. Lloyd and Graumlich 1997; Lloyd and Fastie 2002; Wilmking et al. 2004), evidencing that even at the ATL trees can be negatively affected when temperatures warm beyond a physiological threshold (D'Arrigo et al. 2004). Moreover, ATL stability and differential tree line responses (e.g. Dalen and Hofgaard 2005; Payette 2007; Green 2009; Bogaert et al. 2011; Liang et al. 2011) have been reported under the global warming context. ATL heterogeneity and factors controlling its position and structure are highly scale-dependent (Sveinbjörnsson et al. 2002), and regional to local-scale factors other than temperature (e.g. wind regime, topographic and geomorphologic features, biotic interactions) can be responsible of ATL stability, abrupt changes and reversible responses at the tree line (Noble 1993; Holtmeier and Broll 2010; Kharuk et al. 2010; Stueve et al. 2011). This may preclude general projections of global change assuming simple, linear ecological responses. Overall, despite over a century of research at the tree line, the varying sensitivity of ATL responses to climate among sites, the degree to which ATLs dynamic may lag climate change, as well as the reasons of such site-dependent responsiveness remain largely unknown (but see Harsh and Bader 2011).

Climate warming is only one aspect of environmental change that affects the location and dynamics of ATL. Among the other factors, anthropogenic disturbances (i.e. land-use) may be the most important in shaping tree line dynamics in many regions across the globe, determining and constraining actual ATL advances under a warming climate (Bogaert et al. 2011). However, in spite of the fact that responses due to human-induced changes can initially mask or even override climatic controls (e.g. land-use abandonment, Hofgaard 1997), after disturbance cessation tree lines formerly disturbed by anthropogenic use become sensitive to climate under sustained warming conditions (Holtmeier and Broll 2005). For instance, unfavourable site conditions constrain the invasibility of subalpine grasslands by trees more than in low elevation forests (Dullinger et al. 2004). If such limiting conditions for tree establishment improve due to changing climate, subsequent ATL dynamics (e.g. seedling survival and growth) could be considered as climatically driven even though it actually occurs below the climatic tree line (Gehring-Fasel et al. 2007). This may be the case in many ATLs in the Pyrenean range where, as in many other Eurasian mountain systems (e.g. Alps), local human activities (e.g. seasonal farming) were common at the tree line but they have decreased drastically during the twentieth century (Garcia-Ruiz et al. 1996). ATL advances can be initially associated with a change in height growth of previously

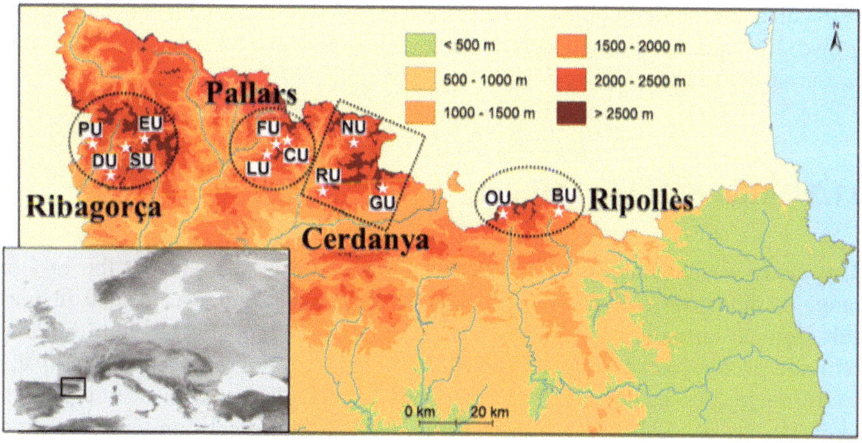

Fig. 11.1 Location of the 12 alpine *Pinus unicnata* tree line ecotones studied in the Catalan Pyrenees, NE Spain, and Andorra (see Table 11.1 for sites and codes). *Dotted lines* depict the four climatic sub-regional areas considered in this study to analyse the relationships between climate and tree growth and recruitment

established individuals (Batllori and Gutiérrez 2008; Aune et al. 2011), but further upward shifts are dependent on new regeneration and survival of recruits beyond the current ATL position. The establishment of new recruits has such a fundamental potential impact on spatial change at the ATL that it has been proposed to be crucial to the study of its dynamics (Malanson et al. 2007).

In this study we describe the spatial patterns of tree recruitment, examine the age frequency distribution of recruits and evaluate the influence of climate factors on both recruitment and radial growth of *Pinus uncinata* ATLs at a regional scale in the Pyrenees. Our study focuses on 12 sites close to the potential ATL altitude, including sites whose dynamics is climatically driven and sites influenced by the after-effects of human-induced perturbations (see Batllori and Gutiérrez 2008, for further details). Despite land-use changes have played a major role in driving the dynamics of *P. uncinata* forests over its distributional range in the Pyrenees, especially at low altitudes (Améztegui et al. 2010), without a reactivation of pastoral use and under future climate warming predictions (IPCC 2007), changing environmental conditions may become the most important factor in determining the dynamics of the studied ATLs (Holtmeier and Broll 2005; Gehring-Fasel et al. 2007). Moreover, similar recruitment processes have been reported between the study sites irrespective of past disturbance regime (Batllori et al. 2010).

A preliminary analysis of the climatic influence on tree growth and recruitment at each ATL precluded a site-specific approach (see Sects. 11.3 and 11.4). Thus, we evaluated the climatic influence on ATL in the Pyrenees at a sub-regional scale (Fig. 11.1). We specifically address the following questions: (1) Which is the spatiotemporal pattern of recent recruitment at the study sites? (2) Are synchronous responses in tree growth and recruitment pulses present over the studied range?

(3) Which are the main climatic factors related to seedling establishment and tree growth? We aim towards a deeper understanding on which climate conditions may favour or limit future ATL responses.

11.2 Materials and Methods

This study was conducted on *P. uncinata* (Ramond ex DC.) ATLs at the eastern range of the Spanish Pyrenees (Catalan Pyrenees) and Andorra, located in four sub-regional climatic areas (Fig. 11.1, Table 11.1). Homogeneous sub-regional climatic areas were defined on the basis of major topographic features at the study range (e.g. catchments) and of previous ecological studies of tree growth (e.g. Macias et al. 2006). *P. uncinata* is a shade-intolerant conifer presenting several attributes typical of good invaders (e.g. light-winged seeds, short pre-reproductive period) which dominates most of the subalpine forests in the Pyrenees (Ninot et al. 2007); its potential tree line elevation ranges between 2,200 and 2,450 m a.s.l. depending on continentality, exposure and landform (Carreras et al. 1996). A combination of field visits and interpretation of aerial photographs was used to select 12 sites where a rectangular plot, its longer side parallel to the elevational gradient, was set up from the closed forest to the treeless alpine communities. Plot area varied from 940 to 7,600 m^2. The study sites were located on uniform north-facing slopes out of avalanche paths, and no evidence of recent fires or current human use were found in any of the stands (i.e. they were no longer intensively grazed). However, the 12 sites can be classified in two groups according to the influence of recent (ca. 50 years) human-induced disturbances as inferred from tree height and age and their transition patterns across the ecotone (Batllori and Gutiérrez 2008). Seven of the plots were representative of climatic ATLs whereas the other five were regarded as sites influenced by the after effects of land-use changes (Table 11.1). Important densification trends have been reported at all sites starting at 1950s, in parallel with land-use changes and warming trends. Since the mid-1970s warming has accelerated in the study area (Fig. 11.2).

The Pyrenees constitute a transitional area between more humid conditions in their northern margin and drier conditions southwards (Vigo and Ninot 1987). A second longitudinal gradient is caused by the location of the range between the Atlantic Ocean westwards and the Mediterranean Sea eastwards. The climate in the study area is continental with oceanic (western sites) or Mediterranean (eastern sites) influences. The Mediterranean influence is characterized by higher summer temperatures, and plausibly water deficit, than westwards.

11.2.1 Field Sampling

The location of all tree stems, using rectangular coordinate axes, was noted in each plot. Positions of all living trees were recorded in Cartesian coordinates using tapes

Table 11.1 Characteristics of the 12 alpine *Pinus uncinata* tree line ecotones studied in the Catalan Pyrenees, NE Spain, and Andorra

Climate sub-region	Site	Latitude (north)	Longitude (east)	Plot size (m^2)	Altitude (m a.s.l.)	Slope (°)	Aspect	Bedrock	Transition pattern: Tree age	Transition pattern: Tree size
Ribagorça	PU	42° 31′	0° 45′	6,000	2,199–2,268	28	N	Lime	Step-like	Gradual
	DU	42° 28′	0° 49′	1,200	2,010–2,069	30	N	Lime	Gradual	Gradual
	SU	42° 32′	0° 53′	1,500	2,270–2,338	27	NW	Lime	Gradual	Gradual
	EU	42° 33′	0° 56′	940	2,299–2,339	24	W	Granodiorite	Step-like	Gradual
Pallars	LU	42° 31′	1° 21′	1,180	2,363–2,405	21	N	Slate	Gradual	Gradual
	FU	42° 33′	1° 23′	7,600	2,352–2,435	24	NW	Slate	Gradual	Gradual
	CU	42° 33′	1° 25′	1,160	2,250–2,297	26	NW	Lime	Step-like	Gradual
Cerdanya	RU	42° 26′	1° 32′	1,650	2,236–2,271	24	W	Slate	Gradual	Gradual
	NU	42° 34′	1° 37′	980	2,266–2,314	29	N	Lime	Gradual	Gradual
	GU	42° 27′	1° 44′	1,500	2,312–2,365	16	NE	Slate	Gradual	Gradual
Ripollès	OU	42° 23′	2° 08′	1,500	2,241–2,308	27	N	Slate	Step-like	Step-like
	BU	42° 24′	2° 19′	1,500	2,184–2,241	22	N	Slate	Step-like	Step-like

Sites recently affected (since the 1950s) by anthropogenic disturbance (logging, grazing) were regarded as those with step-like transition patterns in age and/or size along the tree line ecotone, whereas sites with gradual age and height transitions were considered as undisturbed

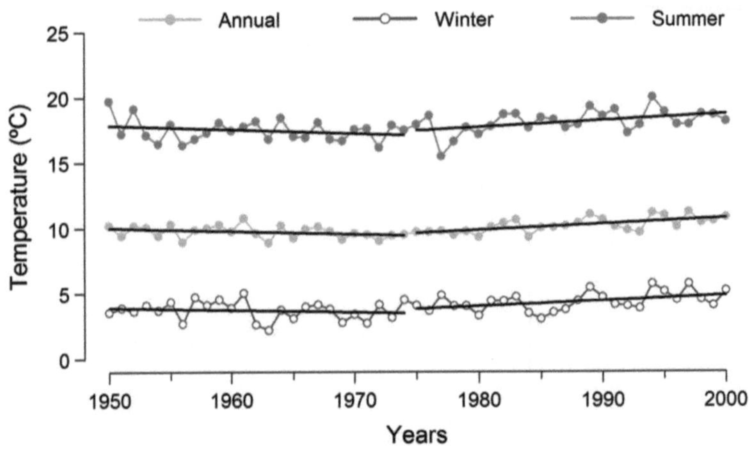

Fig. 11.2 Climatic trends during the second half of the twentieth century in the eastern Spanish Pyrenees and Andorra. Meteorological data were taken from the CRU TS 2.0 0.5°-gridded data set (Mitchell et al. 2004, 2005) considering the region encompassed by coordinates 42° 30′–43° 00′ N and 0° 30′–2° 30′ E. Linear models (*black lines*) for the periods 1950–1974 and 1975–2000 are depicted to highlight temperature trends

along the shorter (x-axis) and longer (y-axis) sides of the plot. Point $(x,y) = (0,0)$ was located in the lower left corner looking upslope. Additional tapes were placed every 5 m to subdivide the plot and measure the coordinates of the centre of each tree to the nearest 0.1 m. Basic biometric measurements (e.g. diameter at breast height; d.b.h., total tree height; h.t.) and age determination by means of core extraction or counting of bud scars or internodes was performed for 3,639 *P. uncinata* trees (see Batllori and Gutiérrez 2008, for details). Trees were classified by size as follows: large upright trees (d.b.h. > 7.5 cm), saplings (d.b.h. < 7.5 cm and h.t. > 0.5 m) and seedlings (h.t. < 0.5 m). All individuals big enough (d.b.h. > 10 cm) were cored at ca. 1.30 m to develop ATL ring-width chronologies, and natural regeneration was carefully examined in each plot for the demographic analysis of recruitment. Field sampling was conducted during the summers of the period 2003–2006.

11.2.2 Laboratory Methods

Cores were prepared following the standard dendrochronological techniques (Stokes and Smiley 1968). All samples were visually cross-dated under a stereomicroscope. Light rings, frost and narrow rings of each core were used as signature rings (Yamaguchi 1991). In accordance with Tardif et al. (2003), false or incomplete rings were rarely encountered. Afterwards, ring widths were measured to a precision of 0.01 mm using the Lintab measuring device (Frank Rinn Heidelberg, Germany) and the program TSAP-Win (Rinn 2003). Finally, cross-dating quality and measurement errors were further validated using the program COFECHA (Holmes 1983).

A ring-width chronology was developed for each site (Table 11.2). The measured and crossdated ring-width series from each tree were detrended and standardized using the "dplR package" in R.2.13 (Bunn 2008, R Development Core Team 2011). Given the short length of the measured series, we used a conservative detrending method to remove age-related growth trends while preserving as much low-frequency variation possible. Detrending involved fitting a negative exponential curve to each series, which were then standardized by transforming ring-width values into a dimensionless index (RWI) by dividing the observed by the expected values given by the negative exponential function. Site chronologies were obtained by averaging the RWI value for each year across all trees. Then, autoregressive modelling was performed on each chronology to remove temporal autocorrelation. RWI series were averaged by site (site ATL chronologies) and by each of the four sub-regional climatic areas included in the study region (sub-regional ATL chronologies). Comparison of tree growth patterns among sites were made by calculating Pearson's product–moment correlation coefficients among all pairs of ATL chronologies for the period 1955–2000 (Table 11.3). Despite significant and positive correlations ($P<0.05$) were observed among many of the tree line chronologies, correlation coefficients were on average higher between sites in the same climate sub-region than between sites from different sub-regions. Accordingly, sub-regional tree line composite chronologies were computed by averaging the RWI value for each year across all trees in each sub-region (DU growth series were excluded from this analysis because they were poorly related to the other series in the same region). The sub-regional chronologies were well replicated and showed strong common signal for the period 1951–2000 (see Table 11.2 for details).

11.2.3 Data Analysis

11.2.3.1 Spatiotemporal Patterns of Recruitment

We used spatially corrected Spearman correlation coefficients to examine the spatial relationship, at plot scale, between densities of two regeneration periods (1950–1974 vs. 1975–1999). Such approach may indicate whether recent recruitment dynamics reflects ATL shifts or densification processes (ecotone infilling). To perform this analysis we used 5-m distance gridded density data to compute Spearman rank correlation coefficients between the two recruitment periods. Correlation coefficients were calculated using the program Spatial Analysis in Macroecology (SAM) (Rangel et al. 2006). We evaluated the spatial pattern of consecutive life stages (seedlings, saplings and large trees) and assessed changes from one class to another (univariate point-pattern analyses). Further, we analysed the spatial interaction between life stages (bivariate point-pattern analyses) to determine if they are primarily positive (aggregation) or negative (segregation) (see Batllori et al. 2010, for further details).

Table 11.2 Chronology characteristics of the 12 tree lines studied

Sub-region	Site	Time span	N° trees	N° radii	Mean series length	Mean tree-ring width (mm ± 1SD)	m_{sx}[a]	r_{bt}[a]	EPS[a]
Ribagorça	PU	1955–2000	17	24	86.5	1.60 ± 0.62	0.132	0.470	0.865
	DU	1955–2000	12	18	54.2	1.97 ± 0.76	0.141	0.612	0.872
	SU	1955–2000	16	28	56.5	1.57 ± 0.61	0.152	0.542	0.852
	EU	1955–2000	15	23	61.3	1.49 ± 0.53	0.125	0.616	0.902
	LU	1955–2000	10	18	57.0	1.42 ± 0.56	0.149	0.687	0.896
Pallars	FU	1955–2000	24	39	54.0	1.56 ± 0.67	0.180	0.554	0.910
	CU	1960–2000	13	16	50.0	1.91 ± 0.68	0.154	0.547	0.850
	RU	1955–2000	6	8	38.1	2.36 ± 0.77	0.166	0.685	0.853
Cerdanya	NU	1955–2000	9	20	83.4	1.47 ± 0.55	0.176	0.682	0.872
	GU	1955–2000	18	21	48.2	1.50 ± 0.52	0.267	0.552	0.889
Ripollès	OU	1960–2000	7	8	34.7	2.04 ± 0.77	0.217	0.730	0.868
	BU	1955–2000	20	25	49.6	1.78 ± 0.66	0.191	0.746	0.961

[a]Dendrochronological statistics (see Fritts 1976): mean sensitivity (m_{sx}), a measure of the year-to-year variability in width of consecutive tree rings; mean between-trees correlation (r_{bt}), a measure of the similarity in growth among trees; expressed population signal (EPS), a measure of the statistical quality of the mean site chronology as compared with a perfect infinitely replicated chronology (a chronology with EPS > 0.85 is regarded as statistically reliable; see Wigley et al. 1984)

Table 11.3 Pearson product–moment correlation coefficients between the 12 tree line chronologies

	DU	SU	EU	LU	FU	CU	RU	NU	GU	OU	BU
PU	0.018	0.572*	0.557*	0.638*	0.601*	0.471*	0.194	0.316*	0.616*	0.572*	0.537*
DU		0.053	0.192	0.065	0.049	0.165	0.296*	0.183	0.192	–0.052	0.231
SU			0.738*	0.773*	0.802*	0.597*	0.268	0.474*	0.688*	0.603*	0.076
EU				0.693*	0.758*	0.253*	0.512*	0.586*	0.831*	0.483*	0.198
LU					0.909*	0.716*	0.385*	0.548*	0.693*	0.626*	0.209
FU						0.581*	0.365*	0.436*	0.674*	0.690*	0.341*
CU							0.476*	0.576*	0.703*	0.409*	0.113
RU								0.510*	0.568*	0.088	0.124
NU									0.644*	0.195	0.080
GU										0.437*	0.216
OU											0.571*

*Significant correlation coefficients at $P<0.05$ level are highlighted in bold

11.2.3.2 Recruitment Age Frequency Distributions

Based on the dating error of 270 collected small individuals at the study sites (Batllori and Gutiérrez 2008), and of a recent evaluation of growth forms and age estimation of tree line species conducted by Caccianiga and Compostella (2011), we developed 5-year age-class frequency distributions. Although we intensively examined all area inside each plot, field sampling may have underestimated the frequency of newly established recruits due to the difficulty of detecting very small pine seedlings (1–3 years old). Thus, we restricted age structures and spatiotemporal analyses of recruitment to the period 1950–1999, which covers the period of increasing ATL density reported at the study sites. Spatiotemporal patterns of recruitment were evaluated for each site, whereas the climatic influence (due to climate data limitations) was evaluated at a sub-regional scale. Temporal patterns of recent recruitment (age frequency distributions in 5-year classes) were compared between sites using paired Kolmogorov–Smirnov tests (with Bonferroni adjustment of the significance level; $\alpha=0.05$/number of comparisons; Gamache and Payette 2005). Despite maximum recruitment pulses varied between sites, and that demographic age structures were visually different between the studied ATLs, Kolmogorov–Smirnov tests yielded no significant differences between the temporal patterns of recruitment between any of the study sites. Thus, we computed sub-regional demographic structures of recent recruitment at the ATL, which represent the total number of recruits per hectare recruited over time in each of the four climatic sub-regions.

11.2.3.3 Relationships with Climate

We assessed growth and recruitment relationships with climate for ATL sub-regional composite chronologies and age frequency distributions using correlation analyses (Fritts 1976). We used two sets of climatic data: (a) sub-regional climatic series developed by Macias et al. (2006) based on 18 meteorological stations located throughout the eastern Pyrenees, and (b) meteorological data from the CRU TS 2.0 gridded (0.5° resolution) data set (Mitchell et al. 2004, 2005).

Relationships between RWIs and monthly average temperature and cumulated precipitation were examined over the period common to the chronology and the instrumental climatic record (1955–2000 period for site-specific chronologies and 1951–2000 period for composite ATL chronologies). For these analyses temperatures and precipitation from October prior to the year of growth ($t-1$) to October of the year of tree-ring formation (t) were used. We extended the period to October since latewood development and the synthesis of carbohydrates in *P. uncinata* and other conifers growing in cold environments also occur during fall (Camarero et al. 1998; Rossi et al. 2008). To evaluate the stationarity of the growth–climate associations over time, we computed moving correlations; we adopted a 25-year interval progressively shifted over time every year to compute the correlation coefficients for the same monthly variables. Both correlations and moving correlations were computed using the "bootRes package" (Zang 2011) with 1,000 bootstrap iterations; their significance was tested using the 95% percentile range method (Dixon 2001).

Given the characteristics of the generated age frequency distributions (encompassing only 10 age-classes), recruitment–climate relationships were computed by correlation analyses between recruits density and seasonal climate. Following Lloyd and Fastie (2002) we created the following seasonal variables: winter temperature (average November–March), spring temperature (average April–May), summer temperature (average June–August), and fall temperature (average September–October). By using this approach, we aimed to avoid a reduced robustness of the regression analysis, and thus of incorrect inferences, as a result of a too large number of independent variables in the models. Since climate also influences post-establishment survival and growth of recruits (e.g. Gamache and Payette 2005) we also performed correlations between recruitment and 5-years forwarded climate data.

11.3 Results

11.3.1 Radial-Growth Patterns

Overall, the dendrochronological statistics revealed a high common growth signal among trees sampled in each ATL (for instance the mean correlation between trees ranged 0.47–0.75; see Table 11.2). The mean radial-growth rate (1.72 mm) was high, probably because of the youthfulness of most trees, whereas the mean sensitivity was relatively low (range 0.12–0.27) and increased eastwards significantly ($r=0.71$, $P=0.008$). The mean correlation between trees' growth series in each ATL also increased significantly eastwards ($r=0.70$, $P=0.01$).

P. uncinata growth year to year variations for the period 1955–2000 are held in common by most of the study sites as illustrated by the significant ($P<0.05$) correlation coefficients between site chronologies (Table 11.3). DU and BU are the sites that present a more dissimilar growth pattern in relation to the other ATLs. Despite such similarity in growth variations among ATLs at regional scale, correlation coefficients were on average higher between sites in the same climate sub-region than between sites from different sub-regions. Sites in the Pallars sub-region presented the highest correlation coefficient between them ($r=0.73$ compared to $r=0.54$ with the other sub-regions), followed by sites at the Ribagorça, Cerdanya and Ripollès ($r=0.62$, $r=0.57$, $r=0.57$ respectively). Correlation coefficients with the other sub-regions for the latter were $r=0.53$, $r=0.43$ and $r=0.33$, respectively.

11.3.2 Spatiotemporal Patterns of Recruitment

No significant spatial segregation between recruitment periods (1950–1974 and 1975–1999) was detected in any of the ATL studied (Fig. 11.3). Five of the sites had slightly negative but non-significant spatially corrected Spearman correlation coefficients between recruitment periods, whereas values for the other seven sites were positive and mostly non-significant. The analysis of small-scale distribution patterns at the study sites (Fig. 11.4; Batllori et al. 2010) revealed that recruitment was characterized by a strong clustering of seedlings at small distances (0–3 m). Saplings showed clustering at the same scale range and strength as seedlings did, but to a lesser degree, and large trees' distribution was not significantly different from a random distribution. Both saplings and seedlings showed spatial segregation from large trees in the vast majority of studied ATLs. Small-scale distribution patterns were very similar between climatically driven ATLs and the ones more influenced by the cessation of human-induced perturbations.

Despite temporal patterns of recruitment were more variable between sites than growth year-to-year variations, paired Kolmogorov–Smirnov tests yielded no significant differences in the age frequency distribution of recruits between any of the studied ATLs for the period 1950–1999. Sub-regional demographic age structures in 5-year classes are presented in Fig. 11.5. Increased recruitment after 1970s is observed in all sub-regions. Maximum recruitment rates occurred during the second half of the 1980s and first half of the 1990s, followed by a reduced number of recently recruited individuals (1995–1999 period) in all sub-regions (Fig. 11.5). Significant differences in the temporal patterns of recruitment between sub-regions were only observed between Ribagorça and Pallars sub-regions ($P<0.025$ Kolmogorov–Smirnov test).

11.3.3 Climatic Drivers of Recruitment

Analysis of recruitment–climate relationships was done between sub-regional patterns of establishment and mean seasonal temperature using the CRU TS 2.0

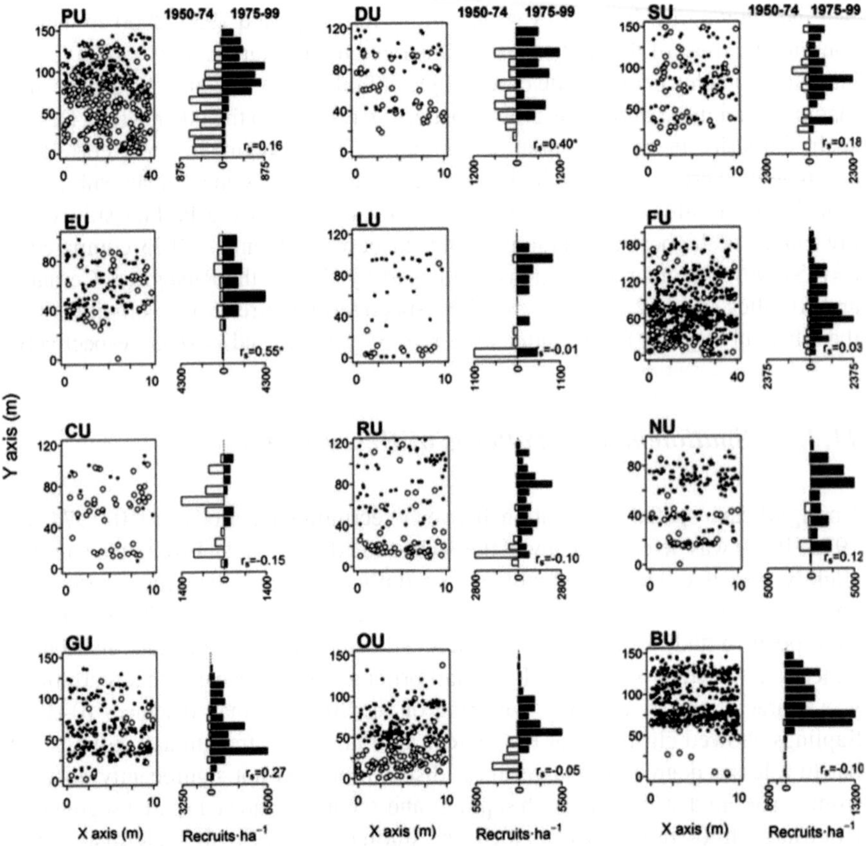

Fig. 11.3 Mapped point patterns (*left scatter plots*) and density by 10 m altitudinal bands (*right bar plots*) of all *Pinus uncinata* individuals recruited during the period 1950–1999 in the 12 tree line ecotones studied. Two recruitment periods are illustrated: 1950–1974 (*white dots*; *white bars*) and 1975–1999 (*black dots, black bars*). The spatially corrected Spearman's correlation coefficient (r_s) between the two recruitment periods is noted for each site; symbol *asterisk* denotes significant ($P<0.05$) r_s coefficients. Note that scales and plot sizes vary between sites (modified from Batllori et al. 2010)

climate dataset (Fig. 11.5). Summer temperatures were positively and significantly related with the established number of recruits in all sub-regions, except for Ribagorça were summer temperature had a positive but non-significant relationship. Spring and fall temperatures showed weaker (non-significant) and more variable relationship with the abundance of recruits. Winter temperatures showed a positive significant relationship with recruitment only in Cerdanya. Forwarded climate analysis evidenced also a positive and significant relationship between density of recruits and summer temperatures, as well as with winter temperatures, in all the sub-regions and also with spring temperature in Pallars.

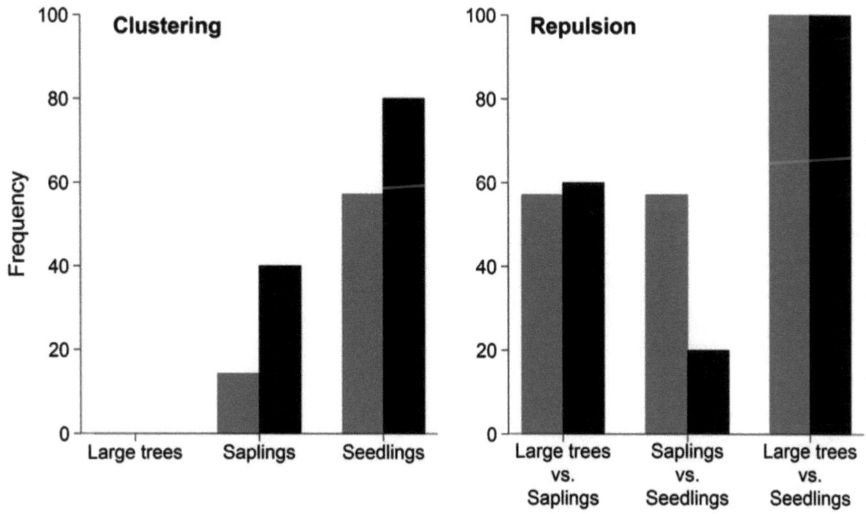

Fig. 11.4 Summary of the univariate and bivariate point pattern analyses in the 12 *Pinus uncinata* tree line ecotones studied. The frequencies of significant clustering (*left plot*) in at least some analysed distance for each tree class, and the negative spatial relationship (repulsion; *right plot*) between tree classes are illustrated. *Light grey bars* correspond to the undisturbed sites and *dark grey bars* correspond to the disturbed sites (see Table 11.1 for details; figure modified from Batllori et al. 2010)

11.3.4 Climatic Drivers of Radial Growth

The preliminary analyses of growth–climate relationships for site ATLs yielded very weak, non-consistent and only few significant relationships (results not presented). Correlation analyses revealed important differences in the growth–climate relationships obtained when using specific climate datasets (i.e. CRU TS 2.0 and sub-regional meteorological datasets described in Macias et al. 2006). Differences were striking, and of concern, when evaluating the precipitation effects on tree growth; both direction and significance of the obtained correlation coefficients were largely inconsistent between the two data sets. In light of these results, and adopting a conservative approach, we decided not to interpret the relationships between radial growth and precipitation. Growth–temperature relationships presented much more consistent results between the two climate datasets, being the CRU TS 2.0 dataset the one presenting higher and more significant correlation coefficients with tree growth year-to-year variations. Results from the four sub-regional ATL composite chronologies and monthly temperature are presented in Fig. 11.6.

Tree growth showed significant and mostly positive responses to the climate conditions of the year of growth, mainly spring and summer months (May through August), and to the late fall (November) prior to the growing year (Fig. 11.6). A significant and positive relationship between growth and temperatures of the

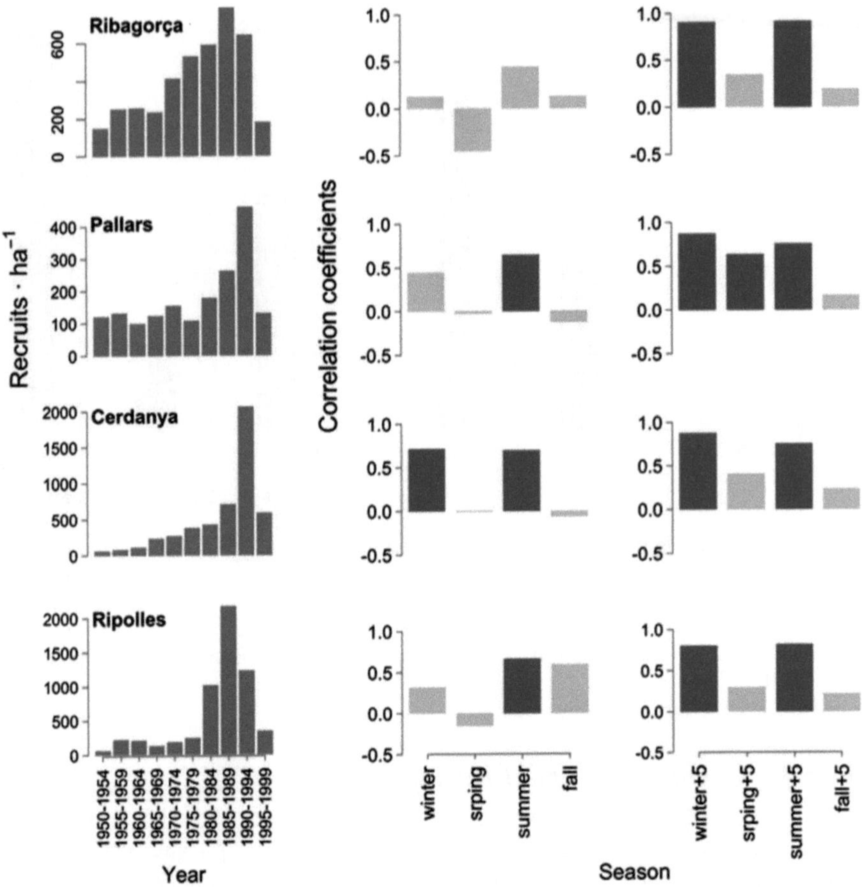

Fig. 11.5 Age frequency distributions (*left*) and bootstrapped correlation coefficients performed between regional recruits density and current and forwarded seasonal temperatures (*right*). Significant correlation function coefficients ($P < 0.05$) are indicated by *dark grey bars*

current August and prior November was detected in all sub-regions, whereas current May temperature related positively and significantly in three of the four sub-regions. Current April, June, July and September temperatures showed a more sub-region specific relationship with tree growth, having only significant coefficients in one of the sub-regional ATL chronologies.

The moving correlations temperature patterns for the growing season (May to October) generally confirm the results obtained from correlation analyses (Fig. 11.6). In many cases, growth–temperature correlations show a relatively stationary response over time (e.g. current June and July months in Pallars, current July and September in Cerdanya, previous November in Ripollès). In contrast, monthly temperatures to which the sub-regional ATL chronologies exhibit a significant response display in some cases more dynamic or unstable responses with important fluctuations over time (e.g. current May and August months in Ribagorça, current

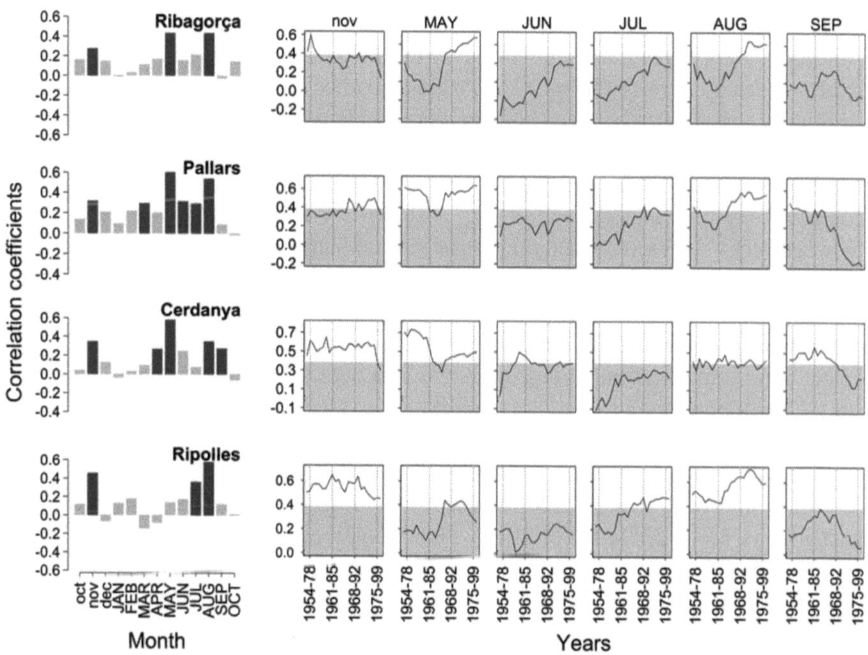

Fig. 11.6 Bootstrapped correlations coefficients (*left bar plots*) and moving correlation coefficients (*right line plots*) performed between the indexed radial growth of the four regional tree line chronologies (period 1951–2000) with monthly averaged temperature from prior October (year *t*–1) to October of the year (*t*) of tree-ring formation. Moving correlation coefficients are only depicted for months with a climatic variable showing a significant influence on tree growth in the studied sub-regions. Lowercase: prior-year months; capitals: current-year months. Significant correlation coefficients ($P<0.05$) are indicated by *dark grey bars* (*left*), and by values falling outside the shaded area (*right*)

September in Cerdanya). September temperatures of the growing season showed, despite being non-significant, a clear decrease trend in all the sub-regional ATLs after ca. the 1970s.

11.4 Discussion

We expected large variation in radial growth and patterns of recruitment establishment between ATLs due to the influence of topography on local climate regimes in the studied high-elevation forests. However, the tree lines studied mostly exhibit synchronous growth year-to-year variations at a regional scale for the period 1955–2000, and surprisingly, the abundance of recruitment and its fluctuations over time were not significantly different between sites. Our results are in accordance with those of Tardif et al. (2003), who reported the influence of macroclimatic environmental conditions on tree ring growth of *P. uncinata* stands from the Central Pyrenees. The coupling of upright trees' to general atmospheric conditions may explain the common growth fluctuations observed at the ATL over the Pyrenees mountains (Körner 2007). On the other hand, we attribute the lack of significant differences between sites in recruits' density over time to the resolution of the analysis (5-year classes). When analysing fluctuations of recruitment at a yearly resolution, 50% of paired Kolmogorov–Smirnov tests (with Bonferroni correction of the significance level, $P<0.001$) yielded significant differences between ATLs. However, the dating error in establishment dates needs careful consideration when defining the width of age-classes, since it can significantly change the shape of age frequency distributions (Wong and Lertzman 2001). This justifies our approach based on 5-year classes, which at the same time reduces the ability to detect important variations in recruitment trends over time, both within and between sites. The importance of microsite conditions (e.g. microtopography), of the interplay between abiotic and biotic controls, and of mechanisms operating at small-spatial scales on regeneration dynamics has been largely documented at the ATL (Smith et al. 2003; Resler 2006; Körner 2007; Batllori et al. 2009; Bekker et al. 2009; Aune et al. 2011; Elliot 2011). Hence, we believe the demographic trends presented in this study are representative of general *P. uncinata* recent dynamics in ATL environments in the Pyrenees rather than indicative of site-specific responses.

As expected, due to the shade intolerance of *P. uncinata* (Bosch and Gutiérrez 1999), the spatial characterization of the study sites evidences significant segregation between large trees and seedlings and saplings (Batllori et al. 2010). *P. uncinata* recruits establishes predominantly above the forest limit where favourable microsites for regeneration occur (Camarero et al. 2000; Batllori et al. 2009), and, in common with other tree line species (e.g. Srutek et al. 2002; Alftine et al. 2003), seedlings and saplings—to a lesser degree—appear strongly clustered at small distances (0–3 m). Despite a remarkable increase in tree recruitment after the 1970s observed in all climatic sub-regions (see also Batllori and Gutiérrez 2008), as summer temperatures rose, all ATL sites are characterized by an absence of spatial segregation between the 1950–1974 and 1975–1999 recruitment periods. This suggests that a continuous infilling process of the ecotone (densification), rather than upward encroachment, has characterized ATL dynamics in the Pyrenees during the last 50 years.

The small-scale clustering, together with remarkable increases in tree recruitment observed in all sub-regions since the 1970s (acceleration period) followed by years of limited establishment (period 1995–1999), may be indicative of positive feedback processes coupled with accelerated warming rates (Alftine et al. 2003). Tree clumping at the ATL promotes a more favourable microenvironment by enhancing snow retention, which increases recruits survival (Batllori et al. 2009; Smith et al. 2003, 2009). However, climatic variability over time disrupts establishment and survival of recruits, limits feedback processes, and may explain the recent recruitment drop observed in the study sites (Alftine et al. 2003; Camarero and Gutiérrez 2004).

In agreement with other studies from cold environments, summer temperatures were the most important factor related to tree growth (e.g. Pfeifer et al. 2005;

Wieser and Tausz 2007; Carrer et al. 2007; Mathisen and Hofgaard 2011) and recruits abundance (e.g. Camarero and Gutiérrez 2004, 2007; Vittoz et al. 2008). Warm Novembers in the year before growth, and warm Mays and Augusts during the year of tree ring formation dominated the growth–climate association at *P. uncinata* tree lines. Moving correlation functions point to a general increase in the sensitivity of May and August temperatures (except in the Cerdanya sub-region) and the relative stability of November temperatures over the last ca. 50 years. This finding is in agreement with the summer warming trend observed during the last 2 decades of the twentieth century which might enhance cambial activity through an extension of the earlywood growing season at the ATL. Furthermore, the eastward increase in mean sensitivity and correlation of growth series within sites indicates a greater importance of summer conditions for growth as we moved towards the Mediterranean Sea. Such relevance may not only refer to temperature because eastern sites, particularly in the Cerdanya sub-region, were also subject to stronger drought conditions than western sites. The low responsiveness of Cerdanya trees to summer warming trends may also be caused by long-term adaptive growth patterns of these ATL populations to warm and dry summer conditions. The lack of adequate local precipitation data at high-elevation sites precludes testing this idea throughout most of the Pyrenean range.

The general positive climate–growth relationship during fall may indicate the negative influence of cold temperatures during this season in the synthesis and accumulation of carbohydrates for following year radial growth (Rolland and Schueller 1994; Kagawa et al. 2006). The described positive response to May temperatures is consistent with the xylogenesis of *P. uncinata* in the Pyrenees (Camarero et al. 1998), and with the fact that cambial reactivation of evergreen conifers is triggered by temperature rise in spring (Tranquillini 1979). Furthermore, May temperatures show an increased influence during the second half of the twentieth century (Tardif et al. 2003), as we generally detected at the ATL, suggesting that trees begin to grow earlier due to an enlargement of the growing period (Menzel et al. 2006). Similarly, despite being non-significant, we observed a decrease in the positive influence of September temperatures on tree growth at all sub-regions. This may illustrate a trend towards potentially increased drought stress in early fall (Macias et al. 2006). Oberhuber et al. (2008) reported that the strongest long-term stable climatic signal on tree growth at the ATL is a direct response to July temperature. However, that was not the case at the study sites where August temperature exhibited a strongest relationship with tree growth. This may be due to age-dependent responses to climate. Carrer and Urbinati (2004) described a strongest influence of July temperature on growth in old trees and of August temperature in young trees, thus the relatively young age of the trees sampled at the study sites may explain the observed growth–climate relationships. Moreover, we suggest the divergence of radial growth and July temperature since the mid 1980s in ATL environments (Oberhuber et al. 2008), which might be due to the impact of temporary drought stress during the growing season, may have a varying influence in young and old trees. Unfortunately, we could not evaluate the influence of precipitation on *P. uncinata* radial growth at the study sites due to climatic data constraints. While a local and high-elevation climate station can be highly representative of year-to-year variability in temperature across the study region (Agustí-Panareda et al. 2000), precipitation presents much variability at local scales due to the topographic complexity in mountain regions and to site-specific factors as slope or exposure to wind (Barry 1992; Wilmking and Juday 2005). Additional studies must be performed (e.g. with other proxies for precipitation) to determine the influence of precipitation on ATL tree growth at the eastern range of the Spanish Pyrenees.

Recruits–climate relationship evidenced that temperature is among the most important factors in determining recruits establishment and survival at the ATL (Körner 1998). Recruitment density appeared positively correlated with winter and summer temperatures both for current and forwarded climate as has been found in nearby ATLs (Camarero and Gutiérrez 2004), and in isolated *P. uncinata* populations in the Iberian range growing under continental conditions (Camarero and Gutiérrez 2007). Camarero and Gutiérrez (2004) also attributed low levels of recruitment and tree line advance due to summer droughts. We cannot rule out an equivalent response at the study sites. First, we could not evaluate the influence of precipitation on the temporal variation of recruitment. Second, the resolution (age class width) and seasonalized climate is not comparable with the previous studies. Our analysis shows that forwarded winter temperatures are also consistently related to recruits density, they have a positive and significant relationship with recruitment fluctuations in all the sub-regions. This confirms that winter conditions in the following years after establishment may be among the most limiting factors for seedling survival at the ATL (Batllori et al. 2009). The analysis presented here is not suitable to perform an accurate evaluation on the effects of climatic conditions and their variability on *P. uncinata* recruitment dynamics at local scales. Thus, as recently highlighted by Malanson et al. (2011), we emphasize the need of explicitly assess the role of local climatic conditions and its temporal variability in tree establishment and survival at the tree line.

11.5 Conclusions

Results presented here evidence that sub-regional response patterns of growth and recruitment to temperature are rather homogeneous at the Pyrenean ATL ecotone. Both radial growth and recruitment have had similar responses to climate variability over the second half of the twentieth century. In light of the positive effects of warm temperature at the study range, one could expect little inertia to climatically induced shifts of the Pyrenean ATLs to recent and future warming trends. However, while environmental conditions during the second half of the twentieth century may have enhanced radial and height growth of already established trees and favoured increased recruitment within the ecotone, tree line shifts have not been ubiquitous (Batllori and Gutiérrez 2008; Batllori et al. 2010). Thus, even if the described sub-regional trends could suggest potential conspicuous future changes in ATL position (Malanson et al. 2007; Aune et al. 2011) varying recent site-specific dynamics preclude such generalization. Our results highlight the importance of spatial and temporal scale in linking process and pattern at the ATL (Levin 1992; Resler 2006), and evidence the importance of small-scale detailed studies to detect and gain understanding on contingent factors that may distort expected regional ATL responses at local (e.g. microsite conditions for regeneration) and regional (e.g. extreme climatic events for growth) scales.

Acknowledgements We are grateful to all the people who helped us in the field work, and to PN Alt Pirineu, PN Cadí-Moixeró and PN Aigüestortes i Estany de Sant Maurici for their collaboration. Special thanks to J.M. Ninot, E. Carrillo, J. Carreras and A. Ferré for their permanent support. This study was supported by the project REN2002-04268 (Spanish Ministry of Research). E.B. and J.J.C. also acknowledge the support from MEC-FPU grant and ARAID, respectively.

References

Agustí-Panareda A, Thompson R, Livingstone DM (2000) Reconstructing temperature variations at high elevation lake sites in Europe during the instrumental period. Verhandlungen der Internationalen Vereinigung Limnologie 27:479–483

Alftine KJ, Malanson GP, Fagre DB (2003) Feedback-driven response to multidecadal climatic variability at an alpine treeline. Phys Geogr 24(6):520–535

Améztegui A, Brotons L, Coll L (2010) Land-use changes as major drivers of mountain pine (*Pinus uncinata* Ram.) expansion in the Pyrenees. Glob Ecol Biogeogr 19(5):632–641

Aune S, Hofgaard A, Söderström L (2011) Contrasting climate- and land-use-driven tree encroachment patterns of subarctic tundra in northern Norway and the Kola Peninsula. Can J For Res 41(3):437–449

Barry RG (1992) Mountain weather and climate, 2nd edn. Routledge, London

Batllori E, Gutiérrez E (2008) Regional tree line dynamics in response to global change in the Pyrenees. J Ecol 96(6):1275–1288

Batllori E, Camarero JJ, Ninot JM, Gutiérrez E (2009) Seedling recruitment, survival and facilitation in alpine *Pinus uncinata* tree line ecotones. Implications and potential responses to climate warming. Glob Ecol Biogeogr 18(4):460–472

Batllori E, Camarero JJ, Gutiérrez E (2010) Current regeneration patterns at the tree line in the Pyrenees indicate similar recruitment processes irrespective of the past disturbance regime. J Biogeogr 37(10):1938–1950

Bekker MF, Clark JT, Jackson MW (2009) Landscape metrics indicate differences in patterns and dominant controls of ribbon forests in the Rocky Mountains, USA. Appl Veg Sci 12(2): 237–249

Beniston M (2003) Climatic change in mountain regions: a review of possible impacts. Clim Chang 59(1–2):5–31

Bogaert RV, Haneca K, Hoogesteger J, Jonasson MD, Callaghan TV (2011) A century of tree line changes in sub-Arctic Sweden shows local and regional variability and only a minor influence of 20th century climate warming. J Biogeogr 38(5):907–921

Bosch O, Gutiérrez E (1999) La sucesión en los bosques de *Pinus uncinata* del Pirineo. De los anillos de crecimiento a la historia del bosque. Ecología 13:133–171

Bunn R (2008) A dendrochronology program library in R (dplR). Dendrochronologia 26: 115–124

Caccianiga M, Compostella C (2011) Growth forms and age estimation of treeline species. Trees. doi:10.1007/s00468-011-0595-1

Camarero JJ, Gutiérrez E (2004) Pace and pattern of recent treeline dynamics: response of ecotones to climatic variability in the Spanish Pyrenees. Clim Chang 63(1–2):181–200

Camarero JJ, Gutiérrez E (2007) Response of *Pinus uncinata* Recruitment to Climate Warming and Changes in Grazing Pressure in an Isolated Population of the Iberian System (NE Spain). Arctic Antarctic Alpine Res 39(2):210–217

Camarero JJ, Guerrero-Campo J, Gutiérrez E (1998) Tree-ring growth and structure of *Pinus uncinata* and *Pinus sylvestris* in the central Spanish Pyrenees. Arctic Antarctic Alpine Res 30(1):1–10

Camarero JJ, Gutiérrez E, Fortin M-J (2000) Spatial pattern of subalpine forest-alpine grassland ecotones in the Spanish Central Pyrenees. For Ecol Manage 134:1–16

Carrer M, Urbinati C (2004) Age-dependent tree-ring growth responses to climate in *Larix decidua* and *Pinus cembra*. Ecology 85(3):730–740

Carrer M, Nola P, Eduard JL, Motta R, Urbinati C (2007) Regional variability of climate–growth relationships in *Pinus cembra* high elevation forests in the Alps. J Ecol 95(5):1072–1083

Carreras J, Carrillo E, Masalles R, Ninot J, Soriano I, Vigo J (1996) Delimitation of the supra-forest zone in the Catalan Pyrenees. Bulletin de la Societé linnéenne de Provence 47:27–36

D'Arrigo RD, Kaufmann RK, Davi N, Jacoby GC, Laskowski C, Myneni RB, Cherubini P (2004) Thresholds for warming-induced growth decline at elevational tree line in the Yukon Territory, Canada. Global Biogeochem Cycles 18. doi:10.1029/2004GB002249

Dalen L, Hofgaard A (2005) Differential regional treeline dynamics in the Scandes Mountains. Arctic Antarctic Alpine Res 37(3):284–296

Danby RK, Hik DS (2007) Variability, contingency and rapid change in recent subarctic alpine tree line dynamics. J Ecol 95(2):352–363

R Development Core Team (2011) R: a language and environment for statistical computing. R Foundation for Statistical Computing, Vienna, Austria, ISBN 3-900051-07-0

Dixon PM (2001) Bootstrap resampling. In: El-Shaarawi AH, Piegorsch WW (eds) The encyclopedia of environmetrics. Wiley, New York

Dullinger S, Dirnböck T, Grabherr G (2004) Modelling climate change-driven treeline shifts: relative effects of temperature increase, dispersal and invasibility. J Ecol 92(2):241–252

Elliot GP (2011) Influences of 20th-century warming at the upper tree line contingent on local-scale interactions: evidence from a latitudinal gradient in the Rocky Mountains, USA. Glob Ecol Biogeogr 20(1):46–57

Ettinger AK, Ford KR, HilleRisLambers J (2011) Climate determines upper, but not lower, altitudinal range limits of Pacific Northwest conifers. Ecology 92(6):1323–1331

Fang K, Gou X, Chen F, Peng J, D'Arrigo R, Wright W, Li M-H (2009) Response of regional tree-line forests to climate change: evidence from the northeastern Tibetan Plateau. Trees 23(6):1321–1329

Fritts HC (1976) Tree rings and climate. Academic, New York

Gamache I, Payette S (2005) Latitudinal response of subarctic tree lines to recent climate change in eastern Canada. J Biogeogr 32(5):849–862

Garcia-Ruiz J, Lasanta T, Ruiz-Flano P, Ortigosa L, White S, González C, Martí C (1996) Land-use changes and sustainable development in mountain areas: a case study in the Spanish Pyrenees. Landsc Ecol 11(5):267–277

Gehring-Fasel J, Guisan A, Zimmermann NE (2007) Tree line shifts in the Swiss Alps: climate change or land abandonment? J Veg Sci 18(4):571–582

Grace J, Berninger F, Nagy L (2002) Impacts of climate change on the tree line. Ann Bot 90(4):537–544

Green K (2009) Causes of stability in the alpine treeline in the Snowy Mountains of Australia—a natural experiment. Aust J Bot 57(3):171–179

Harsh MA, Bader MY (2011) Treeline form—a potential key to understanding treeline dynamics. Glob Ecol Biogeogr 20(4):582–596

Harsh MA, Hulme PE, McGlone MS, Duncan RP (2009) Are treelines advancing? A global meta-analysis of treeline response to climate warming. Ecol Lett 12(10):1040–1049

Hofgaard A (1997) Inter-relationships between treeline position, species diversity, land use and climate change in the Central Scandes Mountains of Norway. Glob Ecol Biogeogr Lett 6(6):419–429

Holmes RL (1983) Computer-assisted quality control in tree-ring dating measurement. Tree Ring Bull 43:68–78

Holtmeier F-K (2009) Mountain timberlines: ecology, patchiness, and dynamics. Springer, New York

Holtmeier F-K, Broll G (2005) Sensitivity and response of northern hemisphere altitudinal and polar treelines to environmental change at landscape and local scales. Glob Ecol Biogeogr 14(5):395–410

Holtmeier F-K, Broll G (2010) Wind as an ecological agent at treelines in North America, the Alps, and the European Subarctic. Phys Geogr 31(3):203–233

IPCC (2007) Climate change 2007: the physical science basis. In: Solomon S, Qin D, Manning M et al (eds) Contribution of Working Group I to the Forth Assessment Report of the Intergovernmental Panel on Climate Change. Cambridge University Press, Cambridge

Kagawa A, Sugimoto A, Maximov TC (2006) $^{13}CO_2$ pulse-labelling of photoassimilates reveals carbon allocation within and between tree rings. Plant Cell Environ 29(8):1571–1584

Kharuk VI, Im ST, Dvinskaya ML, Ranson KJ (2010) Climate-induced mountain tree-line evolution in southern Siberia. Scand J For Res 25:446–454

Körner C (1998) A re-assessment of high elevation treeline positions and their explanation. Oecologia 115(4):445–459

Körner C (2007) Climatic treelines: conventions, global patterns, causes. Erdkunde 61(4): 316–324

Körner C, Paulsen J (2004) A world-wide study of high altitude treeline temperatures. J Biogeogr 31(5):713–732

Kullman L (2005) Pine (*Pinus sylvestris*) treeline dynamics during the past millennium—a population study in west-central Sweden. Ann Bot Fennici 42:95–106

Lenoir J, Gégout JC, Marquet PA, Ruffray P, Brisse H (2008) A significant upward shift in plant species optimum elevation during the 20th century. Science 320(5884):1768–1771

Levin SA (1992) The problem of pattern and scale in ecology. Ecology 73(6):1943–1967

Liang E, Wang Y, Eckstein D, Luo T (2011) Little change in the fir tree-line position on the southeastern Tibetan Plateau after 200 years of warming. New Phytol 190(3):760–769

Lloyd AH, Fastie CL (2002) Spatial and temporal variability in the growth and climate response of treeline trees in Alaska. Clim Chang 52(4):481–509

Lloyd AH, Graumlich LJ (1997) Holocene dynamics of treeline forests in the Sierra Nevada. Ecology 78(4):1199–1210

Macias M, Andreu L, Bosch O, Camarero JJ, Gutiérrez E (2006) Increasing aridity is enhancing silver fir (*Abies alba* Mill.) water stress in its south-western distribution limit. Clim Chang 79(3–4):289–313

Malanson GP, Butler DR, Fagre DB, Walsh SJ, Tomback DF, Daniels LD, Resler LM, Smith WK, Weiss DL, Peterson DL, Bunn AG, Hiemstra CA, Liptzin D, Bourgeon PS, Shen Z, Millar CI (2007) Alpine treeline of western North America: linking organism-to-landscape dynamics. Phys Geogr 28(5):378–396

Malanson GP, Resler LM, Bader MY, Holtmeier F-K, Butler DR, Weiss DJ, Daniels LD, Fagre DB (2011) Mountain treelines: a roadmap for research orientation. Arctic Antarctic Alpine Res 43(2):167–177

Mathisen IE, Hofgaard A (2011) Recent height and diameter growth variation in Scots pine (*Pinus sylvestris* L.) along the Arctic margin: the importance of growing season versus non-growing season climate factors. Plant Ecol Divers 4(1):1–11

Menzel A, Sparks TH, Estrella N, Koch E, Aasa A, Ahas R, Alm-Kübler K, Bissolli P et al (2006) European phenological response to climate change matches the warming pattern. Glob Chang Biol 12(1):1969–1976

Mitchell, TD, Carter TR, Jones, PD, Hulme, M, New, M (2004) A comprehensive set of high-resolution grids of monthly climate of Europe and the globe: the observed record (1901-2000) and 16 scenarios (2001-2100)

Mitchell, TD, Jones PD (2005) An improved method for constructing a database of monthly climate observations and associated highresolution grids. Int J Climatol 25: 693-712

Ninot JM, Carrillo E, Font X, Carreras J, Ferré A, Masalles RM, Soriano I, Vigo J (2007) Altitude zonation in the Pyrenees. A geobotanic interpretation. Phytocoenologia 37(3–4):371–398
Noble IR (1993) A model of the responses of ecotones to climate change. Ecol Appl 3(3): 396–403
Oberhuber W, Kofler W, Pfeifer K, Seeber A, Gruber A, Wieser G (2008) Long-term changes in tree-ring-climate relationships at Mt. Patscherkofel (Tyrol, Austria) since the mid-1980s. Trees Struct Funct 22(1):31–40
Payette S (2007) Contrasted dynamics of northern Labrador tree lines caused by climate change and migrational lag. Ecology 88(3):770–780
Pfeifer K, Kofler W, Oberhuber W (2005) Climate related causes of distinct radial growth reductions in *Pinus cembra* during the last 200 years. Veg Hist Archaeobot 14(3):211–220
Rangel TFLVB, Diniz-Filho JAF, Bini LM (2006) Towards an integrated computational tool for spatial analysis in macroecology and biogeography. Glob Ecol Biogeogr 15(4):321–327
Resler LM (2006) Geomorphic controls of spatial pattern and process at alpine treeline. Prof Geogr 58(2):124–138
Rinn F (2003) TSAP-Win—time series analysis and presentation dendrochronology and related applications. Frank Rinn, Heidelberg
Rolland C, Schueller JF (1994) Relationship between mountain pine and climate in the French Pyrenees (Font-Romeu) studied using the radiodensitometrical method. Pirineos 143–144: 55–70
Rossi S, Deslauriers A, Gričar J, Seo J-W, Rathgeber CBK, Anfodillo T, Morin H, Levanic T, Oven P, Jalkanen R (2008) Critical temperatures for xylogenesis in conifers of cold climates. Glob Ecol Biogeogr 17(6):696–707
Smith WK, Germino MJ, Hancock TE, Johnson DM (2003) Another perspective on altitudinal limits of alpine timberlines. Tree Physiol 23(16):1101–1112
Smith WK, Germino MJ, Johnson DM, Reinhardt K (2009) The altitude of alpine treeline: a bellwether of climate change effects. Bot Rev 75:163–190
Srutek M, Dolezal J, Hara T (2002) Spatial structue and associations in a *Pinus canariensis* population at the treeline, Pico del Teide, Tenerife, Canary Islands. Arctic Antarctic Alpine Res 34(2):201–210
Stokes MA, Smiley TL (1968) An introduction to tree-ring dating. University of Chicago Press, Chicago
Stueve KM, Isaacs RE, Tyrrell LE, Densmore RV (2011) Spatial variability of biotic and abiotic tree establishment constraints across a treeline ecotone in the Alaska Range. Ecology 92(2):496–506
Sveinbjörnsson B, Hofgaard A, Lloyd A (2002) Natural causes of the Tundra-Taiga boundary. Ambio Spec Rep 12:23–29
Tardif J, Camarero JJ, Ribas M, Gutiérrez E (2003) Spatiotemporal variability in tree growth in the central Pyrenees: climatic and site influences. Ecol Monogr 73(2):241–257
Tranquillini W (1979) Physiological ecology of the alpine timberline: tree existence at high altitudes with special reference to the European Alps. Springer, Berlin
Vigo J, Ninot JM (1987) Los Pirineos. In: Peinado-Lorca M, Rivas-Martínez S (eds) La vegetación de España. Publicaciones Universidad Alcalá de Henares, Alcalá de Henares, pp 351–384
Vittoz P, Rulence B, Largey T, Freléchoux F (2008) Effects of climate and land-use change on the establishment and growth of Cembran Pine (*Pinus cembra* L.) over the altitudinal treeline ecotone in the Central Swiss Alps. Arctic Antractic Alpine Res 40(1):225–232
Wieser G, Tausz M (2007) Trees at their upper limit. Treelife limitation at the alpine timberline. Plant Ecophysiol 5. Springer, Dordrecht
Wigley TML, Briffa KR, Jones PD (1984) On the average value of correlated time series, with applications in dendroclimatology and hydrometeorology. J Clim Appl Meteorol 23(2):201–213
Wilmking M, Juday GP (2005) Longitudinal variation of radial growth at Alaska's northern treeline-recent changes and possible scenarios for the 21st century. Glob Planet Change 47(2–4):282–300

Wilmking M, Juday GP, Barber VA, Zald HJ (2004) Recent climate warming forces contrasting growth responses of white spruce at treeline in Alaska through temperature thresholds. Glob Chang Biol 10(10):1724–1736

Wong CM, Lertzman KP (2001) Errors in estimating tree age: implications for studies of stand dynamics. Can J For Res 31(7):1262–1271

Yamaguchi DK (1991) A simple method for cross-dating increment cores from living trees. Can J For Res 21(3):414–416

Zang C (2011). bootRes: The bootRes package for bootstrapped response and correlation functions. URL:http://cran.r-project.org/web/packages/bootRes/bootRes.pdf

Chapter 12
The Alpine Treeline Ecotone in the Southernmost Swedish Scandes: Dynamism on Different Scales

Leif Kullman

12.1 Introduction

Alpine (altitudinal) treeline ecotones are elusive, spatially heterogeneous and dynamic transitional zones between closed mountain forest and treeless alpine tundra. From a distance the may look sharp, but a closer view usually reveals a highly complex structural pattern (Kullman 1979). Thermal growth limitation is considered as the fundamental part of the mechanism that creates and maintains the treeline ecotone (Grace et al. 2002; Hoch and Körner 2003; Holtmeier 2003; Kullman 1998, 2007a, 2010a; Diaz et al. 2003; Lloyd and Fastie 2002). At finer scales, the straight thermal forcing is modulated by other agents, e.g., topography, geomorphology, wind, soil depth, species interactions, fire, herbivory, human impacts, and site history (Walsh et al. 1994; Holtmeier and Broll 2005; Gehrig-Fasel et al. 2007; Kullman and Öberg 2009; Leonelli et al. 2011; Aune et al. 2011). Among these, wind appears to have a superior role (cf. Seppälä 2004; Holtmeier and Broll 2010; Kullman 2010a).

In recent years, treeline ecotones have received increased attention in their postulated roles as sensitive ecological indicators of climate change (Gosz 1993; Kullman 1998; Skre et al. 2002; Holtmeier 2003; Nagy 2006). However, from a theoretical perspective, there has been some doubt and discussion as to the responsiveness and usefulness of treeline ecotones for the purpose of early detection and monitoring of climate-mediated ecological change in high-mountain regions (Noble 1993; Baker and Weisberg 1995; Kupfer and Cairns 1996; Körner 1999; Bekker 2005). On the other hand, paleoecological studies strongly support the contention of perceivable elevational dynamics of alpine treelines on centennial and longer timescales, mainly as a response to climate change (e.g., Tinner and Kaltenrieder 2005; Kullman and Kjällgren 2006; Paus 2010). Also on shorter scales,

L. Kullman (✉)
Department of Ecology and Environmental Science, Umeå University, 901 87 Umeå, Sweden
e-mail: leif.kullman@emg.umu.se

R.W. Myster (ed.), *Ecotones Between Forest and Grassland*,
DOI 10.1007/978-1-4614-3797-0_12,

treelines are found to respond by position, structure, and composition to climate variability. For example, during the past century treelines of different species in the Swedish Scandes have shifted upslope by a common maximum of little more than 200 m, in balance with recorded air temperature change (Kullman and Öberg 2009).

Recent treeline rise has become a circumboreal phenomenon, ultimately conditioned by predominant climate warming during the past century (cf. Shiyatov 2003; Esper and Schweingruber 2004; Lloyd 2005; Tape et al. 2006; Kapralov et al. 2006; Danby and Hik 2007; Holtmeier and Broll 2007; Harsch et al. 2009). A more detailed mechanistic understanding of ecotonal control and performance on decadal to centennial scales is urgently needed as a basis for dynamic modeling and accurate projection of the transformation of the treeline landscape under the auspices of hypothetical future climate change (cf. Malanson 2001; Butler et al. 2007; Elliot and Kipfmueller 2010; Leonelli et al. 2011). For that purpose, the importance of real-world data on treeline performance, representing multiple temporal and spatial scales cannot be overstated (Körner 1999; Fagre et al. 2003).

Alpine treeline ecotones may be relictual (lacking reproductive capacity) or newly shaped (ongoing establishment). The present chapter strives to explore this issue by reconciling observed/reconstructed regional scale (10^5 m) and local (10^2–10^3 m) positional dynamics of the alpine treeline ecotone on different timescales. This effort synthesizes and reviews experiences from a regional treeline monitoring program in the southern Swedish Scandes (TREENET), comprising multisite data on treeline performance over the past 100 years (Kullman and Öberg 2009; Kullman 2010a) and paleoecological (megafossil) treeline data for the same region and the entire postglacial period (Kullman 2004a; Kullman and Kjällgren 2006; Öberg and Kullman 2011a, b).

The "treeline ecotone" is taken as the elevational belt between the uppermost closed forest (irrespective of species) and the *treeline* of the species which reaches the highest elevation. The concept *treeline*, as distinguished from the *treeline ecotone* (cf. Hustich 1979), is narrowly defined as the elevation of the uppermost individual of a particular tree species with a minimum height of 2 m (e.g., Smith 1920; Miehe and Miehe 2000; Kullman 2010a). This very precise definition allows adequate comparisons in time and space and is useful as an index and expression of prevailing growth conditions for trees and other plant species at high elevations in the mountains.

In the Scandes, the altitude of the treeline coincides with more or less abrupt changes in certain vegetation and flora qualities in the understory layer (Smith 1920; Hofgaard and Wilmann 2002; Sundqvist et al. 2008). Furthermore, several studies have demonstrated that elevational treeline shifts (progression or regression) are paralleled with corresponding transformations of tree and ground cover vegetation in the entire treeline ecotone as well as in the alpine zone (Hustich 1978; Kullman 2001a, 2005a, 2007b, c, 2010c; Sundqvist et al. 2008; Hallinger et al. 2010). For example, species which previously did not ascend above the birch treeline, e.g., *Maianthemum bifolium* and *Cornus suecica*, have recently shifted upslope in concert with the treeline. Thus, the treeline, as defined above, integrates positional and structural trajectories in subalpine/alpine vegetation. As clearly realized long ago by experienced field botanists (e.g., Enquist 1933; Nägeli 1969), the treeline rather than any "forest line" provides the most clear-cut and direct relationship to the regional climate (cf. Körner 2007). In addition it can be objectively assessed, which is not the case for the forest line, irrespective of its definition (Samuelsson 1914; Hustich 1937; Kullman 2010a). Thus, the treeline appears as a most appropriate tool and target for climate impact monitoring (Kullman 1998).

Each tree species has its own *tree species line*, at a highly variable distance above the respective treeline. This is the highest occurrence irrespective of size and age. Scientifically, this feature has received less attention than it actually deserves, as part of an "early warning system" with focus on climate-mediated ecological change.

The nomenclature of vascular plant taxa is according to Mossberg and Stenberg (2003). When not otherwise stated, the photographs are taken by Leif Kullman.

12.2 Research Region

The Swedish Scandes (the Caledonides) trend north–south from 69 to 61 °N lat. In the north, the highest mountains reach about 2,000 m a.s.l., with the upper treeline stopping about 700 m a.s.l. In the south corresponding figures are 1,200 and 1,100 m a.s.l., respectively. This study concerns the latter region (province of Dalarna), approximately 61–62 °N lat., where southwards ascending treelines approach the highest mountain tops. This regional-scale landscape pattern takes the form of an "archipelago" of widely scattered and southwards progressively smaller caps of open alpine tundra in a matrix of closed boreal forest (Figs. 12.1 and 12.2).

In the following, these objects will be termed "low fells." On the local scale, each of them possesses a structurally unique alpine treeline ecotone as a mosaic of tree stands and open patches of alpine tundra (Kullman 1979, 2010a).

Given the small alpine caps, which characterize most of the low fells, it appears likely that even relatively insignificant climate shifts would rapidly and profoundly alter the vegetation landscape as the local ecotones adjust to new climatic conditions. It is commonly speculated, although with little empirical support from treeline history, that future hypothetical climate warming (IPCC 2007) would cause the replacement of most of the alpine tundra with forest in the Scandes (e.g., Bernes 1996; Moen et al. 2004). This prospect appears realistic for the low fell region concerned here, but may be questioned for most of the Scandes, where much higher and windier mountains prevail (Kullman and Öberg 2009).

Fig. 12.1 Mt. Storfjället. (**a**) Characteristically, most low fells have a smooth and rounded topography. The treeline ecotone extends diffusively over more than a kilometer. 15 June 2010. (**b**) During the past 30–35 years, scattered trees of mountain birch, spruce, and pine have emerged right at the highest point of the mountain. 10 July 2007

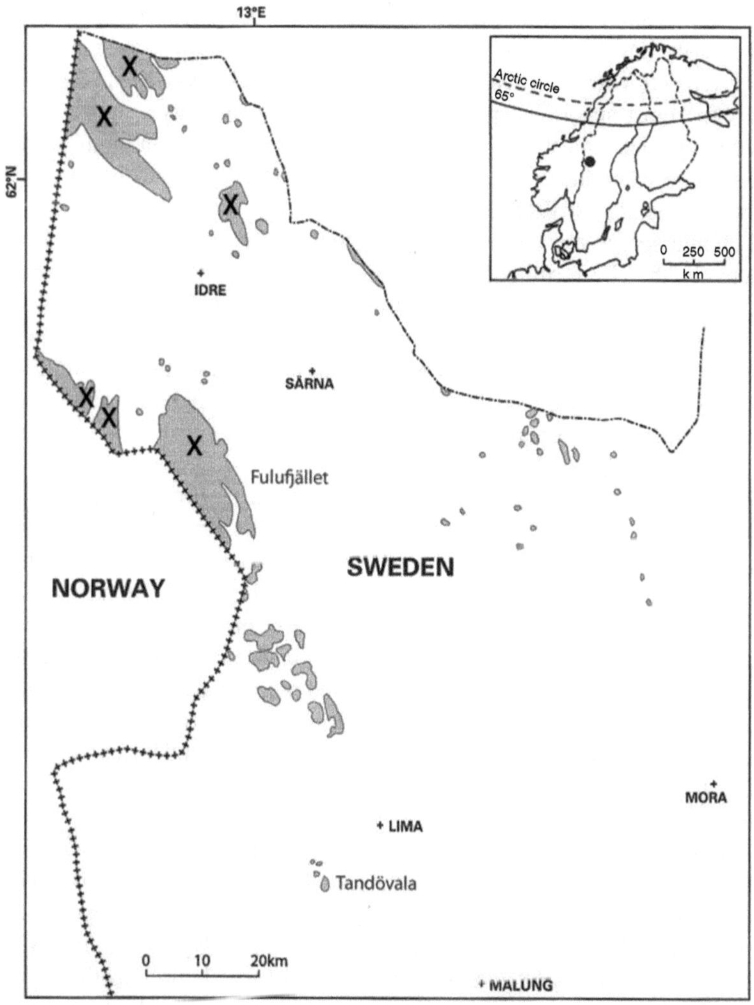

Fig. 12.2 Map showing the study area in the county of Dalarna. All low fells that existed by the early twentieth century (shaded) were mapped by Samuelsson (1917). Today, only those objects marked with an X remain as low fells, i.e., without tree-sized individuals of any species growing at the highest peak

12.3 The Alpine Treeline Ecotone: General Features

As a rule, the alpine treeline ecotone in this region is formed as a relatively broad, sparse, and indistinct belt with variable proportions of mountain birch (*Betula pubescens* ssp. *czerepanovii*), Norway spruce (*Picea abies*), and Scots pine (*Pinus sylvestris*). In contrast to more northerly parts of the Scandes, a discrete and monospecific birch belt, as the major part of the ecotone, is mostly lacking or fragmentarily developed. Only particularly steep and snow-rich sites support an upper birch belt, with a vertical extension of 50 m or less. As a rule, mountain birch is the uppermost species displaying tree form. In some cases, however, outpost individuals of pine and spruce reach even higher elevations than birch, which owes to the relatively dry and continental climate with virtually no persistent snow in the summer (cf. Lundqvist 1948; Kullman 2004a).

Birch and spruce reproduce mainly vegetatively by basal sprouts and layering, respectively, thereby forming clones of infinite age (spruce in particular) and variable form and size, concurrent with shifting climate trends and variations. Accordingly, treeline shifts for these species are predominantly phenotypical (Öberg and Kullman 2011a). Pine, on the other hand, lacks these abilities and therefore has to rely exclusively on seed reproduction in order to maintain its presence and to raise its treeline. A detailed account of these species and their high-mountain ecologies is given by Kullman (2010a).

As indicated above, the local spatial structure of the treeline ecotone is extremely complex. The transition from closed forest to treeless alpine tundra may be relatively abrupt or broadly diffuse. In many cases it takes the form of distinct patches or "fingers," surrounded by open alpine heath or mire (cf. Kullman 1979, 2010a; Moen et al. 2008). Complexity and heterogeneity also concern the tree growth forms. Mountain birch displays great variation between monocormic and polycormic types, which reflects snow cover, water and nutrient availability, and herbivory (Carlsson et al. 1999). Spruce and pine perform with a broad spectrum of forms, ranging from erect-growing individuals to krummholz (stunted forms) (cf. Holtmeier 1981). Ultimately, these modes originate from wind- and winter stress conditions (Holtmeier 2003).

Oligotrophic dwarf-shrub heaths with predominant *Calluna vulgaris* and *Empetrum hermaphroditum* prevail in the ecotone and slightly above. The most wind-exposed sites support a sparse cover of species with true alpine affinities, *Loiseleuria procumbens*, *Arctostaphylos alpinus*, *Carex bigelowii*, *Juncus trifidus*, *Hieracium alpinum*, *Diphasiastrum alpinum*, and variable proportions of reindeer lichens (*Cladina* spp.). The actual abundance of lichens is largely dependent on the intensity of reindeer grazing. Perennial snow patches do not exist, but scattered and relatively species-poor (low herbs and bryophytes) snow bed plant communities prevail in the alpine belt, where snow may stay until the mid-summer. Sites with relatively deep and persistent snow cover are characterized with *Vaccinium myrtillus* as the dominant species. The cover of mires is exceptionally small.

Widely scattered old-growth and vegetatively regenerating trees and krummholz individuals of spruce and birch form the core of a savanna-like landscape and add conservative aspects to the treeline ecotone. Particularly the spruces constitute an ancient structural component in this and many other low fell regions of northern Fennoscandia (e.g., Kihlman 1890; Hustich 1958; Kullman 2001b). As revealed by ^{14}C-dating of wood remnants preserved in the soil beneath the canopies, certain spruce clones are relics from the early and mid-Holocene (Kullman 2005a; Öberg and Kullman 2011a) (Fig. 12.3). The "eternal" nature of many krummholz spruces presupposes a harsh and open landscape, which prevents spruces from attaining

Fig. 12.3 Old-growth layering spruce, which attained tree size during the past century. Radiocarbon-dated wood pieces, preserved in the peat underneath the canopy, indicate that the individual spruce existed 7,940 cal year BP. Mt. Köarskärsfjället (825 m a.s.l.). *Source*: Öberg and Kullman (2011a)

senescence (cf. Öberg and Kullman 2011a). Thus the prevalence of ancient spruce clones indicates that the treeline ecotone contained open spaces even during the early Holocene, when treelines potentially reached higher than the highest mountains in the region.

The influence of human activities in the past has been virtually negligible for the current position of the treelines. This contention is based on the fact that there is no significant positional difference between treelines at sites with different degrees of past human utilization, e.g., livestock grazing, haymaking, reindeer husbandry (Kullman 1979; Kjällgren and Kullman 1998).

12.4 Holocene Retreat and Transformation of the Alpine Treeline Ecotone

A continuous Holocene chronology of elevational pine treeline change has been constructed for the Swedish Scandes. The basis is a large set of radiocarbon-dated tree megafossils (stems, roots, cones) (Fig. 12.4), retrieved from numerous sites at

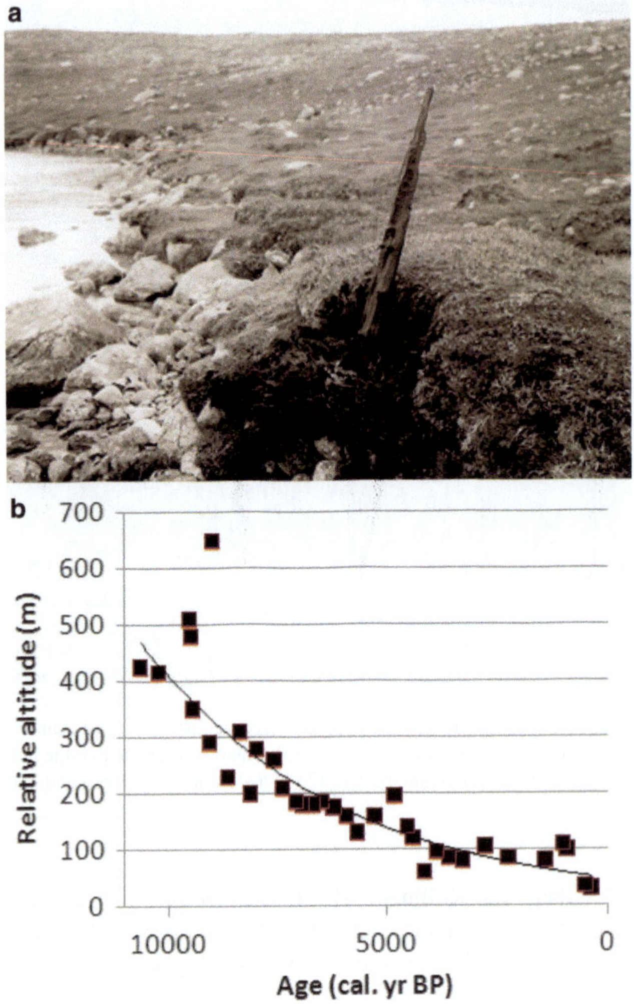

Fig. 12.4 (**a**) Mt. Storvätteshågna (1,180 m a.s.l.). Unearthed megafossil pine, which has been preserved in peat close to the highest mountain in the region. Radiocarbon dating yielded 10,390 cal year BP. 15 Aug 2002. (**b**) Treeline chronology valid for the Swedish Scandes. The graph is based on the highest megafossil pine for each 250-year period and relative to the early twentieth century position of the pine treeline at the respective site. *Source*: Kullman and Kjällgren (2006); Öberg and Kullman (2011b)

different elevations above the modern treeline along the entire Swedish Scandes (Kullman and Kjällgren 2006; Öberg and Kullman 2011a, b). It appears that the pine treeline has descended virtually steadily by about 50 m per millennium since the Holocene Thermal Maximum, 10,000–9,000 years ago. An analogous course of treeline retraction is obtained from southern Norwegian mountains (Paus 2010). Ultimately, this process has been a response to orbitally driven insolation decline and associated summer temperature cooling (cf. COHMAP 1988; Renssen et al. 2009). Drawing of this rate, the first emergence of all low fells and implicit, an alpine treeline ecotone, may be tentatively reconstructed in the study region, by back-counting from the current treeline elevation.

For example, the highest mountain in the region Mt. Storvätteshågna rises to 1,204 m a.s.l., which is 230 m higher than the current (2004) treeline of pine. This implies that the inception of alpine tundra took place about 4,500 years ago. Thereafter the width of the treeless cap has increased gradually until about a century ago, when its Holocene maximum was reached (Öberg and Kullman 2011a). At this point of time, the majority of the low fells in the region were mapped by Samuelsson (1917) (Fig. 12.2). The majority extended less than 100 m above the contemporary treeline and consequently they had gradually evolved during the past 2,000 years and in many cases over the past millennium (Kullman 2004a; Kullman and Öberg 2009). As a consequence, a plethora of new low fells emerged and their latitudinal distribution limit was shifted by several tens of kilometers southwards. This progressive "alpinization" of the concerned region after the mid-Holocene is consistent with the Neoglacial cooling, which has stressed high-altitude ecosystems over major parts of Scandinavia (cf. Karlén 1976; Caseldine and Matthews 1987; Paus 2010).

Another Neoglacial phenomenon concerns the spruce, which in line with its role as a late interglacial successor (Robertsson 1997), has gained its present-day codominant position in the ecotonal tree species mixture during the past 3,000 years. Concurrently, broadleaved thermophilous tree species, e.g., *Quercus robur*, *Corylus avellana*, and *Ulmus glabra*, which immigrated to the southern Scandes during the early Holocene (Kullman 2004a; Bang-Andersen 2006), were extirpated from the levels of the modern treeline ecotone during the cooling late Holocene (Lundqvist 1951; Giesecke 2005; Kullman 2005a, 2008; Öberg and Kullman 2011a). In addition, the minute alpine and northern floral elements in this mountain region seem to have spread to the low fells in parallel with the Neoglacial cooling and associated treeline descent (cf. Samuelsson 1910; Smith 1920; Hustich 1937; Almquist 1949). In particular, snow bed plant communities in and slightly above the treeline ecotone appear to have evolved mainly in response to this cooling phase (cf. Velle et al. 2005; Kullman 2003).

Most of the alpine ground cover species, which have spread to the region during the Neoglacial, are such which tolerate and regenerate on acidic forest soils with a raw humus layer, e.g., *Pedicularis lapponica*, *Phyllodoce caerulea*, *H. alpinum*, and *Athyrium distentifolium*. To a large extent, their low fell habitats may be characterized as "forest without trees," since the ground cover and top soils are residuals from the relatively warm early Holocene and have not yet fully adjusted to the Neoglacial climate (cf. Kullman 1997, 2005d; Hofgaard and Wilmann 2002). To some extent, this circumstance may explain the poverty of the alpine flora, as most genuine alpine plant species are less competitive on these soils.

Many low fells owe their existence to a particularly cool Neoglacial subperiod, the so-called "Little Ice-Age," approximately 1300–1900 AD. (Grove 1988; Kullman 2005c; Ljungqvist 2009). This contention is supported by exploratory studies by botanists and geographers active at the end of this period. They witnessed a miserable

Fig. 12.5 Mt. Storvätteshågna, 900 m a.s.l. About a century ago, the victims of prolonged Little Ice Age cold could be witnessed somewhat above the treeline, which then had retracted to its lowest position ever during the postglacial period. Photo: Karl-Erik Forsslund 1912. Reproduced by permission of Dalarnas Museum, Falun

state of forest vitality in the alpine treeline ecotone, with predominance of dead and dying trees and virtually no regeneration at all (e.g., Kellgren 1891; Forsslund 1921; Kullman 2000 and literature cited therein). Photographs from the final part of the Little Ice Age highlight a common situation with dead standing pine snags remaining in an open alpine landscape (Fig. 12.5), which was previously a part of the treeline ecotone. From adjacent parts of Norway, contemporary eyewitness reports tell of previously entirely forest clothed mountains, which were transformed to untreed low fells (Grove 1988). Possibly, low fells came into existence in Sweden even somewhat south of Mt. Tandövala during the Little Ice Age. Today, some high mountains here display a summit area with sparse, even-aged, and fairly young tree stands and a flora with a few alpine/subalpine plant species, e.g., *A. alpinus* and *B. pubescens* ssp. *czerepanovii* (Kullman 2004a and literature cited therein). Presence of more alpine plant species in the surrounding boreal forest than on the low fells (Samuelsson 1910; Almquist 1949) may further support the contention of a substantial and still remaining biogeographic impact of the Little Ice Age.

The national park, Sonfjället, in the southern Swedish Scandes was established in 1909, with the outspoken purpose to preserve a suitable site for scientific study of the ongoing "battle" between alpine and forest, which was felt as a great problem to forest authorities (Öberg 2009).

To some extent, disturbance by fire and cattle grazing may have interacted with late Holocene climate cooling as drivers of this course of landscape evolution (Kullman 1980; Kardell et al. 1982; Oldhammer 2005). Possibly, fire (natural or intentional), in combination with different geological substrates and positions along regional climatic continentality-maritimity gradients, has some role for the local variations in the relative proportions of spruce and pine in the treeline ecotone. Prevalence of fire-sensitive spruce clones in the treeline ecotone may signal a little (if any) fire-impacted environment.

12.5 Post-little Ice Age Evolution of the Alpine Treeline Ecotone

12.5.1 Meteorological Background

Instrumental meteorological records (1876–2007), representative of the study region, show a positive linear trend of 1.4 °C for the mean annual temperature and approximately the same for summer and winter periods. Precipitation has increased continuously over the past century (Alexandersson 2006; Kullman and Öberg 2009). Despite this fact, the treeline ecotone has become progressively drier during the summer period, mainly as a consequence of increased evapotranspiration and earlier disappearance of late-lying snow patches, which previously irrigated the slopes for most of the summer (Smith 1957; Kullman 2004a, 2007b, c). A decreasing trend of snow cover duration has been recorded over the past century (Moberg et al. 2005).

12.5.2 From Alpine Tundra to Forest: A Case Study on Mt. Tandövala

The mountain Tandövala (60°50′N, 13°11′E) reaches 774 m a.s.l., and for long it was considered to be the southernmost low fell in Sweden (Samuelsson 1917; Forsslund 1921; Lundqvist 1951). The last-mentioned author made a photographic survey of the summit area. Recent repeat photography of some of these images is presented below and shows a trend of tree and forest encroachment over the entire mountain cap (755–774 m a.s.l.), which therefore cannot be characterized as a low fell today (Figs. 12.6, 12.7, and 12.8).

The summit area has become covered with a sparse and patchy, 4–7 m high forest of mixed spruce, pine, and mountain birch. Some spruces and birches are old-growth individuals, which existed as prostrate krummholz prior to the twentieth century, as evidenced by ground observations and the historical photographs displayed above. Many trees have broken tops, as a consequence of heavy snow load in combination with strong winds. In accordance with a more common pattern in the southern Swedish Scandes, some trees suffered from frost desiccation injuries (needle loss) during the 1980s (cf. Kullman 1997), further indicating the marginal position of this new forest.

The static age structure of spruce (Fig. 12.9) was assessed in a plot within the newly established summit forest. Coring was carried out close to the root collar of the largest and most ancient-looking stem belonging to each individual, most of which were multistemmed. Since this kind of retrospective analysis expresses both natality and subsequent mortality, it should be interpreted with great caution. In the present case, absence of remains of dead saplings and young trees suggests that

Fig. 12.6 Mt. Tandövala, previously the southernmost low fell in Sweden. (**a**) By 1919, the summit area was devoid of trees. A few krummholz individuals of birch and spruce are strewn over the peak area. Photo: Karl-Erik Forsslund. (**b**) Throughout the twentieth century and up to the present day, a tree cover of spruce and birch has gradually emerged. 16 June 2010

Fig. 12.7 Mt. Tandövala. (a) View westwards from the summit. The landscape is virtually treeless. Photo: Karl-Erik Forsslund 1919. (b) By 1974, a tree cover (predominantly pine) has become established and the ground cover has increased in density and height. (c) The progressive trends of the tree and ground cover have proceeded further. Up-growing birches in the foreground preclude retake of the photo from the same point as previously. 16 June 2010

Fig. 12.8 Mt. Tandövala. (**a**) Boulder fringing open alpine tundra over the summit plateau. Photo: Karl-Erik Forslund 1919. (**b**) Today, this landscape supports fairly dense tree stands of pine. 16 June 2010

the obtained age spectrum reproduces the major variations in reproductive output, which peaked during the particularly warm decades 1935–1944 and 1995–2004.

It is confirmed by the age distribution that a few spruces grew here during the early nineteenth century, likely as krummholz. This situation still prevailed in the early twentieth century, as Forsslund (1921) explicitly reports that a few "dwarfed" spruces and "low growing birches" existed close to the summit. During the past century, and particularly after the 1920s, the number of emerging spruce stems has

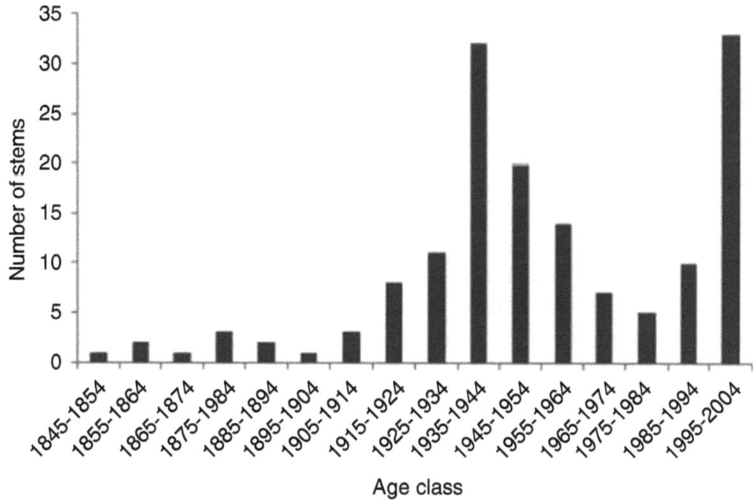

Fig. 12.9 Age frequency distribution of all spruces growing in a plot of 80 × 80 m, which covers a major part of the view depicted in Fig. 12.6. In order to minimize the uncertainty caused by coring slightly above the true root–stem interface, the datings are grouped into 10-year classes. *Source*: Kullman (2005b)

increased substantially, as emphatically evidenced by the re-photo record. Also the pine age structure (not shown) supports that the build-up of the present-day population is exclusively an affair of the twentieth century (Kullman 2005b). Tree ring count on a large and multistemmed birch at the top plateau shows that at least one specimen existed as a low-growing shrub by the mid-1880s.

Overall, the field-layer in the summit area is dominated by *C. vulgaris,* while *E. hermaphroditum* occupies more exposed sites in the microtopography. *V. myrtillus* prevails in minor depressions with ample snow cover and wind shelter. A sparse shrub layer of prostrate *Juniperus communis* is spread over the summit plateau. Some plant species, characteristic of natural low alpine heaths in the region, exist in the understory. These are *L. procumbens*, *A. alpinus*, *C. bigelowii*, *D. alpinum*, and the lichen *Flavocetraria nivalis*. In perspective of the ongoing tree cover closure, the persistence of these vascular plant species reasonably relates to their clonal and mat-forming growth modes, which do not invite the establishment of colonizers. No indications of substantial human impacts in the past are perceived.

Based on the photo record from the early twentieth century (Fig. 12.7), it is evident that the field-layer has grown denser, higher, and lusher. Minor boulders and stones, which were clearly visible some decades ago, are currently more or less overgrown with vegetation.

Relative to a species list compiled in 1915 for the summit plateau (Samuelsson 1917), the plant species richness has increased and not a single species has disappeared (Kullman 2005b). Most newcomers are genuine "forest species," some with relatively high thermal demands, e.g., *Acer platanoides* and *Betula pendula*. *Pinus*

Table 12.1 The magnitude of treeline rise (1915–2007) for different species in the study area (county of Dalarna)

Species	Mean (m)	Range (m)	Number of sites
Birch	55	0–125	17
Spruce	70	0–185	14
Pine	110	40–180	15

Source: Kullman and Öberg (2009)

contorta is an escape from forestry plantations at lower elevations in the region. This course of biodiversity transformation is consistent with current trends in other high mountain regions of the Scandes (Klanderud and Birks 2003; Kullman 2007b, c, 2010b; Odland et al. 2010).

In addition to Tandövala, there are some other, less intensively studied low fells in this region, which signals corresponding pending trajectories and have experienced a fundamental ecosystem shift in the same way (cf. Kullman 2005b, 2010a; Oldhammer 2005).

12.5.3 Treeline Change: The Regional Perspective

Recent repetition of historical altimeter measurements, in combination with dendrochronological analyses, forms the basis for assessments of elevational treeline change. This endeavor concerns birch, spruce, and pine over the period 1915–2007 at a large number of sites (belt transects) in the study region (Kullman and Öberg 2009). Practically all sites experienced an upslope shift and in no case has the treeline descended (Table 12.1). The maximum rises of all species concur with a pattern recorded along the entire Swedish Scandes, showing values around 200 m (Kullman and Öberg 2009; Öberg and Kullman 2011b). Advances by this order should be expected for a case with treelines in balance with ambient air temperature, assuming a common lapse rate 0.6 °C per 100 m altitude (Laaksonen 1976) and recorded secular warming by 1.4 °C. Thus, the largest extents of upward treeline movement represent treelines predominantly controlled by temperature. Localities with relatively lesser upshifts, i.e., the majority, are more or less constrained by local factors, e.g., geology/geomorphology, wind, etc. (Kullman and Öberg 2009).

A noteworthy implication of the results outlined above concerns the possibility to use treeline dynamics for the purpose of climate change monitoring (cf. Baker and Weisberg 1995). Treeline surveillance at selected localities with the largest advances during the past century and where the treeline has performed in near-perfect equilibrium with climate may serve as a complement to and a proxy check of instrumental climate trend records.

Pine displays the most consistent elevational trend, with an average rise by more than 100 m (Fig. 12.10). That is double the figure expressed by birch, although birch responded most readily during the early phase of warming, prior to the mid-twentieth century (Kullman and Öberg 2009). Age structure analyses of pine treeline populations

Fig. 12.10 Young pine (30–35 years old), representing treeline rise by 145 m since the early twentieth century. Mt. Fulufjället, 07 Aug 2004

show discrete recruitment pulses coincident with intervals with relatively high temperatures during the past century (Kullman 2000, 2004a, 2005b). Obviously, climate evolution over the entire past century has favored population expansion of the drought-tolerant pine more than birch and spruce, which both prefer a higher level of soil water availability (Tallantire 1977; Kullman 1986b). Increasing soil drought, correlated with recent warming and earlier and more complete annual melt-out of late-lying snow patches (Kullman 2004a, 2005a), is likely to have constrained the establishment of both birch and spruce in this region (Kullman and Kjällgren 2006). Already by the early twentieth century, Samuelsson (1917) inferred that climatic dryness (cf. Lundqvist 1951) prevents birch from attaining its upper thermal limit in these mountains, a situation which has obviously become further aggravated during the past century (cf. Smith 1957). Field-experimental results have evidenced the pivotal role of summer soil moisture for the growth and survival of birch seedlings in the treeline ecotone (Kullman 1986a). A supporting and more large-scale manifestation of this circumstance is provided on Mt. Storvätteshågna, where birch stands reach about 100 m higher on the moister north facing slope than on the relatively dry south slope. Likewise, the exceptionally dry Mt. Städjan displays treelines of pine, spruce, and birch at 1,045, 1,130, and 960 m a.s.l., respectively. During the past century, these treelines were displaced upslope by 180, 135, and 60 m, respectively (Kullman and Öberg 2009).

The constraint of further altitudinal birch expansion during recent decades may to some extent also depend on the fact that birch, as usually the highest ascending tree species, rapidly reached an elevation where wind velocity rather than temperature limits growth and survival and thwarts further elevational advance (Kullman and Öberg 2009; Kullman 2010a). As wind stress in general increases with altitude (e.g., Baker and Weisberg 1995), this is an important insight with relevance for projective treeline modeling.

Age structure analyses indicate that a majority of pines in the treeline ecotone has become established after 1990 (Kullman 2000, 2004a, 2005b). In general they have grown rapidly (10–30 cm/year) during this period, as winter-desiccation injuries, which in the past frequently checked the height increment, have been rare (cf. Kullman 2007a). However, the winters 2009/2010 and 2010/2011 were unusually cold in the perspective of the past 2 decades, which has caused severe damage to many young pine saplings and small trees in the treeline ecotone (Kullman and Öberg 2009, updated; Kullman 2010d). Also if this is a passing short-term extreme, the effects may remain for several years. Evidently, the new and rising pine treeline and tree species line are still in a vulnerable state, which should caution against taking projections too definitely, particularly as climate models for the future are unable to account for the balance of natural vs. human secular forcings (e.g., Karlén 2008).

The obtained coherence in maximum displacement magnitudes among species, despite widely different regeneration strategies, indicates that the treeline phenomenon may relate primarily to constraints operating well after the seed and establishment stages (cf. Kullman 1993, 2010a; Aune et al. 2011).

12.5.4 Recent Treeline Rise in a Long-Term Perspective

The current displacement and restructuring of the alpine treeline ecotone tend to resemble the situation prevailing by the earliest (and warmest) part of the Holocene, 10,000–9,000 years ago. During this interval, pine constituted the core of the treeline ecotone and covered most present-day low fells in the region up to the summits. Pure birch stands were strictly confined to isolated spots and slope sections with wind-accumulated and late-lying snow patches (Lundqvist 1951; Kullman and Kjällgren 2006; Öberg and Kullman 2011b). In the warm and dry climate, scattered spruces were restricted to north facing slopes and sites adjacent to wells (Kullman 2001b).

By projecting regional maximum treeline pine line rise by 180 m during the past century on the postglacial treeline chronology displayed in Fig. 12.4, it appears that the recently raised pine treeline position and associated climate forcing may be unsurpassed during at least the past 5,000 years (cf. Kullman and Kjällgren 2006; Velle et al. 2005; Bakke et al. 2008; Öberg and Kullman 2011a, b). However, any such projection should be judged cautiously, since in some cases the new and higher treeline is still formed by rather tiny trees (Fig. 12.10), sharply contrasting with the stout trees which formed the treeline about a century ago (Fig. 12.11). Such young treeline markers might easily succumb to short-term cooling episodes, possibly leaving few traces in the paleorecord (cf. Kullman 1997). Thus, the existence of one or a few prior short-term advances of similar magnitude as that of the twentieth century cannot be entirely ruled out, e.g., during the warm Medieval period

Fig. 12.11 The pine treeline by the early twentieth century is represented by this more than 500-year-old tree. Mt. Fulufjället, 25 May 2011

(cf. Kullman 1998, 2000, 2010a; Ljungqvist 2009). A further complicating aspect is that evolutionary processes throughout the Holocene may have changed the genetic structure and climate requirements of pine (cf. Rehfeldt et al. 2002). Moreover, the precise interpretation in degrees Celsius of the long-term Holocene treeline descent is complicated by the fact that the pine treeline integrates both long-term trends and short-term climatic extremes of summer and winter temperatures (Kullman 2007a, 2010a).

12.5.5 Local Spatial Aspects of Recent Treeline Evolution

It is a common observation that treeline markers of all species, birch in particular, seek wind shelter in close association with large boulders and other topographic irregularities. Likewise, mutual protection and facilitation by other specimens and various, still poorly understood seed and seedling mechanisms contribute to the structuring and maintaining of the treeline ecotone, as often discussed (Germino et al. 2002; Alftine and Malanson 2004). A certain clustering of pine seedlings around old seed bearers at the treeline is likely explained merely by the simple fact that most seeds hit the ground underneath their canopies (Kullman 2007a).

Typically, seed regeneration by spruce and birch in open patches is currently rare in the treeline ecotone. In great contrast, pine has reproduced prolifically during the past 3 decades, even far beyond and above putative "mother trees." Pine seedlings and saplings of all sizes are currently strewn over subalpine and lower alpine heaths and the pine tree species line is definitely on the rise. Even the highest mountains in the region harbor some pine saplings in their summit areas (Kullman 2000; Öberg 2008).

12.5.6 Recent Changes in a Broader Biogeographic Context, with Some Bearing on the Future

A continuation of the current climate and treeline trends over some coming decades would further alter and latitudinally contract this low fell landscape, both upslope and northward. With reference to the standard definition of a fell or alpine tundra, i.e., land above the highest treeline (e.g., Smith 1920; Hustich 1937; Bernes 1996; Nagy 2006), a major part of the alpine world at the southern periphery of the Scandes has already been lost, as young pine, spruce, and birch trees now grow on practically all summits up to about 950 m a.s.l. Thereby, the southern limit of low fells in Sweden has been displaced about 80 km northwards, from Mt. Tandövala to Mt. Fulufjället (Fig. 12.2). Merely lack of seed delays the infilling of the remaining gaps in the treeline ecotone and the final complete conversion to continuous forest (cf. Hustich 1948; Shiyatov 2003; Holtmeier 2003; Kullman 2007a), as described explicitly for Mt. Tandövala.

The progressive secular trend for pine in particular is a phenomenon which may herald the advent of a new and distinct pine belt, largely replacing the subalpine birch belt and thereby forming a major part of the treeline ecotone as a subalpine pine belt (Kullman 2004a, 2007b, 2010a, b). It appears that this is a fairly common process, which has progressed at variable rates on low fells over much of northern Fennoscandia since the 1930s (Hustich 1948, 1958; Wistrand 1981; Holtmeier and Broll 2007).

The likelihood of a major pine advance stands out even more distinctly as mountain birch has lost some of its vigor and competitive power and spruce reproduces poorly. In fact, on many of the low fells, birch displays a conspicuous geriatric age structure, with a preponderance of old-age trees from the mid-nineteenth century, and a conspicuously low regeneration rate for most of the twentieth century (Figs. 12.12 and 12.13) (Kullman 1987, 2004a). In more northerly regions of the Scandes, with a less continental climate, a putative takeover by pine is likely to procrastinate, since pine has to penetrate and compete with a more or less dense and still relatively vital birch forest belt (cf. Kullman 2010a). These regions are less marginal with respect to summer soil moisture availability, and therefore birch forest demise is conditional upon relatively severe warming-induced drought (cf. Young et al. 2011).

A further indication of the overriding role of climate change for the current and future reorganization and upward displacement of the treeline ecotone is provided by the fact that many low fells have received some new and particularly warmth

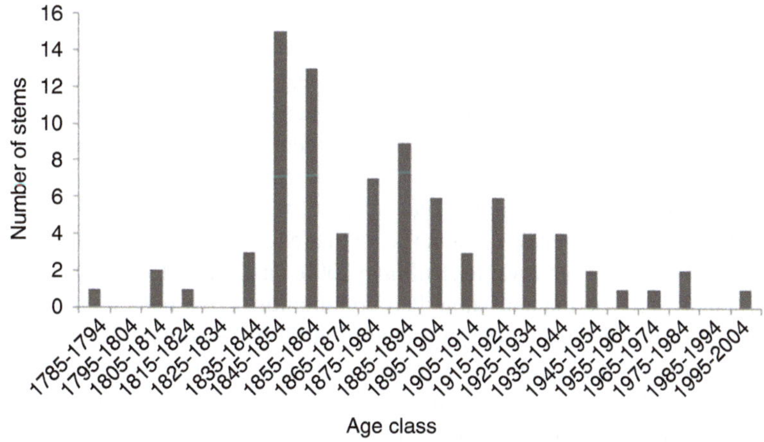

Fig. 12.12 Age frequency distribution of all occurring birches, sampled close to the ground in a plot of 200×200 m in the upper treeline ecotone on Mt. Köarskärsfjället (850–865). For multi-stemmed individuals (the majority), the seemingly oldest-looking stem was cored. In many cases, the ages do not represent establishment, but rather the inception of accelerated vertical growth of old-established krummholz birches. *Source*: Kullman (2005b)

Fig. 12.13 A major proportion of the mountain birches are of strikingly old age. This individual existed by the early nineteenth century, presumably as low-growing krummholz, which has attained tree-size over the past century. Mt. Köarskärsfjället, 18 July 2010

demanding and exotic plant species from lower regions during the past few decades. Striking examples are e.g., *A. platanoides*, *B. pendula*, *Larix sibirica*, *P. contorta*, *Pteridium aquilinum*, and *Lupinus polyphyllus*. This complies with results from more northerly parts of the southern Scandes, showing recent establishment of thermophiles such as *Q. robur* and *U. glabra* close to the birch treeline (Kullman 2008). Some of these newcomers may be "cultural fugitives," while others have a more uncertain origin (Kullman 2004b). Given that the concerned mountain region is the part of the Swedish Scandes, which is nearest to large human population centra, this kind of enrichment and floral adjustment to a new climatic situation will proceed most readily here in the case of future warming.

Repeat photography highlights that in tracts without a continuous history of pasturing by reindeer herds, tendencies for increasing stature and density of the dwarf-shrub layer are perceivable (Kullman 2000, unpublished; Oldhammer 2005). In particular, semi-permanent snow bed communities have been commonly converted to *V. myrtillus*-heaths (cf. Virtanen et al. 2003; Kullman 2004a, b, 2005a, 2007b). In the northern part of the region, reindeer grazing and trampling have substantially reduced the mats of reindeer lichens (*Cladina* spp., *F. nivalis*) over the past 5–6 decades. Comparing landscape photos from the mid-1940s (Lundqvist 1948, 1951) with the present-day situation reveals marked lichen decline and transformation of a prior whitish/yellowish landscape into a darker phase, leaving mats of dwarf-shrubs and spots with lichen fragments or bare mineral soil (Allard 2003; Kullman 2004a). Possibly, this course of change has promoted seed germination and early plant development and thereby having some role in the current pine expansion (cf. Sedia and Ehrenfeld 2003; Tømmervik et al. 2009). However, pine regenerates abundantly also in tracts virtually without reindeer presence during the past century and with a preponderance of thick lichen mats (Kullman 2004a, 2005a). In contrast to low fells in continental regions of northernmost Fennoscandia (Tømmervik et al. 2009), birch reproduction has not gained from reduced reindeer lichen mats in this region.

Current changes of the treeline ecotone fit a more widespread and ongoing reorganization of Fennoscandian biogeography. One aspect of this dynamism is the transformation and upward shifts of the principal alpine vegetation belts (Kullman 2004b, 2010b, c; Odland et al. 2010). For example, the mid-alpine belt in the Scandes is currently colonized by subalpine and low alpine species. The cover of *J. trifidus* is growing denser, which facilitates the establishment of other species, as *Juncus* has an excellent ability to collect snow leeward of the tufts, thereby ameliorating the microenvironment (Kullman 2004b). The high alpine belt, previously virtually devoid of higher vegetation, except for some scattered specimens, experiences primary succession of snow bed plant communities in the wake of receding permanent snow and ice (Kullman 2010c).

Another facet of recent climate-driven biogeographical disorder concerns the tendencies for poleward migration of the main phytogeographical border (regional ecotone) between southern (boreo-nemoral) and northern (boreal) Sweden, the so-called *Limes norrlandicus* (Sjörs 1999). This interface, which is defined from presence/absence of certain thermophilic tree species, retreated to its early twentieth century position in response to climatic stress during the Little Ice Age (e.g., Erkamo 1956; Huldén 2001; Molinari et al. 2005). Currently, it appears to be on the move northwards, spontaneously or somewhat aided by human activity (Erkamo 1956; Mascher 2007; Kullman 2008 and literature cited therein; Lidberg and Lindström 2010).

12.6 Conclusions

1. At broad spatial and temporal scales, alpine treelines in the Swedish Scandes perform in near-equilibrium with climate change.
2. At fine spatial and temporal scales, the magnitude of treeline responses to climate forcings is species-specifically constrained and modulated by various site characteristics (present and past).
3. Since the Holocene Thermal Maximum (10,000–9,000 years before the present) and until about a century ago, treelines descended in altitude more or less constantly, following orbitally driven insolation changes and correlated decrease of temperature and seasonality.
4. Treeline rise by 200 m over the past century has reversed the long-term trend of treeline descent and southward spread of alpine tundra. This course of biogeographic evolution appears anomalous in the perspective of several millennia.
5. *P. sylvestris* seems to be the tree species most favored by the new climatic situation. In case of continued warming it may regain the dominant position it held by the warm and dry early Holocene. The present-day ecotonal dominant *B. pubescens* ssp. *czerepanovii* is likely to regress substantially as the supply of meltwater tends to diminish in the mid- and late summer.
6. Given the responsiveness recorded for the near and distant past and considering the importance of local site characteristics, treelines appear well suited for monitoring and projecting climate change impacts of high-elevation ecosystems.
7. A multisite approach based on a narrow and precise treeline definition is essential to capture the entire spectrum of responses occurring in the heterogeneous mountain landscape.

References

Alexandersson H (2006) Klimat i förändring. En jämförelse av temperatur och nederbörd 1991–2005 med 1961–1990. SMHI Faktablad 29:1–8

Alftine K, Malanson GP (2004) Directional positive feedback and pattern at an alpine treeline. J Veg Sci 15:3–12

Allard A (2003) Vegetation changes in mountainous areas. PhD-thesis. Department of Physical Geography and Quaternary Geology. Stockholm University

Almquist E (1949) Dalarnas flora i växtgeografisk belysning. In: Natur i Dalarna. K-H, Forsslund C, Curry-Lindahl C (eds). Bokförlaget Svensk Natur, Göteborg, pp 55–7

Aune S, Hofgaard A, Söderström L (2011) Contrasting climate- and land-use-driven tree encroachment patterns of subarctic tundra in northern Norway and the Kola Peninsula. Can J For Res 41:437–449

Baker WL, Weisberg PJ (1995) Landscape analysis of the forest-tundra ecotone in Rocky Mountain National Park, Colorado. Prof Geogr 47:361–375

Bakke J, Lie Ø, Dahl SO, Nesje A, Bjune AE (2008) Strength and spatial patterns of the Holocene wintertime westerlies in the NE Atlantic region. Glob Planet Chang 60:28–41

Bang-Andersen S (2006) Charcoal in hearths: a clue to the reconstruction of palaeo-environments of Mesolithic dwelling sites. Archaeol Environ 12:5–16

Bekker MF (2005) Positive feedback between tree establishment and patterns of subalpine forest advancement, Glacier National Park, Montana, USA. Arct Antarct Alp Res 37:97–107
Bernes C (1996) Arktisk miljö i Norden—orörd, exploaterad, förorenad. Nordiska ministerrådet, Stockholm
Butler DR, Malanson GP, Walsh SJ, Fagre DB (2007) Influences of geomorphology and geology on alpine treeline in the American west—more important than climatic influences? Phys Geogr 28:434–450
Carlsson BÅ, Karlsson PS, Svensson BM (1999) Alpine and subalpine vegetation. Acta Phytogeogr Suec 84:75–89
Caseldine CJ, Matthews JA (1987) Podzol development, vegetation change and glacier variations at Haugabreen, southern Norway. Boreas 16:215–230
COHMAP members (1988) Climatic changes of the last 18.000 years: observations and simulations. Science 241:1043–1052
Danby RK, Hik DK (2007) Variability, contingency and rapid change in recent subarctic alpine treeline dynamics. J Ecol 95:352–363
Diaz HF, Grosjean M, Graumlich L (2003) Climate variability and change in high elevation regions: past, present and future. Clim Chang 59:1–4
Elliot GP, Kipfmueller KF (2010) Multi-scale influences of slope aspect and spatial pattern on ecotonal dynamics at upper treeline in the Southern Rocky Mountains, U.S.A. Arct Antarct Alp Res 42:45–56
Enquist F (1933) Trädgränsundersökningar. Svenska Skogsvårdsför Tidskr 31:145–191
Erkamo V (1956) Untersuchungen über die pflanzenbiologischen und einige andere Folgerscheinungen der neuzeitlichen Klimaschwankung in Finnland. Ann Bot Soc Zool Bot Fenn Vanamo 28:1–283
Esper J, Schweingruber FH (2004) Large-scale tree-line changes recorded in Siberia. Geophys Res Lett 31:1–5
Fagre DB, Peterson DL, Hessle AE (2003) Taking the pulse of mountains: ecosystem responses to climatic variability. Clim Chang 59:263–282
Forsslund K-E (1921) Omkring Storvaln. Sveriges sydligatse fjäll. Sveriges Natur 1921:33–42
Gehrig-Fasel J, Guisan A, Zimmermann NE (2007) Tree line shifts in the Swiss Alps. Climate change and land abandonment. J Veg Sci 18:571–582
Germino MJ, Smith WK, Resor AC (2002) Conifer seedling distribution and survival in an alpine-tundra ecotone. Plant Ecol 162:157–168
Giesecke T (2005) Holocene forest development in the central Scandes Mountains, Sweden. Veg Hist Archaeobot 14:133–147
Gosz JR (1993) Ecotone hierarchies. Ecol Appl 3:369–373
Grace J, Berninger F, Nagy L (2002) Impacts of climate change on the tree line. Ann Bot 90:537–544
Grove JM (1988) The little Ice Age. Methuen, New York
Hallinger M, Manthey M, Wilmking M (2010) Establishing a missing link: warm summers and winter snow cover promote shrub expansion into alpine tundra in Scandinavia. New Phytol 186:890–899
Harsch MA, Hulme PE, McGlone MS, Duncan RP (2009) Are treelines advancing? A global meta-analysis of treeline response to climate warming. Ecol Lett 12:1040–1049
Hoch G, Körner C (2003) The carbon charging of pines at the climatic treeline: a global comparison. Oecologia 135:10–21
Hofgaard A, Wilmann B (2002) Plant distribution across the forest-tundra ecotone: the importance of treeline position. Ecoscience 9:375–385
Holtmeier F-K (1981) What does the term "krummholz" really mean? Observations with special reference to the Alps and the Colorado Front Range. Mt Res Dev 1:253–260
Holtmeier F-K (2003) Mountain timberlines: ecology, patchiness and dynamics. Kluwer, Dordrecht
Holtmeier F-K, Broll G (2005) Sensitivity and response of northern hemisphere altitudinal and polar treelines to environmental change at landscape and local scales. Glob Ecol Biogeogr 14:395–410

Holtmeier F-K, Broll G (2007) Treeline advance—driving processes and adverse factors. Landsc Online 1:1–33
Holtmeier F-K, Broll G (2010) Wind as an ecological agent at treeline in North America, the Alps and the European Subarctic. Phys Geogr 31:203–233
Huldén L (2001) Ektunnor och den medeltida värmeprioden i Satakunta. Terra 113:171–178
Hustich I (1937) Pflanzengeographische Studien im Gebiet der niederen Fjelde im westlichen Finnischen Lappland. Acta Bot Fenn 19:1–156
Hustich I (1948) The Scotch pine in northernmost Finland and its dependence on the climate in the last decades. Acta Bot Fenn 42:1–75
Hustich I (1958) On the recent expansion of the Scotch pine in northern Europe. Fennia 82:1–25
Hustich I (1978) The growth of Scots pine in northern Lapland, 1928–77. Ann Bot Soc Zool Bot Fenn Vanamo 15:241–252
Hustich I (1979) Ecological concepts and biogeographical zonation in the north: the need for a generally accepted terminology. Holarct Ecol 2:208–217
IPCC (2007) Climate change 2007. The physical scientific basis. Cambridge University Press, Cambridge
Kapralov DS, Shiyatov SG, Moiseev PA, Fomin VV (2006) Changes in composition, structure, and altitudinal distribution of low forests at the upper limit of their growth in the Northern Ural Mountains. Russ J Ecol 37:367–372
Kardell L, Arvidsson B, Nilsson E (1982) Tandövala—vårt sydligaste lågfjäll. Sveriges Lantbruksuniv Avd För Landskapsvård Rapport 24:1–157
Karlén W (1976) Lacustrine sedimenst and tree-limit variations as indicators of Holocene climatic fluctuations in Lappland: northern Sweden. Geogr Ann 58A:1–34
Karlén W (2008) Recent changes in climate: natural or forced by human activity. Ambio Special Report 14:483–488
Kellgren AG (1891) Om de skogbildande trädens utbredning i Dalarnes fjältrakter. Botaniska Notiser 1891:182–186
Kihlman AO (1890) Pflanzenbiologische Studien aus Russich Lappland. Acta Soc Fauna Flora Fenn 68(3):1–263
Kjällgren L, Kullman L (1998) Spatial patterns and structure of the mountain birch tree-limit in the southern Swedish Scandes—a regional perspective. Geogr Ann 80A:1–16
Klanderud K, Birks HJ (2003) Recent increases in species richness and shifts in altitudinal distributions of Norwegian mountain plants. Holocene 13:1–6
Körner C (1999) Alpine plant life. Springer, Berlin
Körner C (2007) Climatic treelines: conventions, global patterns, causes. Erdkunde 61:316–324
Kullman L (1979) Change and stability in the altitude of the birch tree-limit in the southern Swedish Scandes 1915–1975. Acta Phytogeogr Suec 65:1–121
Kullman L (1980) Radiocarbon dating of subfossil Scots pine (*Pinus sylvestris* L.) in the southern Swedish Scandes. Boreas 9:101–106
Kullman L (1986a) Demography of *Betula pubescens* ssp. *tortuosa* sown in contrasting habitats close to the birch limit in Central Sweden. Vegetatio 65:13–20
Kullman L (1986b) Recent tree-limit history of *Picea abies* in the southern Swedish Scandes. Can J For Res 16:761–771
Kullman L (1987) A decade of tree-limit monitoring in the southern Swedish Scandes. UNGI Rapport 65:191–202
Kullman L (1993) Tree limit dynamics of *Betula pubescens* ssp. *tortuosa* in relation to climate variability: evidence from central Sweden. J Veg Sci 4:765–772
Kullman L (1997) Tree-limit stress and disturbance. A 25-year survey of geoecological change in the Scandes Mountains of Sweden. Geogr Ann 79A:139–165
Kullman L (1998) Tree-limits and montane forests in the Swedish Scandes: sensitive biomonitors of climate change and variability. Ambio 27:312–321
Kullman L (2000) Tree-limit rise and recent warming: a geoecological case study from the Swedish Scandes. Norw J Geogr 54:49–59

Kullman L (2001a) 20th century climate warming and tree-limit rise in the southern Scandes. Ambio 30:72–80
Kullman L (2001b) Immigration of *Picea abies* into North-Central Sweden. New evidence of regional expansion and tree-limit evolution. Nordic J Bot 21:39–54
Kullman L (2003) Recent reversal of Neoglacial climate cooling trend in the Swedish Scandes as evidenced by mountain birch tree-limit rise. Glob Planet Chang 36:77–88
Kullman L (2004a) Tree-limit and landscape evolution at the southern fringe of the Swedish Scandes (Dalarna province)—Holocene and 20th century perspectives. Fennia 182:73–94
Kullman L (2004b) A face of global warming—"ice birches" and a changing alpine plant cover. GeoÖko 25:181–202
Kullman L (2005a) Gamla och nya träd på Fulufjället—vegetationshistoria på hög nivå. Sven Bot Tidskr 99:315–329
Kullman L (2005b) Trädgränsen i Dalafjällen. Del. 2. Tandövala—försvinnande sydlig fjällvärld. Länsstyrelsen Dalarnas Län Miljövårdsenheten Rapport 10:1–25
Kullman L (2005c) Pine (*Pinus sylvestris*) treeline dynamics during the past millennium a population study in west-central Sweden. Ann Bot Soc Zool Bot Fenn Vanamo 42:95–106
Kullman L (2005d) Wind-conditioned 20th century decline of birch treeline vegetation in the Swedish Scandes. Arctic 58:286–294
Kullman L (2007a) Tree line population monitoring of *Pinus sylvestris* in the Swedish Scandes, 1973–2005: implications for tree line theory and climate change ecology. J Ecol 95:41–52
Kullman L (2007b) Modern climate change and shifting ecological states of the subalpine/alpine landscape in the Swedish Scandes. GeoÖko 28:187–221
Kullman L (2007c) Long-term geobotanical observations of climate change impacts in the Scandes of West-Central Sweden. Nord J Bot 24:445–467
Kullman L (2008) Thermophilic tree species reinvade subalpine Sweden—early responses to anomalous Late Holocene warming. Arct Antarct Alp Res 40:104–110
Kullman L (2010a) One century of treeline change and stability—experiences from the Swedish Scandes. Landsc Online 17:1–31
Kullman L (2010b) Alpine flora dynamics—a critical review of responses to climate change in the Swedish Scandes since the early 1950s. Nord J Bot 28:398–408
Kullman L (2010c) A richer, greener and smaller alpine world: review and projection of warming-induced plant cover change in the Swedish Scandes. Ambio 39:159–169
Kullman L (2010) Fjällens nya ansikte. Rikare, grönare, vackrare. Lustgården 90:13–24. In Swedish with English summary
Kullman L, Kjällgren L (2006) Holocene treeline evolution in the Swedish Scandes: recent treeline rise and climate change in a long-term perspective. Boreas 35:159–168
Kullman L, Öberg L (2009) Post-Little Ice Age tree line rise and climate warming in the Swedish Scandes: a landscape ecological perspective. J Ecol 97:415–429
Kupfer JA, Cairns DM (1996) The suitability of montane ecotones as indicators of global climatic change. Prog Phys Geogr 20:253–272
Laaksonen K (1976) The dependence of mean air temperatures upon latitude and altitude in Fennoscandia. Ann Acad Sci Fenn Ser A3 199:1–19
Leonelli G, Pelfini M, Morra di Cella U, Garavaglia V (2011) Climate warming and the recent treeline shift in the European Alps: the role of geomorphological factors in high-altitude sites. Ambio 40:264–273
Lidberg R, Lindström H (2010) Medelpads flora. SBF-förlaget, Uppsala
Ljungqvist FC (2009) Temperature proxy records covering the last two millennia: a tabular and visual overview. Geogr Ann 91A:11–29
Lloyd AH (2005) Ecological histories from Alaskan treelines provide insight into future change. Ecology 86:1687–1695
Lloyd AH, Fastie CL (2002) Spatial and temporal variability in the growth and climate response of treelines in Alaska. Clim Chang 52:481–509
Lundqvist G (1948) De svenska fjällens natur. Svenska Turistföreningens Förlag, Stockholm

Lundqvist G (1951) Beskrivning till jordartskarta över Kopparbergs län. Sver Geol Und Ca 21:1–213
Malanson GP (2001) Complex responses to global change at alpine treeline. Phys Geogr 22:333–342
Mascher JW (2007) Nya fynd i Ångermanlands flora. Sven Bot Tidskr 101:321–346
Miehe G, Miehe S (2000) Comparative high mountain research on the treeline ecotone under human impact. Erdkunde 54:34–50
Moberg A, Tuomenvirta AH, Nordli Ø (2005) Recent climatic trends. In: Seppälä M (ed) The physical geography of Fennoscandia. Oxford University Press, Oxford, pp 112–133
Moen J, Aune K, Edenius L, Agerbjörn A (2004) Potential effects of climate change on treeline position in the Swedish mountains. Ecol Soc 16:1–10
Moen J, Cairns DM, Lafon CW (2008) Factors structuring the treeline ecotone in Fennoscandia. Plant Ecol Divers 1:77–87
Molinari C, Bradshaw RHW, Risbøl O, Lie H, Ohlson M (2005) Long-term vegetational history of a *Picea abies* stand in south-eastern Norway: implications for the conservation of biological values. Biol Conserv 126:155–165
Mossberg B, Stenberg L (2003) Den nya nordiska floran. Wahlström & Wistrand, Stockholm
Nägeli W (1969) Waldgrenze und Kampfzone in den Alpen. Hespa Mitt 19:1–44
Nagy L (2006) European high mountain (alpine) vegetation and its suitability for indicating climate change impacts. Biol Environ Proc R Ir Acad 106B(3):335–341
Noble IR (1993) A model of the responses of ecotones to climate change. Ecol Appl 3:396–403
Öberg L (2008) Trädgränsen som indikator för ekologiska klimateffekter i fjällen. Länsstyrelsen i Jämtlands län. Miljö/Fiske Miljöövervakning. Rapport, 2008,01
Öberg L (2009) The heart of Härjedalen. Sonfjället. National Park since 1909. Jamtli Förlag, Östersund
Öberg L, Kullman L (2011a) Ancient subalpine clonal spruces (*Picea abies*): sources of postglacial vegetation history in the Swedish Scandes. Arctic 64:183–196
Öberg L, Kullman L (2011b) Recent glacier recession—a new source of postglacial treeline and climate history in the Swedish Scandes. Landsc Online 26:1–38
Odland A, Høitomt T, Olsen SL (2010) Increasing vascular plant richness on 13 high mountain summits in southern Norway since the early 1970s. Arct Antarct Alp Res 42:458–470
Oldhammer B (2005) Varder—en försvinnande naturtyp. Sven Bot Tidskr 99:330–331
Paus A (2010) Vegetation and environment of the Rødalen alpine area, Central Norway, with emphasis on the early Holocene. Veget Hist Archeobot 19:29–51
Rehfeldt GE, Tchebakova NM, Parfenova YI, Wykoff WR, Kuzmina NA, Milyutin LI (2002) Intraspecific responses to climate in *Pinus sylvestris*. Glob Chang Biol 8:912–929
Renssen H, Seppä H, Heiri O, Roche DM, Goosse H, Fichefet T (2009) The spatial and temporal complexity of the Holocene thermal maximum. Nat Geosci 2:411–414
Robertsson AM (1997) Reinvestigation of the interglacial pollen flora at Leveäniemi, Swedish Lapland. Boreas 26:81–89
Samuelsson G (1910) Regionförskjutningar inom Dalarne. Sven Bot Tidskr 4:1–57
Samuelsson G (1914) Om Dalafjällen. Ymer 1914:331–345
Samuelsson G (1917) Studien über die Vegetation der Hochgebirgsgegenden von Dalarne. N Acta Reg Soc Sci Upsal 4(8):1–253
Sedia EG, Ehrenfeld JG (2003) Lichens and mosses promote alternate stable plant communities in the New Jersey pinelands. Oikos 100:447–458
Seppälä M (2004) Wind as a geomorphic agent in cold climates. Cambridge University Press, Cambridge
Shiyatov SG (2003) Rates of change in the upper treeline ecotone in the Polar Urals. Pages News 11:6–8
Sjörs H (1999) The background: geology, climate and zonation. Acta Phytogeogr Suec 84:5–14
Skre O, Baxter R, Crawford RMM, Callaghan TV, Fedorkov A (2002) How will the tundra-taiga interface respond to climate change? Ambio Special Report 12:37–46

Smith H (1920) Vegetationen och dess utvecklingshistoria i det centralsvenska högfjällsområdet. Almqvist & Wiksell, Uppsala

Smith H (1957) En botanisk undersökning av Neans dalgång. K Svenska Vetenskapsakad Avhandl Naturskyddsär 16:1–21

Sundqvist MK, Björk RG, Molau U (2008) Establishment of boreal forest species in alpine dwarf-shrub heath in subarctic Sweden. Plant Ecol Divers 1:67–75

Tallantire P (1977) A further contribution to the problem of the spread of spruce (*Picea abies* (L.) Karst.) in Fennoscandia. J Biogeogr 4:219–227

Tape K, Sturm M, Racine C (2006) The evidence for shrub expansion in Northern Alaska and the Pan-Arctic. Glob Chang Biol 12:686–702

Tinner W, Kaltenrieder P (2005) Rapid responses of high-mountain vegetation to early Holocene environmental changes in the Swiss Alps. J Ecol 93:936–947

Tømmervik H, Johansen B, Riseth JÅ, Karlsen SR, Solberg B, Høgda KA (2009) Above ground biomass changes in the mountain birch forests and mountain heaths of Finnmarksvidda, northern Norway, in the period 1957–2006. For Ecol Manag 257:244–257

Velle G, Larsen J, Eide W, Peglar SM, Birks HJB (2005) Holocene environmental history and climate at Råtåsjøen, a low-alpine lake in south-central Norway. J Paleolimnol 33:129–153

Virtanen R, Eskelinen A, Gaare E (2003) Long-term changes in alpine plant communities in Norway and Finland. In: Nagy L, Grabherr G, Körner C, Thompson DBA (eds) Alpine biodiversity in Europe. Springer, Berlin, pp 411–422

Walsh SJ, Butler DR, Allen TR, Malanson GP (1994) Influence of snow patterns and snow avalanches on the alpine treeline ecotone. J Veg Sci 5:657–672

Wistrand G (1981) Bidrag till Pite lappmarks växtgeografi. Växtekol Stud 14:1–99

Young A, Cairns DM, Lafon CW, Moen J, Martin LE (2011) Dendroclimatic relationships and possible implications for mountain birch and Scots pine at treeline in northern Sweden through the 21st century. Can J For Res 41:450–459

Chapter 13
How Climate and Fire Disturbances Influence Contrasted Dynamics of *Picea glauca* Ecotones at Alpine Tree Lines in Atlantic and Continental Eastern North America

Guillaume de Lafontaine and Serge Payette

13.1 Introduction

White spruce (*Picea glauca* (Moench) Voss.) is considered as the tree most characteristic of the North American boreal forest biome (Takhtajan 1986). Yet, in boreal eastern North America (Québec–Labrador Peninsula) it is rather a secondary, companion species (*c.* 10 % of the forest cover) of the balsam fir (*Abies balsamea* (L.) Mill.)–white birch (*Betula paryrifera* Marsh.) dominated forest assemblage (Jurdant 1959; Lafond 1964, 1974), typical of the closed-crown boreal forest south of 49°N (Fig. 13.1). Noticeable exceptions however exist in eastern North America where white spruce is locally the dominant species. On the maritime coast of Hudson Bay and Labrador, monospecific white spruce stands are initiating the colonization (primary succession) of barren surfaces (Payette and Filion 1985; Caccianiga and Payette 2006; Payette 2007; Laliberté and Payette 2008). Other maritime white spruce populations are forming a strip along the Acadian coast where it is a pioneer species of the secondary succession following forest openings by wind blowdowns (Davis 1966). White spruce also readily colonizes abandoned agricultural fields in maritime Canada and New England (Davis 1966; Nienstaedt and Zasada 1990). On Anticosti island in the Gulf of St. Lawrence, selective grazing of balsam fir by overpopulated white-tailed deer (*Odocoileus virginianus* Zimmerman) resulted in a shift from former balsam fir stands to white spruce dominated stands over the last century (Potvin et al. 2003) .

G. de Lafontaine (✉)
Département de biologie and Centre d'études nordiques,
Université Laval, Québec, Canada

UMR Biogeco, INRA and Univ. Bordeaux 1,
Bordeaux, France
e-mail: gdelafon@pierroton.inra.fr

S. Payette
Département de biologie and Centre d'études nordiques,
Université Laval, Québec, Canada

R.W. Myster (ed.), *Ecotones Between Forest and Grassland*,
DOI 10.1007/978-1-4614-3797-0_13, © Springer Science+Business Media New York 2012

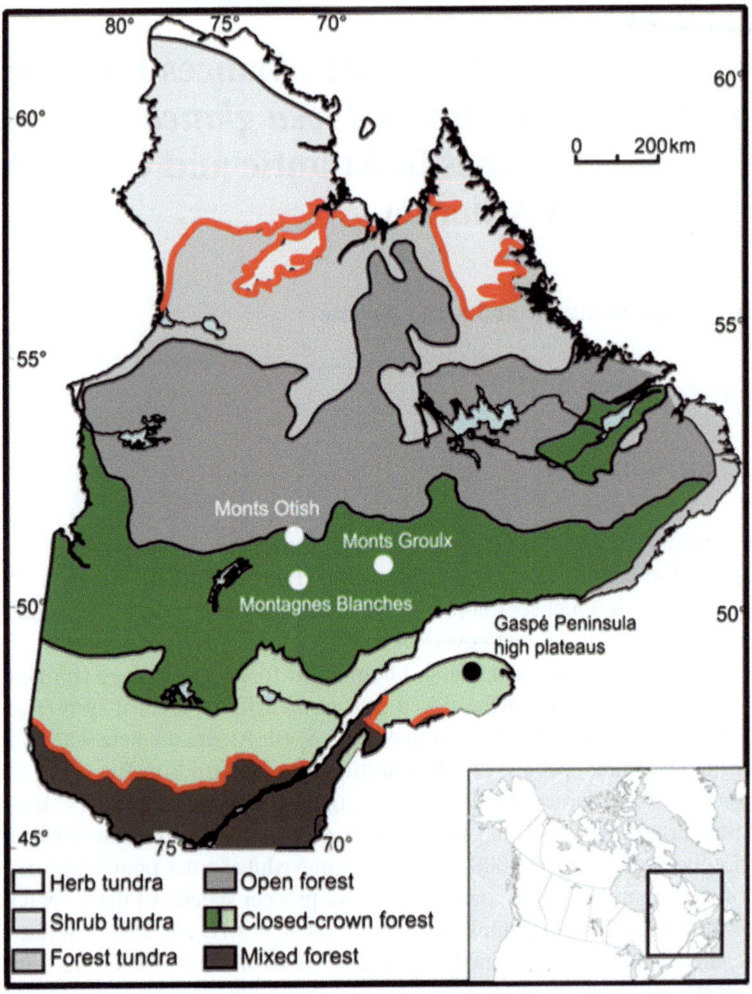

Fig. 13.1 Location of the ecotonal, subalpine white spruce stands in eastern North American boreal forest. The main bioclimatic zones of Québec–Labrador are represented; the zones corresponding to the boreal biome are between the two *red lines*. Note that the closed-crown forest is roughly subdivided along 49°N in two ecological regions represented by different shades of *green*; the *paler green* represents the balsam fir–white birch zone whereas the *darker green* represents the black spruce–feathermoss zone

Other old-growth white spruce stands are found in ecotonal environment, at the altitudinal tree lines of the highest plateaus of the boreal forest in eastern North America. Such subalpine white spruce stands are not found southward in the Appalachian Mountains where the tree line, from north to south is either red spruce (*Picea rubens* Sarg.), or balsam fir or Fraser fir (*Abies fraseri* (Pursh) Poir.), respectively (Cogbill and White 1991). The southernmost subalpine white spruce stands are distributed on extensive high plateaus (*c.* 1,200 m above sea level) within

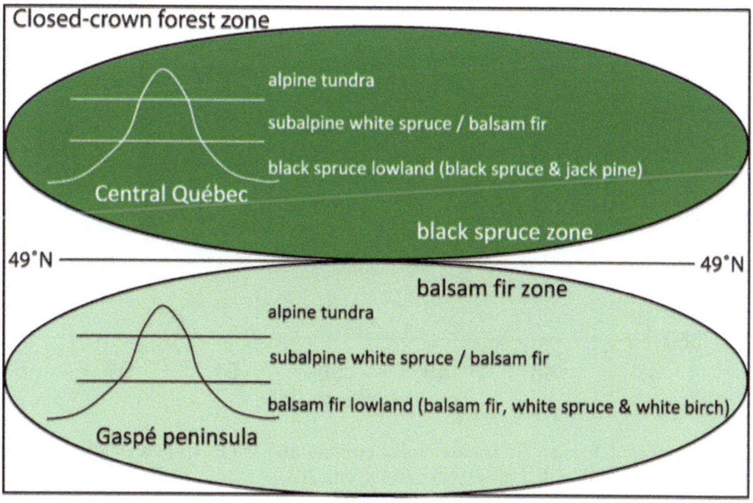

Fig. 13.2 Biogeographical model illustrating subalpine white spruce and balsam fir stands in boreal eastern North America. (From de Lafontaine and Payette 2010)

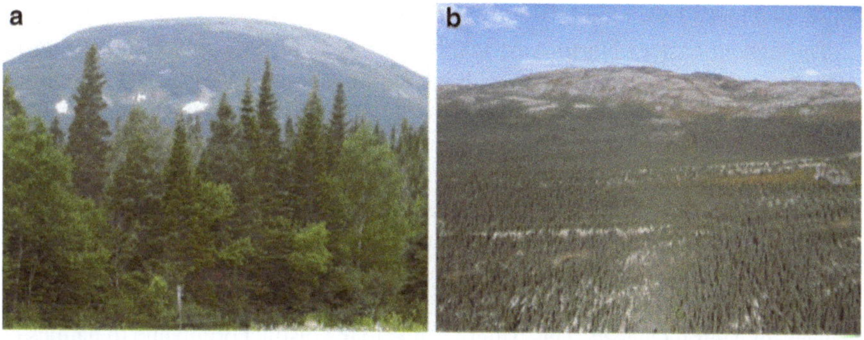

Fig. 13.3 Example of ecotonal subalpine white spruce stands from (**a**) Gaspé Peninsula high plateaus with lowland balsam fir–white birch forest in the foreground; and (**b**) Monts Otish with lowland black spruce–feathermoss forest (interspaced with jack pine stands) in the foreground

the balsam fir–white birch forest assemblage of the (Atlantic) Gaspé Peninsula (Fig. 13.1). These stands form a subalpine ecotonal belt between the balsam fir forest (typical of the lowlands) and the alpine tundra (Payette and Boudreau 1984; Figs. 13.2 and 13.3). The other (northern) subalpine white spruce stands are restricted to three high plateaus (Monts Otish, Monts Groulx, Montagnes Blanches; *c.* 1,000 m above sea level) in continental central Québec. There, the regional lowland matrix is dominated by closed crown black spruce (*Picea mariana* (Mill.) B.S.P.)–feathermoss forests and monospecific jack pine (*Pinus banksiana* Lamb.) stands typical of the boreal landscape north of 49°N (Fig. 13.1). In the subalpine belt of this area, old-growth white spruce stands co-occur with balsam fir stands

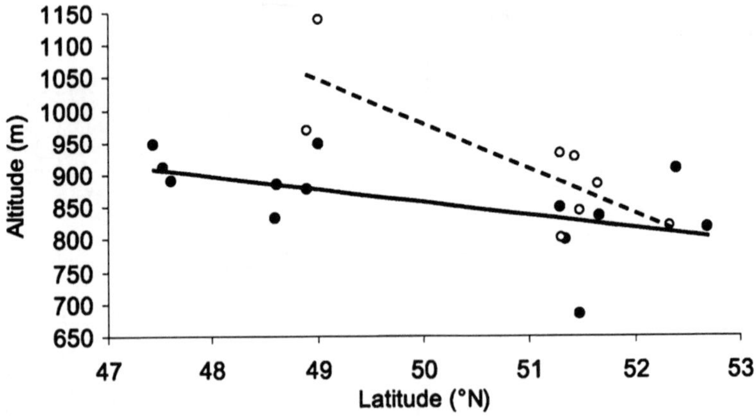

Fig. 13.4 Altitude of balsam fir stands (*solid circles*) and white spruce stands (*open circles*) according to latitude. (From de Lafontaine and Payette 2010)

forming the ecotone between continuous black spruce forest in the lowlands and alpine tundra at higher elevations, a pattern roughly similar to that found on the high plateau of the Gaspé Peninsula, albeit with a different lowland forest community (Figs. 13.2 and 13.3).

Throughout eastern North America, the elevation of both subalpine balsam fir and white spruce stands decreases with increasing latitude. However, the altitudinal position of white spruce stands drops more rapidly with latitude than that of balsam fir stands (Fig. 13.4). This results in the altitude of white spruce stands being significantly higher than balsam fir stands south of 49°N, whereas there is no difference in altitude between white spruce stands and balsam fir stands north of 49°N (Fig. 13.4). In other words, on the high plateaus of the Gaspé Peninsula, white spruce stands form a distinct subalpine belt above the balsam fir forest whereas the two stand types coexist at the same elevation in the ecotonal subalpine belts of the high plateaus of central Québec. This pattern is due to the distinct origin and dynamics of the subalpine old-growth forests north and south of 49°N (de Lafontaine and Payette 2010). Ultimately, all subalpine white spruce stands originated from an ecological sorting of the balsam fir–white birch assemblage allowing local dominance of the companion species (white spruce) over the regionally dominant one (balsam fir). However, the main environmental factors driving the ecological sorting processes are different north and south of 49°N. These environmental factors causing the observed patterns of the ecotonal old-growth subalpine belts are discussed below.

13.2 Origin of the Subalpine Stands North of 49°N

Of all the widespread boreal tree species, white spruce and especially balsam fir are the least adapted to fire, whereas jack pine and black spruce are fire-adapted species (Heinselman 1973; Rowe and Scotter 1973). This suggests that fire might be a major forcing accounting for the distinctiveness of the southern balsam fir–white birch zone (balsam fir zone thereafter) and northern black spruce–feathermoss zone (black spruce zone, thereafter) in eastern North America. Indeed, by studying fire records of subalpine stands of the black spruce and balsam fir zones, de Lafontaine and Payette (2010, 2011) found distinct fire regimes. Subalpine stands from the black spruce zone were all subjected to fire disturbance whereas no fire was found in subalpine stands distributed on the high plateaus of the Gaspé Peninsula. On the one hand, north of 49°N, old-growth stands were all regenerated after a wildfire between 230 and 940 calibrated years BP (cal year BP). On the other hand, south of 49°N, the minimum age of subalpine stands varied between 3,500 and 5,300 cal year BP with no field evidence of past fire activity.

An analysis of macrofossil charcoal buried in the mineral soil allowed direct inferences of the Holocene (i.e., the last 11,000 year) fire regimes (de Lafontaine and Payette 2011, 2012). This analysis showed that subalpine stands from the black spruce zone were subjected to recurrent fires since the second half of the Holocene (5,600 cal year BP) (Fig. 13.5). According to deglaciation data, this area (north of 49°N) was entirely free of ice sometime between 7,800 and 7,450 cal year BP (Dyke 2004). This leaves a conservative 2,200-year window after ice melting during which a there was probably only a few or no fire disturbance in the region. The ecological conditions prevailing during this period were thus favorable for the northward expansion of the fire intolerant balsam fir and its companion species, white spruce. According to published pollen data, postglacial migration of *Abies* may have reached the lowlands surrounding the high plateaus of the black spruce zone sometime between 7,300 and 6,200 cal year BP (King 1986). In all published pollen diagrams from the black spruce zone a decrease in *Abies* pollen was contemporary with an increase in *Picea* pollen (most likely black spruce) (de Lafontaine and Payette 2011 and references therein). These data suggest that a change in wildfire regime around 5,600 cal year BP was responsible for a zonal transition from the early Holocene northern expansion of fire intolerant balsam fir (and perhaps its associated flora) to the current extensive fire-prone black spruce forest north of 49°N (de Lafontaine and Payette 2011). The distinct subalpine flora in this area (with respect to the black spruce lowlands) represents the remains of past balsam fir expansion. The orographic effect of the high plateaus of central Québec increases atmospheric moisture, resulting in a relatively long mean fire-return interval of 542 year over the last 5,600 year allowing the maintenance of fire intolerant taxa despite a regime of recurrent fires (de Lafontaine and Payette 2012). Shorter fire intervals in the black spruce lowlands (between 100 and 270 year [Bouchard et al. 2008; Cogbill 1985; Parisien and Sirois 2003; Payette et al. 1989]) resulted in the regional collapse of the fire intolerant balsam fir forest and its replacement by the fire-prone black spruce forest.

This pattern of isolation of the balsam fir forest left its imprint on the genetic diversity and structure of white spruce chloroplast DNA (cpDNA) (de Lafontaine et al. 2010). Subalpine white spruce populations in central Québec display patterns of reduced genetic diversity with respect to other populations throughout Québec–Labrador. Genetic structure also shows strong differentiation between the subalpine

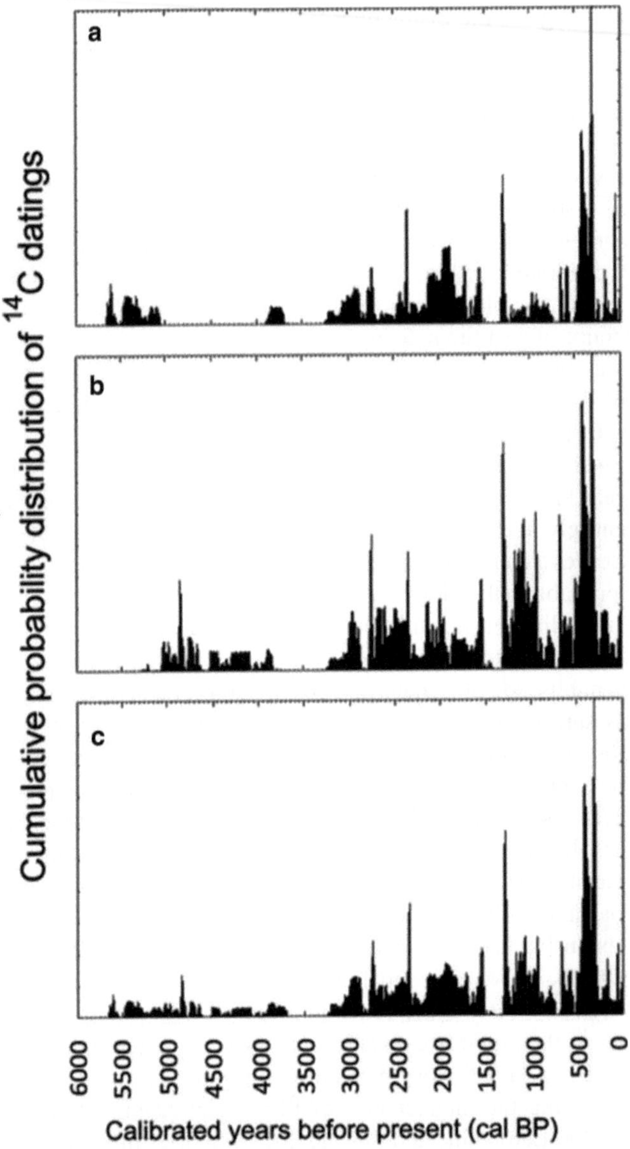

Fig. 13.5 Regional fire histories of the white spruce stands (**a**), the balsam fir stands (**b**), and all subalpine stands (**c**) north of 49°N. The histograms represent cumulated probability of calibrated ^{14}C dates. (From de Lafontaine and Payette 2012)

population and the other populations of Québec–Labrador. Moreover, there is strong genetic differentiation across subalpine populations from each high plateau (Monts Otish, Monts Groulx, and Montagnes Blanches). These genetic patterns (typical of genetic bottlenecks) corroborate the regional demographic decline and local isolation of subalpine white spruce stands associated with the fire driven collapse of the historical northern expansion of the balsam fir forest (de Lafontaine et al. 2010).

13.3 Dynamics of the Subalpine Stands North of 49°N

By analyzing the size structures of subalpine balsam fir and white spruce stands, de Lafontaine and Payette (2010) have shown that old-growth subalpine white spruce stands in central Québec correspond to the late individuals of the ageing first cohorts which initially established after fire disturbance. A chronosequence studying the postfire dynamics in the southern balsam fir forest showed that white spruce is the pioneer species initiating secondary succession on large openings created by wildfire (Galipeau et al. 1997). After a prolonged period without fire, old-growth stands gradually become dominated by balsam fir with white spruce persisting as a secondary species with reduced regeneration (Galipeau et al. 1997). Given the fact that subalpine forests north of 49°N form residual, disjunct stands originating from a former northern expansion of the balsam fir zone, one might expect that the postfire dynamics in these stands should be similar to that found in the southern balsam fir zone.

Indeed de Lafontaine and Payette (2010) showed that the northern old-growth subalpine forests become structurally similar to that of balsam fir stands of the southern balsam fir zone as time since fire increases. The complete successional process of seral replacement of white spruce by balsam fir may take several centuries due to the long life span of white spruce delaying the massive balsam fir infilling into the canopy. Empirical data indicated that old-growth postfire white spruce stands are 370 year on average, whereas the mean age of subalpine balsam fir stands is 480 year (de Lafontaine and Payette 2010). Although the change in stand dominance is gradual, subtle, and probably highly dependent on site-specific ecological factors, a period of 370–480 year after fire is a rough estimate of the time needed to achieve the transition from white spruce stands to balsam fir stands in subalpine belts of central Québec. The process of changing dominance appears somewhat similar to that observed in the subalpine fir (*Abies lasiocarpa* (Hook.) Nutt.)/Engelmann spruce (*Picea engelmanii* (Engelm.) Parry) montane forests (Day 1972; Romme and Knight 1981; Aplet et al. 1988; Antos and Parish 2002). In the Canadian Rockies, Engelmann spruce dominates stand structure from 50 to 250 year after fire before being replaced by subalpine fir (Day 1972; Antos and Parish 2002).

The wildfire regime north of 49°N was instrumental in shaping the landscape of boreal northeastern America (i.e., the zonal replacement of the balsam fir forest by black spruce forest north of 49°N), but it is also the proximal factor explaining the persistence of white spruce stands in subalpine belts of continental central Québec. Fire thus caused a regional decline in the lowlands but also a local concentration of early seral white spruce communities in the subalpine belts.

13.4 Origin and Dynamics of the Subalpine Stands South of 49°N

In sharp contrast with subalpine stands north of 49°N, those of the high plateaus of the Gaspé Peninsula were not submitted to wildfire for at least 3,500 year (de Lafontaine and Payette 2010). Moreover, no charcoal was found in the mineral soils of the subalpine belt in this area. The analysis of macrofossil soil charcoal was used to reconstruct Holocene fire history up to 10,440 cal year BP (Talon et al. 2005) and 9,170 cal year BP (de Lafontaine and Payette 2011) in other parts of Québec–Labrador. Using the same methodological approach, the complete absence of buried soil charcoal in the subalpine belt of the Gaspé Peninsula was interpreted as an absence of fire in these stands over the Holocene period (de Lafontaine and Payette 2011).

Without fire disturbance, the origin and dynamics of subalpine white spruce stands in Gaspé Peninsula are obviously different from those of central Québec. Remember that in this area old-growth subalpine white spruce stands are located at higher elevation than the subalpine balsam fir stands (Fig. 13.4). A study of tree size structure shows a decrease in balsam fir regeneration relative to its abundance in the canopy as well as a decline of overall balsam fir dominance with increasing altitude towards white spruce dominated stands (de Lafontaine and Payette 2010). Conversely, white spruce regeneration is more abundant that in the canopy at higher elevations implying that white spruce dominance in the upper subalpine belt is maintained by high density of seedlings and saplings. Keeping in mind that the lowlands surrounding the high plateaus of the Gaspé Peninsula are dominated by balsam fir forest with increasing altitude, there is a progressive shift from the balsam fir forest (typical of the lowlands, up to the lower subalpine belt) to white spruce dominated forest in the upper subalpine belt. It has been suggested that harsh high-altitudinal climate in the Atlantic (high wind exposure on ridges, deep snow cover in snowbeds, and overall lower temperature in upper subalpine belt) has inhibitory effects on balsam fir which limits its capacity to dominate stands at high elevation (de Lafontaine and Payette 2010). White spruce appears more tolerant to altitudinal climate in the Atlantic and takes advantage of the virtual absence of interspecific competition to dominate the upper subalpine belt on the high plateaus of the Gaspé Peninsula.

13.5 Similar Ecotonal Subalpine Forest, Different Ecological Sorting Processes: Prospects

The balsam fir forest within the black spruce zone (north of 49°N) was maintained in the subalpine belts of this region for several millennia as demonstrated by the continuous presence of *Abies* sp., *Picea* sp. (Fig. 13.6), and *Betula* sp. (de Lafontaine

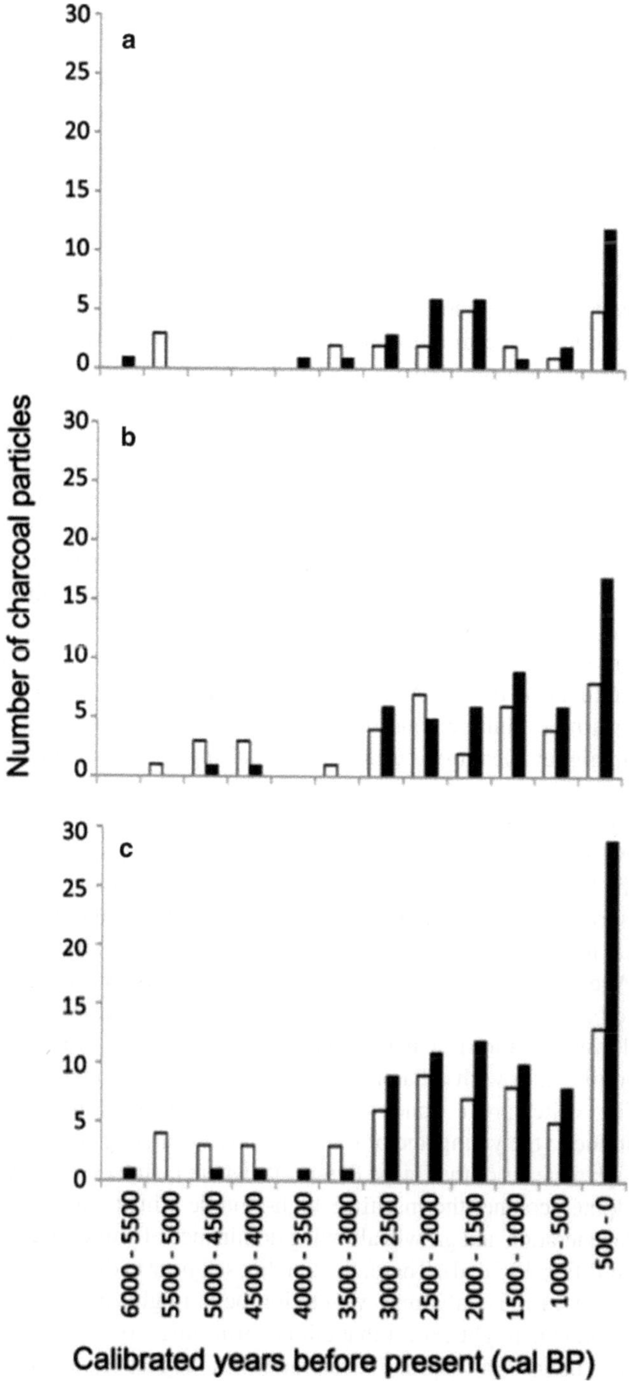

Fig. 13.6 Regional macrofossil charcoal assemblages of the white spruce stands (**a**), the balsam fir stands (**b**), and all subalpine stands (**c**) north of 49°N. *Black* and *white bars* indicate the number of *Picea* and *Abies* charcoal particles recorded in each 500 cal. year BP interval, respectively. (From de Lafontaine and Payette 2012)

and Payette 2011) charcoal since 5,600 cal year BP. The isolated subalpine balsam fir and white spruce stands within the black spruce zone form a distinct species assemblage (with respect to regional vegetation) maintained over the Holocene time scale by fire disturbances occurring on average every *c*. 540 year. In central Québec, the old-growth subalpine forests are thus a mosaic of different postfire successional stages coexisting at similar elevation on the high plateaus. Under current climatic conditions, without the fire regime, it was expected that subalpine stands in central Québec would likely become more similar to the balsam fir forest of the balsam fir zone, thus reducing the number and extent of coexisting white spruce stands (de Lafontaine and Payette 2010).

One can speculate on the fate of the subalpine ecotones under different continental climate change scenarios in central Québec. Under a dryer climate scenario, the fire regime is expected to become more active in the subalpine belt. This would result in the collapse of the residual subalpine balsam fir/white spruce stands and their replacement by fire prone black spruce stands, following the same process inferred earlier in the Holocene in the surrounding lowlands. A wetter, colder climate would allow the maintenance of old-growth white spruce stands in the upper subalpine belt and balsam fir stands in the lower subalpine belt, a pattern similar to that currently observed on the high plateaus of the Gaspé Peninsula influenced by the Atlantic high-altitude climate. A change towards wetter and warmer climatic conditions could allow the expansion of the balsam fir forest at higher elevation, ultimately resulting in the afforestation of the alpine tundra.

The analysis presented here focused exclusively on old-growth stands of the subalpine belts. It should however be acknowledged that above the old-growth stands on the high plateau of central Québec, one will find early successional tree line white spruce stands expanding upslope into the alpine tundra (Fig. 13.7a, b). These stands most likely developed subsequent to climate warming following the Little Ice Age. Whether these tree line stands are transient and will ultimately develop into balsam fir forest (as are the old-growth subalpine white spruce stands located at lower elevation) or if they will remain stable old-growth white spruce stands in equilibrium with a harsher climate in the upper subalpine belt (as are the subalpine white spruce stands of the Gaspé Peninsula) cannot be assessed at the moment due to the young age of these recently expanding stands.

On the high plateaus of the Gaspé Peninsula, where no fire evidence was recorded in the subalpine ecotone, the maritime high-altitude climate alone directly limits balsam fir regeneration and growth allowing dominance of white spruce in the upper subalpine belt (Fig. 13.7c, d). There, the tree lines appear in direct equilibrium with climate and the various subalpine vegetation belts might directly track ongoing climate change. One might expect that climate warming would allow a progression of both balsam fir and white spruce subalpine belts to higher elevation while keeping relatively constant the white spruce and balsam fir belts at upper and lower elevation, respectively.

Fig. 13.7 Example of white spruce stands at the tree line. (**a**) Early successional white spruce stand forming the tree line in the Monts Otish, central Québec. The stunted growth form of the dead stump (*white spruce*) suggests that the tree died during the climatically harsh Little Ice Age. This contrasts with the upright position of the living tree suggesting favorable current climatic conditions. (**b**) Aerial view of the old-growth (*closed crown*) subalpine forest in the Montagnes Blanches, central Québec with example of advancing tree line composed of young white spruce trees (*rectangle*). Note that this colonizing tree line takes the form of an open woodland in which the virtual absence of dead individuals illustrates the young age of this colonizing front. (**c, d**) Old-growth white spruce stand forming the tree line at Mont Jacques-Cartier, Gaspé Peninsula. The stunted growth form of the living trees (*white spruce* and *balsam fir*) exemplifies the current harsh Atlantic high elevation climatic conditions

13.6 Conclusion

Lloyd and Fastie (2003) documented advances of altitudinal white spruce tree lines from 1,800 AD to present at several sites in the Brooks Range in Alaska. Lloyd et al. (2003) also found an upslope expansion in the shrub tundra in several sites of the Seward Peninsula, Alaska. However, Payette and Filion (1985) found no significant upslope shift in white spruce tree line on the coast of Hudson Bay. Instead the response to warming was an increase in cover density. Szeicz and MacDonald (1995) found a pattern similar to that of Payette and Filion (1985) at the white spruce tree lines of several mountain ranges in north-western Canada (Yukon and western Northwest Territories). Other studies have indicated that the response of white spruce to climate change at high elevation was a reduced growth caused by temperature-induced drought stress (Barber et al. 2000; Wilmking et al. 2004). Danby and Hik (2007) and Stueve et al. (2011) have studied interior Yukon and Alaskan white spruce tree lines, respectively. Both studies concluded that the ecological processes controlling tree establishment are spatially variable and thus highly site specific at the tree lines.

We stress that the subalpine white spruce/balsam fir ecotones in eastern Canada might constitute valuable ecosystems to monitor and to model climate change on altitudinal tree lines. Further studying these ecosystems should allow disentangling of the many instrumental ecological factors that could obscure the multiple tree line responses to climate change. For example, assessing the response to climate/fire disturbance interaction can be achieved by comparing alpine tree lines (and subalpine stands) of continental central Québec with that of the maritime Gaspé Peninsula. Monitoring the recent upslope shift can be achieved by further studying the white spruce belt of central Québec. The expanding pattern in this area appears similar to that documented in Alaska by Lloyd and Fastie (2003) and Lloyd et al. (2003). Such recent expansion of white spruce into the tundra was not observed on the high plateaus of the Gaspé Peninsula, suggesting that these tree lines might currently be responding to warmer climate by increasing cover density rather than expanding upslope (similar to the pattern suggested in Payette and Filion (1985) and Szeicz and MacDonald (1995)). Further studying subalpine white spruce/balsam fir ecotones in Québec–Labrador will hopefully help in reconciling the literature on this topic.

References

Aplet GH, Laven RD, Smith FW (1988) Patterns of community dynamics in Colorado Engelmann spruce-subalpine fir forests. Ecology 69:312–319

Antos JA, Parish R (2002) Dynamics of an old-growth, fire-initiated, subalpine forest in southern interior British Columbia: tree size, age, and spatial structure. Can J For Res 32:1935–1946

Barber VA, Juday GP, Finney B (2000) Reduced growth of Alaskan white spruce in the twentieth century from temperature-induced drought stress. Nature 405:668–673

Bouchard M, Pothier D, Gauthier S (2008) Fire return intervals and tree species succession in the North Shore region of eastern Québec. Can J For Res 38:1621–1633

Caccianiga M, Payette S (2006) Recent advance of white spruce (*Picea glauca*) in the coastal tundra of the eastern shore of Hudson Bay (Québec, Canada). J Biogeogr 33:2120–2135

Cogbill CV (1985) Dynamics of the boreal forests of the Laurentian Highlands, Canada. Can J For Res 15:252–261

Cogbill CV, White PS (1991) The latitude-elevation relationship for spruce-fir forest and treeline along the Appalachian mountain chain. Vegetatio 94:153–175

Danby RK, Hik DS (2007) Variability, contingency and rapid change in recent subarctic alpine tree line dynamics. J Ecol 95:352–363

Davis RB (1966) Spruce-fir forests of the coast of Maine. Ecol Monogr 36:79–94

Day RJ (1972) Stand structure, succession, and use of southern Alberta's Rocky Mountain forest. Ecology 53:472–478 de Lafontaine G, Payette S (2010) The origin and dynamics of subalpine white spruce and balsam fir stands in boreal eastern North America. Ecosystems 13:932–947

de Lafontaine G, Payette S (2011) Shifting zonal patterns of the southern boreal forest in eastern Canada associated with changing fire regime during the Holocene. Quat Sci Rev 30:867–875

de Lafontaine G, Payette S (2012) Long-term fire and forest history of subalpine balsam fir (*Abies balsamea*) and white spruce (*Picea glauca*) stands in eastern Canada inferred from soil charcoal analysis. Holocene. 22:191–201

de Lafontaine G, Turgeon J, Payette S (2010) Phylogeography of white spruce (*Picea glauca*) in eastern North America reveals contrasting ecological trajectories. J Biogeogr 37:741–751

Dyke AS (2004) An outline of North American deglaciation with emphasis on central and northern Canada. In: Ehlers J, Gibbard PL (eds) Quaternary glaciations—extent and chronology, part II. Elsevier BV, Amsterdam, pp 373–424

Galipeau C, Kneeshaw D, Bergeron Y (1997) White spruce and balsam fir colonization of a site in the southeastern boreal forest as observed 68 years after fire. Can J For Res 27:139–147

Heinselman ML (1973) Fire in the virgin forests of the Boundary Waters Canoe area, Minnesota. Quat Res 3:329–382

Jurdant M (1959) Étude écologique des associations des forêts résineuses de la région de Québec. MSc Thesis, Université Laval, Québec

King GA (1986) Deglaciation and vegetation history of western Labrador and adjacent Quebec. PhD Thesis, University of Minnesota, Minneapolis

Lafond A (1964) Classification écologique des forêts par la végétation: application à la province de Québec. Faculté d'Arpentage et de Génie forestier, Université Laval, Québec

Lafond A (1974) Forests and forestry in Québec. H.R. MacMillan Forestry Lecture Series. The University of British Columbia, Vancouver

Laliberté A-C, Payette S (2008) Primary succession of subarctic vegetation and soil on the fast-rising coast of Hudson Bay, Canada. J Biogeogr 35:1989–1999

Lloyd AH, Fastie CL (2003) Recent change in treeline forest distribution and structure in interior Alaska. Ecoscience 10:176–185

Lloyd AH, Rupp TS, Fastie CL, Starfield AM (2003) Patterns and dynamics of treeline advance on the Seward Peninsula, Alaska. J Geophys Res 108(D2):8161

Nienstaedt H, Zasada JC (1990) White spruce. In: Burns RM, Honkala BH (eds) Silvics of North America: 1. Conifers. Agriculture Handbook 654. Department of Agriculture, Forest Service, Washington, pp 389–442

Parisien M-A, Sirois L (2003) Distribution and dynamics of tree species across a fire frequency gradient in the James Bay region of Quebec. Can J For Res 33:243–256

Payette S (2007) Contrasted dynamics of northern Labrador tree lines caused by recent climate change and migrational lag. Ecology 88:770–780

Payette S, Boudreau F (1984) Évolution postglaciaire des hauts sommets alpins et subalpins de la Gaspésie. Can J Earth Sci 21:319–335

Payette S, Filion L (1985) White spruce expansion at the tree line and recent climatic change. Can J For Res 15:241–251

Payette S, Morneau C, Sirois L, Desponts M (1989) Recent fire history of the northern Québec biomes. Ecology 70:656–673

Potvin F, Beaupré P, Laprise G (2003) The eradication of balsam fir stands by white-tailed deer on Anticosti Island, Québec: a 150-year process. Écoscience 10:487–495

Romme WH, Knight DH (1981) Fire frequency and subalpine forest succession along a topographic gradient in Wyoming. Ecology 62:319–326

Rowe JS, Scotter GW (1973) Fire in the boreal forest. Quat Res 3:444–464

Stueve KM, Isaacs RE, Tyrrell LE, Densmore RV (2011) Spatial variability of biotic and abiotic tree establishment constraints across a treeline ecotone in the Alaska Range. Ecology 92:496–506

Szeicz JM, MacDonald GM (1995) Recent white spruce dynamics at the subarctic alpine treeline of north-western Canada. J Ecol 83:873–885

Takhtajan AL (1986) Floristic regions of the world. University of California Press, Berkeley
Talon B, Payette S, Filion L, Delwaide A (2005) Reconstruction of the long-term fire history of an old-growth deciduous forest in southern Québec, Canada, from charred wood in mineral soils. Quat Res 64:36–43
Wilmking M, Juday GP, Barber VA, Zald HSJ (2004) Recent climate warming forces contrasting growth responses of white spruce at treeline in Alaska through temperature thresholds. Glob Chang Biol 10:1724–1736

Chapter 14
Conclusion, Synthesis, and Future Directions

Randall W. Myster

14.1 Conclusions

In this section I summarize all previous chapters by reviewing what has been said about the Holocene and last ice age, and also about the current structure, function, and dynamics of each of the three general types of ecotones covered. Specific references are to be found back in the original chapters.

14.1.1 Temperate Forest and Grasslands

Ecotones between deciduous forest and grasslands include prairies (Chaps. 1 and 2), steppes (Chaps. 3 and 4), and pampas (Chap. 5). Natural temperate grasslands encompass vast regions including the North American Great Plains, the Eurasian steppes of Russia, China and Mongolia, and the South American Pampas. These grasslands occur in areas with 250–1,200 mm of annual precipitation, being replaced by forests at higher precipitation. While climatic grassland–forest boundaries have been maintained for millenia, modern grasslands appear to be changing as a result of woody invasions which may be in response to changes in variability and frequency of extreme events rather than to changes in mean conditions.

Temperature and precipitation are major factors controling tree establishment and growth here upon which the effects of other ecological factors—soil type, local vegetation, snow cover, elevation, geomorphology, rainshadow, gravity-induced mass movement, lightning, volcanic eruptions, wildfires caused by meteorite impact or wind shear—interact. Temperate forest–grassland ecotones are also often regions

R.W. Myster (✉)
Department of Biology, Oklahoma State University, Oklahoma City, OK 73107, USA
e-mail: rwmyster@gmail.com

R.W. Myster (ed.), *Ecotones Between Forest and Grassland*,
DOI 10.1007/978-1-4614-3797-0_14,

with rich soils that have been heavily used by humans, and so land-use changes can largely influence its dynamics. Both the historical existence of these ecotones and the present shifts in tree cover can be integrated within a common framework, containing the hierarchical influence of multiple forces acting at different spatiotemporal scales.

In the Cross Timbers of Oklahoma USA, oak invasion into this ecotone is slowed down by mammalian seed dispersal but can be helped by root sprouting if it as common in the ecotone as in the adjoining closed-canopy forest. Likewise resprouting after a tornado may help trees to persist. If tree seeds can be dispersed into grassy patches, they will suffer low seed predation compared to shrub patches and underneath trees, but are likely to germinate anywhere. After germination, those tree seedlings that can escape herbivory have a major advantage. Oak seedlings growing under trees have a low mortality, but suffer high mortality if growing under shrubs and in grass patches, perhaps due to intense blow-ground competition for water. Oak seedlings without below-ground competition survived twice as much as those with it, seedlings in unburned areas survived up to four times more than those in burned areas, reduced below-ground competition lead to twice as much growth, and watering and placement under a shrub also lead to a twofold increase in growth. Taken together then the slow oak invasion, and ecotonal change, that has been observed for decades may be due, in part, to poor oak dispersal and the lack of patches which have high survivorship and growth for both oak seeds and oak seedlings.

Several authors of chapters in this section have believed that the climate of the Holocene and/or the last ice age can act as a mirror, helping us to see much of current ecotone structure, and thus we may use that analysis to infer dynamics and construct models. If so, authors suggest that pollen, charcoal, phytolith, and plant macrofossil data, as well as terrestrial mollusks, may be of use in reconstructing shifts in the lower and upper treeline. Terrestrial mollusks, in particular, imply composition of the vegetation, as well as abiotic ecological parameters such as humidity and temperature. The wind-borne transportation of pollen grains over large distances, unforunately, hampers the identification of small-scale regional fluctuations in the treeline. Plant and animal macrofossils, preserved locally without transportation, can however give reliable information down to a couple of meters, and they are easier to identify than pollen. The current ecotone structure is also the product of intensive human activities present during the past 8,000 years where the original vegetation may be only partially preserved. In just the last 300 years livestock grazing systems have, for example, increased 600% in extent.

The most important climatic parameters favoring the development and sustainment of forest–steppe ecotones are average temperature of the summer months, rainfall, and evapotranspiration. A treeline is a boundary used for marking the edge of the habitat at which trees are capable of growing, where upper and lower treeline can define the transitional ecotone between the referred boundary and the adjacent open vegetation areas. The lower treeland may be a place with a closed canopy of 30–40% and a soil intervowen by roots of arboreal elements, with C_3 grasses are found mostly under the tree canopy and C_4 grasses dominating between trees. There may be a drastic fall in total biomass in the transition zone due to the replacement of trees by smaller bushes and nonarboreal elements. Trees and shrubs can reduce herbaceous species biomass, but it may also increase under tree canopies due to improved nutrient supply and reduced evapotranspiration.

Herbaceous species can increase fine fuel loads, fire frequency and intensity, leading to increased mortality of small woody seedlings that are especially vulnerable to fire. Herbaceous species can reduce woody seedling establishment by effectively competing for light, water, and nutrients. Woody plants may then only establish during wet periods, when competition from herbaceous species are limited. Once woody seedlings establish and grow beyond the height of the herbaceous layer, they become vertically dominant and shade out grasses. This is an important positive feedback mechanism in these ecotones where trees suppress grass through shading and also increased seed input within areas around tree patches. Models predict that woodland distribution would increase in extent in tropical, subtropical, and cool temperate regions of the world under increasing temperature conditions. Similarly, during drought periods woody species are better able to persist than herbaceous species. Periodic droughts, therefore, might be associated with episodic woody plant establishment.

The increasing livestock production systems around the world, at present occupying 25% of the global land surface, suggest that woody–herbaceous–livestock species interactions affect ecotones significantly now and will also into the future. Grazing decreases seed production, seedling establishment, and biomass of palatable herbaceous species, decreases fine fuel accumulation thereby reducing fire frequency, and thereby increasing tree seed germination and woody establishment. Grazed herbs are also less able to compete for resources with woody plants. Livestock disperse woody species seeds across the landscape, which facilitates woody species expansion, but they can also graze on tree seedling and both trample and rub against the bark of seedlings, saplings, and trees.

The structure and function of the Pannonian temperate forest–steppe is fundamentally controlled by local edaphic factors, best visible in the alkaline areas. Their dynamics may be characterized as (1) a shift in the forest–grassland boundary location into the adjacent grassland, resulting from mechanisms and tolerances—especially seed dispersal—which help new trees establish in the adjacent grassland, (2) an increase in tree density at the forest–grassland boundary, or (3) the establishment of a "fairy ring" in the grassland along the edge of the forest when new trees establish as a fringe in the grassland, adjacent to the forest boundary. Fairy rings consist of new stands of densely distributed new stems of similar age where new regeneration—both seed-dispersed and vegetatively reproducing—occurs as pulses advancing the forest boundary into the adjacent grassland. The northern Patagonian transition between one of the most structurally and biological diverse temperate forest and treeless Patagonian steppe develops over only 70 km and has been traditionally directly related to a typical rain shadow effect. Ecotones are structured and controlled by a complex set of natural disturbance regimes with spatially varying roles, different portions of ecotones are unequally affected by climatic variability, and ecotones are areas that are and have been differentially used by humans and thus differently impacted by different land use.

Tree invasion in the Pampa grasslands of Argentina consists of native taxa across natural grassland–forest boundaries and the spread of exotic tree species across habitat mosaics. The latter occurs through "diffuse" ecotones created by anthropogenic landscape fragmentation, and is facilitated by the introduction of preadapted tree genotypes, the addition of dispersal corridors, and the disturbance of resident grasses. Land-use changes associated with agriculture and domestic grazing have caused the widespread loss or fragmentation of large areas of temperate grasslands. The humid Pampas of Argentina are the largest portion of the Río de la Plata grasslands, which also comprise the Campos of Uruguay and southern Brazil. Multicausal constraints to the presence of trees in pristine grassland are key. Facilitative interactions between exotic and native woody taxa could play a key role in accelerating a state shift towards a novel, woody-dominated ecosystem.

14.1.2 Tropical Forest and Grasslands

Ecotones between tropical forest and grasslands include neotropic old fields (Chap. 6), savannas (Chaps. 7 and 8), and llanos (Chap. 10). The dynamics of tropical forest–grassland ecotones may be governed by soil type, fire, hydrology, disturbance (both natural and anthropogenic), as well as climate, in particular the seasonality of precipitation. Determining the underlying control for a forest–savanna transition in a particular location may therefore not be straightforward and may differ markedly among forest–savanna ecotones.

Within the postagricultural ecotones of Puerto Rico, I suggest that the best focus is on how mechanisms and tolerances affect tree seeds and seedlings producing responses, how spaces both within and without these tree individuals change as a result, and how those changes lead to trees replacing (or not replacing) grasses, forbs and other trees. Based on the field experiments and analysis from permanent plot sampling already completed in Puerto Rico, dispersal, seed predation, facilitation, and competition have major roles in controlling tree invasion, establishment and growth in these ecotones, i.e., their dynamics. Facilitation may, however, take many forms. For example, trees can facilitate seed dispersal by supplying woody perches for birds and cover for ground animals. Also patches that are less preferred by predators may facilitate establishment for those seeds found there.

In the southernmost part of Brazil, the southern Campos, grasslands occur in close contact with different forest types. Expansion of woody vegetation throughout the past centuries was driven by changes in the climatic conditions but slowed down as consequence of land use and disturbances. Dynamics in forest–grassland ecotones are thus linked to climatic changes, vegetation management, and disturbance regime. Maintenance of relatively stable forest–grassland boundaries in areas under the influence of fire in grasslands are due to the inability of forest species to become established in grassland, and because grass biomass promotes frequent fires in a positive feedback loop. Under intensive disturbances, forest species can establish successfully only at safe sites or very close to the forest border, leading to a slow process of forest expansion over grassland, corresponding to increasing shrub encroachment in grasslands. Future climate change is expected to lead to an increase in velocity of forest expansion, even under fire.

Paleoecological data can show whether a forest–savanna ecotonal shift constitutes a minor short-term oscillation about an essentially stable ecotone, or instead a much longer, unidirectional ecotonal shift forced by climate change. For example, pre-Columbian earthworks in seasonally flooded landscapes of tropical South America can tell us whether the spatial pattern of forest–savanna mosaics is hydrologically controlled and whether these changes are natural or a function of large-scale landscape modification by ancient construction of raised fields, canals, and/or large habitation mounds.

14.1.3 Boral/Cloud Forest and Grasslands

Ecotones between boreal/cloud forest and grasslands include páramo (Chap. 10) and other alpine grasslands (Chaps. 11, 12, and 13). Temperature is considered as the fundamental factor that creates and maintains the treeline ecotone, modulated by other agents, e.g., topography, geomorphology, wind, soil depth, species interactions, fire, herbivory, human impacts, and site history. Alpine treeline ecotones may lack reproductive capacity or have ongoing establishment. Not surprisingly then, climate warming is expected to cause significant changes in the structure and dynamics of alpine ecotones, which is of concern due to potential loss of biodiversity and ecosystem function as alpine grasslands are replaced by woody species. Treelines are a good choice for monitoring and early warning of climate change impacts on ecological systems in high mountain areas.

In the páramo the most significant transition is from continuous forest to treeless herbaceous vegetation, located at higher elevation. The speed of its altitudinal migrations during Holocene and late Pleistocene times (last 180,000 years) varied significantly: from ~1,400 m (~8 °C) in ~7,000 years from the Late glacial up to mid-Holocene time (at 1°N), frequent oscillations of 300–700 m (2–3.5 °C) at millennial time scales throughout the last two glacial periods, up to rapid shifts of maximally 1,500 m (~10 °C) triggered by temperature and green house gases at glacial–interglacial transitions.

Among other alpine ecotones, the long-term (Holocene) perspective is one of virtually monotonous treeline descent by about 50 m per millennium since the Holocene Thermal Maximum, 10,000–9,000 years before the present. This trend of treeline retreat and compositional change was distinctly broken and reversed about a century ago, correlated with a warming of 1.4 °C. Since then treelines have advanced by a maximum of 185 m, resembling the situation during the warmest part of the Holocene.

Old-growth white spruce stands are found at the altitudinal treelines of the highest plateaus of the boreal forest in eastern North America. In the southern boreal zone, however, subalpine white spruce stands are distributed on extensive high plateaus within the balsam fir–white birch forest assemblage of the (Atlantic) Gaspé Peninsula. *Pinus sylvestris* seems to be the tree species most favored by the new climatic situation, where the present-day ecotonal dominant *Betula pubescens* ssp. *czerepanovii* is likely to regress substantially. In addition exotic plant species from lower regions during the past few decades have ascended, leaving mats of dwarf-shrubs and spots with lichen fragments or bare mineral soil which may promote pine seed germination. Birch and spruce reproduce mainly vegetatively by basal sprouts and layering, while pine relies exclusively on seed reproduction. Also, *Juncus trifidus* is growing denser facilitating the establishment of other species. In addition to climate warming, wind appears to have a superior role.

Finally in these ecotones, low temperatures generally reduce biomass production reducing their competitive abilities. Human-induced changes in land-use can initially mask or even override climatic controls, but after disturbance treelines formerly disturbed by anthropogenic use become sensitive to climate under sustained warming conditions. Recruitment patterns show a strong clustering of recruits at small distances (0–3 m) with both saplings and seedlings segregated from large trees. Synchronous year-to-year growth variations suggest that tree growth sensitivity is greatly affected by macroclimatic conditions, such as summer temperatures affecting tree radial growth and recruitment over time, and winter conditions having a large influence on recruits' survival.

14.2 Synthesis

In the introductory chapters, I defined plant–plant replacements as the fundamental process of community dynamics. Plants respond—given hierarchically as survival [germination if a seed], growth, allocation, and architecture—to various mechanisms and tolerances by creating space which they can physically occupy (called phyto-space) and by influencing, or being influenced by, other space (called neighborhood space). These spaces change as replacements continue. Forest–grassland ecotones are defined by those replacements that involve trees and grasses. If an ecotone does not change over a certain space and time, then those replacements would be dominated by grasses being replaced by other grasses and trees being replaced by other trees. If an ecotone changes, however, and that change moves the grasses move into areas previously occupied by trees, then individual replacements must be dominated by trees being replaced by grasses. Alternately if the ecotone changes and that change causes trees to recruit into areas previously occupied by grasses (happening more and more around the world), then individual replacements must be dominated by grasses being replaced by trees.

Among the three general kinds of forest–grassland ecotones described in this book, the spatial and temporal heterogeneity of temperature and precipitation are primary controlling factors that both grasses and trees respond to. Within that context, human land-uses also set the stage upon which mechanisms and tolerances operate. Among those mechanisms and tolerances, the chapters suggest a key role for (1) seed dispersal, especially when asexual advancement is limited, (2) seed predation, (3) facilitation, and (4) competition. Facilitation may be particularly important in the harsh conditions of alpine ecotones, and competition very important in the species rich environment of tropical ecotones.

14.3 Future Directions

Based on these results, forces at work maintaining (in the short term) and/or changing (in both the short and long term) these forest–grassland ecotones include precipitation and temperature gradients operating along the ecotone (represented by the thick red arrows in Fig. 14.1) and both positive feedback and negative feedback mechanisms and tolerances working within each forest and grassland ecosystems, or also between them (represented by the thin black arrows in Fig. 14.1). These are the major drivers of ecotonal dynamics.

The conceptual frameworks and the studies of forest–grassland ecotones presented in this book hold the following recommendations for future research that needs to be done at each individual site:

1. Sample the evidence for investigation of the Holocene to construct past ecotone structure and dynamics.
2. Set up permanent vegetation plots across the ecotone, spanning the closed-canopy forest on one side and the open grassland on the other, including areas of land-use change, sample them for decades with the aim of investigating the pace and nature of plant changes.
3. Sample soil and environmental factors also across these ecotones (e.g., temperature, precipitation, carbon and energy pathways, biogeochemical cycles).
4. Determine with controlled field experiments how mechanisms and tolerances affect tree seeds and seedlings, how spaces both within and without these tree individuals change as a result of these responses, and how those changes lead to trees replacing (or not replacing) grasses. This entails an exploration of the ecophysiology of these plants, their niches, and other aspects of their biology and ecology.
5. If possible, perform restoration experiments in the field based on the results of these investigations.

I suggest work focusing on plant responses (together niches) and how the local environment (phyto- and neighborhood spaces) determines when and if ecotone trees and grasses replace each other over time and space. There should be no shortcuts, for example in the use of "surrogate" plant parameters instead of direct measurements of plant structure and function. The conceptual framework from Chap. 1, and results from the other chapters, show how to proceed: key species and replacement patterns can be ascertained by using permanent plots, plant niches can be

Fig. 14.1 An artistic representation of the forces at work in ecotones entitled "unstable equilibrium" by the Swiss painter Paul Klee

determined through various kinds of field and greenhouse experiments, and plant–plant replacements can be investigated by sampling local environmental factors and combining them with plant niches. The primary emphasis should be on the ways in which plants respond to and interact with both the living and nonliving constituents of the environment. When these studies are complete they will show how these ecotones are structured, and that they are, I believe, much more individualistic and much more dynamic than are often assumed.

Index

DOI 10.1007/978-1-4614-3797-0, © Springer Science+Business Media New York 2012

MIX
Papier aus verantwortungsvollen Quelle
Paper from responsible sources
FSC® C105338

If you have any concerns about our proc
you can contact us on
ProductSafety@springernature.co

In case Publisher is established outside t
the EU authorized representative is
Springer Nature Customer Service Center
Europaplatz 3, 69115 Heidelberg, Germ

Printed by Libri Plureos GmbH
in Hamburg, Germany